Springer
Proceedings in Physics 4

Springer Proceedings in Physics

Springer Proceedings in Physics is a new series dedicated to the publication of conference proceedings. Each volume is produced on the basis of camera-ready manuscripts prepared by conference contributors. In this way, publication can be achieved very soon after the conference and costs are kept low; the quality of visual presentation is, nevertheless, very high. We believe that such a series is preferable to the method of publishing conference proceedings in journals, where the typesetting requires time and considerable expense, and results in a longer publication period. Springer Proceedings in Physics can be considered as a journal in every other way: it should be cited in publications of research papers as *Springer Proc. Phys.*, followed by the respective volume number, page number and year.

Time-Resolved Vibrational Spectroscopy

Proceedings of the 2nd International Conference
Emil-Warburg Symposium
Bayreuth-Bischofsgrün, Fed. Rep. of Germany
June 3–7, 1985

Editors
A. Laubereau and M. Stockburger

With 187 Figures

Springer-Verlag
Berlin Heidelberg New York Tokyo

Professor Dr. Alfred Laubereau

Universität Bayreuth, Universitätsstraße 30, Postfach 3008,
D-8580 Bayreuth, Fed. Rep. of Germany

Dr. Manfred Stockburger

Max-Planck-Institut für Biophysikalische Chemie, Postfach 2841,
D-3400 Göttingen, Fed. Rep. of Germany

ISBN 978-3-642-47545-0 ISBN 978-3-642-47541-2 (eBook)
DOI 10.1007/978-3-642-47541-2

Softcover reprint of the hardcover 1st edition 1985

2153/3150-543210

Preface

For more than two decades time-resolved vibrational spectroscopy (TRVS) was only part of general conferences on solid-state physics, molecular spectroscopy, photochemistry and photobiology. It was in 1982 when the first meeting on TRVS was organized at Lake Placid. The conference met a strong need among the workers in the field, and it was decided to continue with special conferences on this topic.

The 2nd International Conference on Time-Resolved Vibrational Spectroscopy was held June 3–7, 1985 at Bayreuth-Bischofsgrün, Germany. Scientists from many disciplines came together to discuss their common interest in time-resolved techniques and spectroscopic applications. The high quality of the research presented, the enthusiasm of the participants, and the attractive surroundings combined to an enjoyable atmosphere. Ample time for discussions and the limited number of participants (approximately 100) stimulated the formal and informal exchange of ideas.

Numerous people helped to make the conference run smoothly. Special thanks are due to Mrs. Lenich for making the technical arrangements, and to the program committee for the selection of the scientific presentations. The meeting has benefited from several financial souces. The generous support by the Emil-Warburg-Stiftung was particularly helpful. Financial aid of the "Deutsche Forschungsgemeinschaft" and of the "Bayerisches Staatsministerium für Unterricht und Kultus" is also gratefully acknowledged.

Bayreuth and Göttingen
August 1985

A.Laubereau
M. Stockburger

Contents

Introduction

M. Stockburger

Max-Planck-Institut für Biophysikalische Chemie, D-3400 Göttingen, F. R. G.

A. Laubereau

Physikalisches Institut, Universität Bayreuth, D-8580 Bayreuth, F. R. G.

It has become convenient in recent years to use the term "Time-Resolved Vibrational Spectroscopy" (TRVS) for experiments in which a molecular sample is temporarily perturbed and its consecutive evolution studied, monitoring the time-behaviour of a few vibrational modes or of the total vibrational spectrum. The methods of TRVS have found wide applications and provide information on the vibrational structure and on the dynamical properties of various physical, chemical and biological systems. Mainly two types of TRVS experiments can be distinguished, and will be discussed in the present volume.

In the first kind of measurements a single vibrational mode is excited by a short pump pulse and the subsequent relaxation is followed by delayed probing pulses. Such a "physical" excitation allows to study the dynamics of molecular vibrational modes under various environmental conditions. Since, as a rule, vibrational relaxation of polyatomic molecules occurs in the picosecond and sub-picosecond time-domain, the "physical" TRVS experiment requires sophisticated experimental techniques.

The second kind of TRVS experiments is substantially different in nature. In this case a transient molecular species is formed during the pump process, which is subsequently monitored, observing the time evolution of its vibrational spectrum. The transient (or intermediate) species may be a radical, a radical ion, a conformational isomer or just an excited electronic state of the chemically unmodified sample, which is characterized by its unique set of vibrational modes. Such a "chemical" TRVS-experiment is closely related to conventional spectroscopy. However, it is the great virtue of the TRVS-technique that the information on the frequency position of vibrational modes can be provided in extremely short time intervals, so that structural changes may be readily observed as a function of time. It is often also advantageous to use the intensity of well-resolved vibrational bands as a diagnostic tool to study the kinetic behaviour of chemical intermediates.

A summary of pump and probing schemes is presented in Table 1. The excitation mechanisms are listed in the first column. The first three processes serve for the "physical" excitation of vibrational modes using high intensity Raman or electric-dipole coupling. The other examples are suitable for the initiation of chemical processes via optical absorption (UV,VIS) or high-energy particle ionization. An alternative excitation process is the rapid thermodynamic perturbation by laser pulse heating or concentration changes.

A variety of probing mechanisms is displayed in the second column of the Table. Vibrational spectra are obtained in short time intervals using several kinds of Raman scattering or infrared absorption or emission. Alternatively, vibronic transitions can give information on

Table 1. Excitation and probing processes

stimulated Raman scattering	spontaneous Raman
IR absorption	spontaneous resonance Raman
vibronic excitation	coherent Raman
uv photolysis, ionization, radical formation	coherent resonance Raman
high-energy particle ionization	Raman echo
laser pulse heating	IR absorption
rapid mixing	IR emission (spontaneous or coherent)
	vibronic absorption, photo echo, induced grating

vibrational states by conventional one-photon or multi-photon absorption in the visible or uv because of the involved Franck-Condon factors. As an elegant method to study the dynamics of highly excited vibrational states, the so-called "hot uv absorption spectroscopy" is mentioned in this context. A similar technique allows ultrafast intramolecular thermometry.

Most of the probing schemes apply short probing pulses, the duration of which determines the time resolution of the interrogation process. For these techniques the time-integrated (slow) detection of the probe signal is sufficient. Monitoring the sample by its IR emission (see Table 1) requires sensitive time-resolving detectors. As discussed in one article of the present volume, light gating techniques with nonlinear optical effects can be used to obtain sub-picosecond time resolution in conjunction with conventional detectors. In many experiments, e.g. for resonance Raman studies, one is interested in measuring the dispersed vibrational spectrum of transients or chemical intermediates. In these cases, optical multichannel detectors combined with spectrometers of low straylight level are preferentially applied. Although such detectors are now available in good quality, one can expect further improvements in the near future.

The time-scale of TRVS experiments described in this book extends from a fraction of a picosecond for the population lifetime of vibrationally excited states to several seconds for long-living chemical intermediates. Consequently, quite different methods have to be applied to generate the delay-time between pump and probe event. In the short time-domain ($< 10^{-8}$ s) optical delay lines are used, while in the nanosecond or microsecond regime conventional electronic devices are applicable. An elegant and rather simple technique is described for resonance Raman studies on the photochemical cycle of bacteriorhodopsin. In this case the delay is established by the transit time the liquid sample needs to flow from the pump to the probing laser beam. This technique allows the application of continuous wave lasers for TRVS studies.

In many TRVS experiments the mutual exclusion of high temporal and spectral resolution is not crucial. For example, measurements on the nanosecond time-scale still allow a frequency resolution of 10^{-2} cm^{-1}; this value is considerably smaller than the spectral linewidth in condensed matter of $\sim$ 10 cm^{-1}. For studies in the picosecond and femtosecond range the basic limitations of the uncertainty principle have to be noticed. In the subpicosecond time-domain the inherent instrumental linewidth is several ten cm^{-1}; i.e. a considerable amount of spectroscopic resolution has to be sacrificed.

A special chapter on "Techniques and Instrumentation" of this volume presents papers which are dealing with new techniques or with a detailed description of TRVS instrumentation and its application. We found it also useful to include in this chapter a paper which describes a picosecond transient resonance Raman study with typical applications.

It has been mentioned above that mainly two types of TRVS experiments may be distinguished. This fact is reflected in the present book by corresponding groups of reports with nearly equal numbers. The contributions on "physical" TRVS are presented in two chapters on vibrational dynamics, while the papers on "chemical" TRVS studies are collected in the chapters on photochemical reactions, transient species and biological systems.

In chemical TRVS studies resonance Raman spectroscopy is preferentially used to probe the vibrational spectrum of transient states or chemical intermediates. This method benefits from the fact that a great variety of laser systems is now available in the visible and ultraviolet region for resonance excitation into the molecular chromophore. In this way the Raman scattering cross-section is increased by orders of magnitude for a selected number of vibrational modes with a corresponding simplification of the vibrational spectrum of complex molecular systems. Infrared spectroscopy, on the other hand, is only applied in few cases. However, as is documented by a contribution in the chapter on "Techniques and Instrumentation" Fourier transform infrared spectroscopy can be successfully applied,and one may expect that the role of this technique will increase in the future.

A final chapter is devoted to a few theoretical aspects which are related to time-resolved resonance Raman spectroscopy, to vibrational dynamics, to coherent infrared excitation and to the role of vibrational energy in electron-transfer or in photochemical isomerization processes.

The various reports in this Conference Proceedings represent the enormous progress in the field of TRVS studies,and also reflect the present state of the art. We would be glad if this book could stimulate further work in this fascinating and important research area.

Part I

Techniques and Instrumentation

Energy-Partitioning Measurements in $O(^1D_2)$ Reactions by Time-Resolved Fourier Transform Infrared Emission Spectroscopy

P.M. Aker and J.J. Sloan

Division of Chemistry, National Research Council of Canada,
100 Sussex Drive, Ottawa, Ontario, Canada K1A OR6

1. Introduction

The considerable experimental advantage of Fourier transform spectroscopy and the wealth of detailed kinetic and dynamic information obtainable from time-dependent spectroscopic measurements have led to many different attempts to combine the two. In the most general sense, however, a combination of these techniques is not possible,because of the mathematical requirement that a function which is to be Fourier transformed may depend on only one variable in each domain if a unique result is to be obtained from the transform.

Fourier transform spectroscopy is almost universally implemented using Michelson interferometers. In this context, the above mathematical requirement has the consequence that the time-variable in the measured signal must depend only on the difference between the lengths of the optical paths in the arms of the interferometer. If any other (external) time-dependence is imposed on the measurement, the frequency spectrum generated by the transform will not be a unique function of the signal incident on the interferometer.

It is clear that the two mutually-independent time-variables implied above must be separated for the successful implementation of time-resolved Fourier transform spectroscopy. In practice, this requires that one of the variables be held constant while the measurement is made. The different techniques which have been proposed to achieve this may be classified into two categories depending upon which of the time-variables is held constant. In the "step and sample" method, the time determined by the optical path difference is held constant,while the external time-dependent signal is recorded. The alternative method, which may be termed "timed sampling", requires that the external signal be generated repeatedly and sampled at the same time after each initiation while the optical path difference is varied.

The step and sample technique [1] was the first successful implementation of a time-resolved Fourier transform experiment,and early results from this method and its modifications [2] demonstrated that the throughput and multiplex advantages of Fourier transform spectroscopy could be obtained in a time-resolved experiment. This method of data aquisition ensures that the time-dependence of the external signal is measured with high accuracy. However, it places severe demands on the stability of the interferometer. The timed sampling technique, by contrast, does not make especially stringent demands of the interferometer hardware,and as a result, has been implemented by several commercial instrument manufacturers. Many reports of such work have now appeared in the literature [3]. The first demonstration of the technique [4], however, illustrated that it has a subtle and potentialy dangerous weakness - under certain circumstances it can produce spectral artifacts which are indistinguishable from real spectra. It has been shown that these artifacts are copies of real spectra, shifted by unknown and unpredictable amounts in frequency and phase from the real spectra from which they derive [5]. In particular, it is found that they originate in a sorting process which is a necessary part of the most commom implementation of the timed-sampling technique. In this sorting process, data points collected during the whole time-evolution of the external time-dependent signal are sorted into interferograms which are characteristic of a specific time after the initiation of the signal. If any systematic change in spectral intensity occurs during the observation, this can be converted by the sorting process into a modulation of the interferogram which, in turn, gives rise to spectral artifacts.

If a simplification is made to the timed-sampling data aquisition scheme, the possibility for generation of these undesirable artifacts is removed. While the resulting experiment has less efficient data collection than that describd above, it retains all of the signal-to-noise advantages enjoyed by Fourier transform spectroscopy, combined with the excellent time-resolution of the timed-sampling scheme.

In the following we shall report some initial results obtained using a system of this kind which we have developed in our laboratory. Our system monitors the operation of a commercial Fourier transform spectrometer (which has no intrinsic capability for time-resolution) and operates a time-dependent experiment in such a way that infrared spectra of arbitrary spectral resolution and microsecond time-resolution are obtained. Using this system, we have made the first measurements of the complete OH vibrational distributions formed in the reactions of $O(^1D_2)$ with several small polyatomic hydrides. In particular, we shall report preliminary measurements of the reactions with H_2 and CH_4.

2. Experimental

The experiments were carried out in a large, high-throughput vacuum apparatus which has been described previously [6]. Only those aspects of the apparatus which are important for the present discussion will be outlined here. The reaction chamber was a 25 cm. dia. by 25 cm. high stainless steel cylinder evacuated through a 31 cm. dia. gate valve by an 8000 l/sec diffusion pump (Varian Model VHS 400). Reagents (ozone and the hydride molecule) were introduced via two concentric quartz tubes located on the central axis of the cylinder. Immediately beneath the reagent inlet, the gas flow was intersected at right angles by a KrF excimer laser beam (Lumonics Model TE-860T-4, operated at 350 Hz) which was multipassed through the stream of reagent gas by two dielectric mirrors, coated for maximum reflectivity at 248 nm and mounted facing each other across the cylindrical chamber. The photolysis of the ozone produces $O(^1D_2)$ with about 90% quantum efficiency.

Following the photolysis, vibrationally and rotationally excited OH (OH(v',J')) is formed promptly in collisions between $O(^1D_2)$ and the hydride molecule. For the case of H_2, the reaction rate constant (in units of cm^3 $molec^{-1}$ sec^{-1}) is 3×10^{-10} while for CH_4 it is 4×10^{-10}, hence both reactions occur with about gas kinetic rates [7]. A Welsh cell, mounted with its axis perpendicular to both the axis of the multipass laser mirrors and that of the gas flow, collects the infrared emission from the OH(v',J') and focuses it out through an f/1 CaF_2 lens into a Fourier transform spectrometer, consisting of a Nicolet Model 7199 which had been modified in this laboratory for very-low-light-level emission spectroscopy.

The timing between the excitation event (the excimer laser pulse) and the digitization of a data point by the interferometer was done using an LSI1/23 processor to control a clock and a parallel I/O interface (Data Translation Models DT2769 and DT2768-I respectively). The LSI11/23 processor monitors two signals provided by the Nicolet electronics: a square wave which triggers the A/D convertor on each negative transition and a level which is true when the interferometer is in data aquisition mode. The timing sequence is initiated on the positive transition of the square wave. If the interferometer is in data aquisition mode, the clock is started on this positive transition and runs for a preset time, after which its overflow pulse is detected and used to trigger the excimer laser. The timing is adjusted so that the laser is triggered at a specified time before the next negative transition of the square wave (when a data point is taken). Operation of the experiment in this way has the drawback that only one data point is taken per excitation event. However, it also guarantees that no spectral artifacts of the kind described in the previous section will occur.

The measured reagent inflow rates for these experiments were in the range of $4 - 6 \times 10^{-4}$ mole sec^{-1}. The pumping speed was set by adjusting the gate valve above the diffusion pump in order to maintain the partial pressures of the reagents (ozone and the hydride molecule) in the range of $2 - 10 \times 10^{-2}$ Torr, which was found to provide adequate signal-to-noise. In most experiments, the conditions of pressure and flow were chosen such that the residence time in the photolysis region defined by the multipass laser mirrors was less than 2×10^{-3} Sec in order to ensure that the reaction products were removed between pulses of the photolysis laser.

3. Results

The OH vibrational distributions measured in several experiments on the $O(^1D_2)$ reactions with both H_2 and CH_4 are collected in Table 1, with the relevant experimental conditions. The column headed "Delay" indicates the time-delay between the creation of the $O(^1D_2)$ reagent by O_3 photolysis and the observation of the OH infrared emission spectrum. The vibrational populations were calculated by summing those of the rotational states, obtained from the intensities of the rotationally-resolved emission lines using known Einstein transition probabilities [8]. The sum has been normalized to 1.0 in all cases.

Table 1. Experimental conditions and observed OH vibrational distributions

Run	R-H/[R-H] (Torr×10^{-3})	[O_3]	Delay (μ Sec)	P(v'=1)	P(v'=2)	P(v'=3)	P(v'=4)
1	H_2 / 50	90	400	0.51	0.29	0.15	0.05
2	H_2 / 74	76	190	0.45	0.29	0.19	0.07
3	H_2 / 75	75	50	0.33	0.32	0.21	0.14
4	H_2 / 9	24	40	0.27	0.31	0.26	0.16
5	H_2 / 10	10	80	0.28	0.31	0.29	0.12
6	H_2 / 10	26	150	0.34	0.30	0.23	0.13
7	CH_4 / 30	50	50	0.46	0.27	0.18	0.09
8	CH_4 / 50	20	200	0.59	0.21	0.13	0.07
9	CH_4 / 30	50	400	0.65	0.22	0.11	0.02

Data for both high and low reagent pressures have been shown for the H_2 system. The high pressure runs, (1-3), have vibrationally-cold distributions, which become progressively more-excited at shorter times. The low-pressure data (runs 4-6) show the same effect. The latter runs are shown in Figure 1 without the normalization. This shows that the populations decrease with time, an effect which is partly due to the physical removal of the emitters from the observation zone,and partly the result of vibrational deactivation to the v'=0 state. The higher-pressure data are vibrationally colder than the lower-pressure runs for the same delay times.

After the OH is created in the reaction, it begins to be deactivated in collisions with the unused reagents. This effect is most severe at high reagent pressures and long delay times. Thus the lowest-pressure, shortest-time observation, run 4, represents the distribution which is closest to that created by the reaction. This distribution does not change significantly between the first two observations at 40 and 80 μ Sec, hence it is unlikely to have been deactivated much before 40 μ Sec. We therefore conclude that it is the initial vibrational distribution produced by the H_2 reaction. We estimate the uncertainty for this run to be ±0.04 per vibrational level.

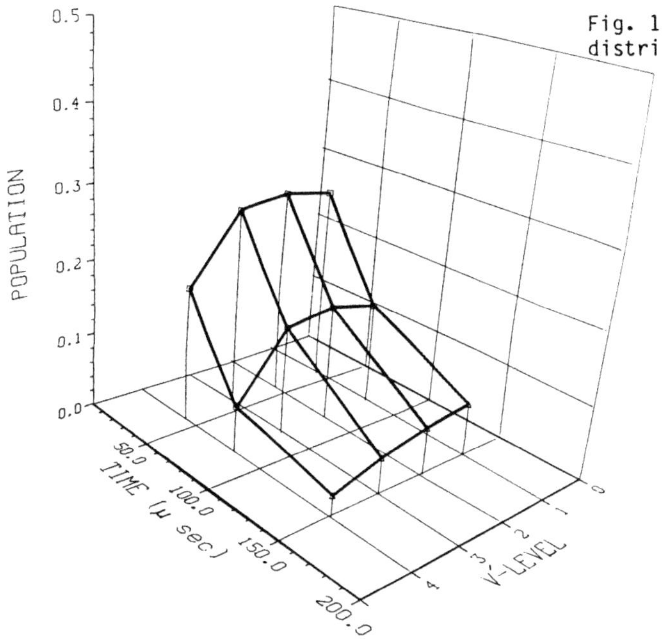

Fig. 1. Time-dependent vibrational distributions for runs 4-6, table 1.

Similar trends are observed in the rotational distributions of these runs. The intensities of the P-branch rotational transitions for OH(v'=3) are shown as a function of time in Figure 2. A substantial nonthermal rotational excitation is seen at the shortest time, 40 μ Sec, but this disappears before the next observation at 80 μ Sec. The same result is observed for the OH(v'=2) rotational distribution, but signal-to-noise constraints prevented the detection of high-J transitions in the other vibrational levels. This implies that the OH is formed

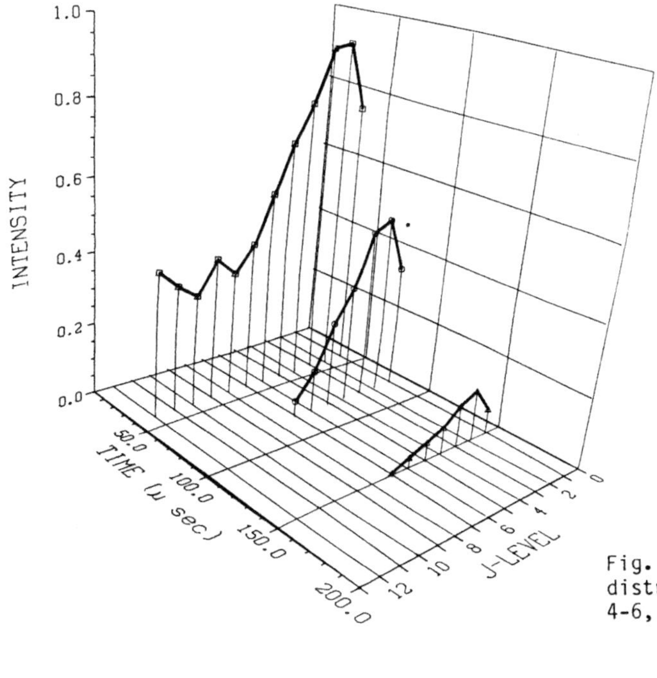

Fig. 2. Time-dependent rotational distributions for OH(v'=3) in runs 4-6, table 1.

with substantial rotational excitation, which is rapidly lost in collisions with unused reagents. In previous laser-induced fluorescence work on this system, very large rotational excitations were observed for the OH(v'=0 and 1) levels as well [9]. We therefore conclude that the reaction produces substantial vibrational and rotational excitation in the OH.

The OH(v') distribution from the CH_4 reaction becomes more vibrationally-excited with decreasing time, as was the case with the H_2 reaction. For the CH_4 case, low-pressure experiments have not been carried out yet, hence the initial vibrational distribution is not known. The higher-pressure experiments, however, have yielded interesting, new information about the overall mechanism. At higher pressures and longer times in this system, emission is observed from internally-excited CO as well as from OH. The CO/OH intensity ratio increases with increasing laser power (increasing $O(^1D_2)$ concentration) and with increasing delay-time. This indicates that the CO originates in a secondary reaction involving $O(^1D_2)$ and a product of the primary reaction, probably CH_3. Such a reaction would likely proceed via the formation of a highly-excited methoxy radical, which would decompose in two steps, forming OCH_2 + H, then CO + H_2 + H. The energy released in the first step is 570 kJ mole^{-1}; the overall exoergicity is 483 kJ mole^{-1}.

4. Conclusion

The reactions of $O(^1D_2)$ with H_2 and CH_4 both produce substantial internal excitation in the OH product. On the basis of the H_2 distributions, which are now reasonably complete, a dynamical model for the reaction which distinguishes between the two mechanistic possibilities - direct abstraction vs. H-O-H intermediate complex formation - should be derivable. The CH_4 reaction is much more complicated; the effects of the two reactive steps must be disentangled before dynamical information can be obtained for either. This work is presently under way using the time-resolved Fourier transform instrument in this laboratory.

References

1. (a) R.E. Murphy and H. Sakai in Proceedings of Aspen International Conference on Fourier Spectroscopy G.A. Vanasse, A.T. Stair and D. Baker, eds. (1971) p. 301.
2. (a) R.E. Murphy, F.H. Cook and H. Sakai, J. Opt. Soc. America 65, 600 (1975). (b) H. Sakai and R.E. Murphy, Appl. Opt. 17, 1342 (1978). (c) G.E. Caledonia, B.D. Green and R.E. Murphy, J. Chem. Phys. 71, 4369 (1979). (c) W. Barowy and H. Sakai, Infrared Phys. 24, 251 (1984).
3. (a) B.D. Moore, M. Poliakof, M.B. Simpson and J.L. Turner, J. Phys. Chem. 89, (1985) 850. (b) J.L. Chao and T. Kumar in Time-Resolved Vibrational Spectroscopy, G.H. Atkinson, ed. Academic Press (1983) p. 97. (c) W.M. Grim, III, J.A. Graham, R.M. Hammaker and W.G. Fateley, American Laboratory 16, 22 (1984).
4. (a) A.W. Mantz, Proc. SPIE 82, 54 (1986). (b) A.W. Mantz, Appl. Opt. 17, 1347 (1978).
5. (a) A.A. Garrison, R.A. Crocombe, G. Mamantov and J.A. de Haseth, Appl. Spectrosc. 34, 399 (1980). (b) J.A. de Haseth, Proc. SPIE 289, 34 (1981). (c) R.A. Crocombe, A.A. Garrison and G. Mamantov, ibid p. 36.
6. (a) D.J. Donaldson and J.J. Sloan, J. Chem. Phys. 82, 1873 (1985). (b) D.J. Donaldson, J.J. Sloan and J.D. Goddard, J. Chem. Phys. (accepted for publication).
7. G. Paraskevopoulos and R.J. Cvetanovic, J. Am. Chem. Soc. 91, 7572 (1969).
8. (a) J.R. Gillis and A. Goldman, J. Quant. Spectrosc. Radiat. Transfer 26, 23 (1981). (b) A. Goldman Appl. Optics 21, 2100 (1982).
9. (a) A.C. Luntz, J. Chem. Phys. 73, 1144 (1980). (b) G.K. Smith and J.E. Butler, J. Chem. Phys. 73, 2243 (1980).

High-Resolution Fourier Transform Raman Spectroscopy with Ultrashort Laser Pulses

H. Graener and A. Laubereau

Physikalisches Institut, Universität Bayreuth, D-8580 Bayreuth, F.R.G.

In the past, considerable progress has been achieved in high resolution Raman spectroscopy /1/. Using several versions of the stimulated Raman scattering process and narrow band laser sources, the limitations of conventional spectrometers that determine the frequency resolution of spontaneous Raman spectroscopy could be avoided. These frequency domain techniques necessitate a considerable technical effort towards frequency stabilisation at the high power level inherent to nonlinear Raman processes. In this letter a novel alternative approach is demonstrated applying time rather than frequency spectroscopy. In fact, a frequency resolution of $\sim 10^{-3}$ cm^{-1} corresponds to time-domain observations on the nanosecond time-scale which are experimentally feasible with sub-nanosecond or even shorter laser pulses /2/. We report here on a first demonstration of FT-CARS measurements /3,4/. It is straighforward to extend our technique to other nonlinear Raman effects, e.g. coherent Stokes Raman scattering or stimulated Raman gain (or loss) spectroscopy.

Our method consists of two steps: coherent anti-Stokes Raman scattering of delayed probing pulses is measured after short pulse stimulated Raman excitation; subsequent numerical Fourier transformation of the signal transient provides the desired spectroscopic information. In a generalized two-level approach with a distribution of neighbouring molecular transitions of relative weight f_j and frequency position ω_j, the time-integrated scattering signal of the probing pulse is given by

$$S^{coh}(t_D) \propto \Phi^2(t_D)\left[\sum_j f_j^2 + 2\sum_{i<j} f_i f_j \cos(\omega_i-\omega_j)t_D\right] \qquad (1)$$

The probe delay is denoted by t_D. $\Phi(t_D)$ represents the phase correlation function, which accounts for the damping of the molecular excitation after the excitation process. The above equation describes a complex modulation with the frequency differences $(\omega_i-\omega_j)$ which decays proportional to Φ^2. Measuring $S^{coh}(t_D)$ the frequency differences $\omega_i-\omega_j$ can be determined by numerical Fourier transformation. The resolution of the difference frequency spectrum is dominantly given by the total delay-time interval during which the signal transient is observed.

As a first demonstration of the FT-Raman technique, we present experimental data on the Q-branch of the ν_1-vibrational mode of methane. Using the supersonic expansion, the temperature of CH_4 is effectively lowered leading to a reduced number and width of lines in the Q-band /3,5/. In our measurements, a 1:2 mixture of CH_4 with argon buffer gas was expanded at approximately 10 bar pressure and room temperature through a pulsed nozzle into vacuum (nozzle diameter 250 μm, background pressure 10^{-5} bar). Details of our experimental apparatus have been reported recently /3/. Single pulses of a modelocked Nd-YAG laser system ($\omega_L/2\pi c$ = 9398 cm^{-1}, duration 23 ps) and of a parametric generator-amplifier device ($\omega_S/2\pi c$ = 6482 cm^{-1}, duration 12 ps) are directed into the sample and serve for the excitation of the Q-lines via stimulated Raman amplification. The second harmonic of the laser pulse ($\omega_{2L}/2\pi c$ = 18796 cm^{-1}) is used for the coherent Raman probing process with collinear optical beams and perpendicular to the supersonic stream. A multiple-pass delay set-up is incorporated in the experimental apparatus, which achieves a maximum delay of the green probe pulse of t_{obs} = 12 ns with sub-picosecond precision.

An example for the experimental data obtained approximately 2 mm behind the nozzle is presented in Fig. 1. The coherent anti-Stokes Raman signal $S^{coh}(t_D)$ is plotted versus delay time. The data points extend over 4 orders of magnitude, and represent an average of 20 individual measurements. A complicated beating structure and the decay of the signal envelope are readily seen.

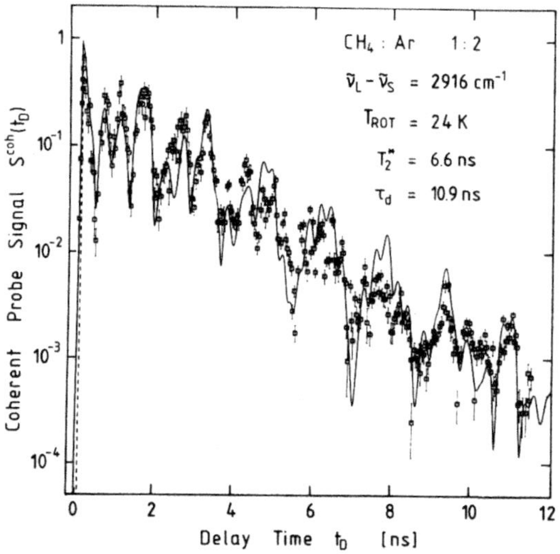

Fig. 1: Coherent probe signal $S^{coh}(t_D)$ versus delay time for the ν_1-Q band of CH_4 ($T_{ROT} \simeq$ 24 K); experimental points; theoretical (solid) curve. Broken line: instrumental response with picosecond decay

The solid line in the Figure is a theoretical curve which agrees well with the experimental points; it represents a computer simulation using the data of Ref. 6 for the frequency positions ω_j and level degeneracies effecting the weight factors f_j. A Boltzmann distribution with same rotational temperature T_{rot} is assumed for the rotational levels of the three nuclear spin subensembles of CH_4 /1/. The value of T_{rot} governs the line strength parameters f_j and is treated as a fitting parameter. From the data of Fig. 1 we obtain $T_{rot} \simeq$ 24 K in good agreement with earlier results /3/. As decay function Φ we use

$$\Phi = \exp\left[-(t'-t)/T_2^* - (t'-t)^2/\tau_D^2\right] \quad (2)$$

with the fitted parameter values T_2^* = 6.6 ns and τ_D = 10.9 ns.

The desired spectroscopic information is obtained by numerical Fourier transformation of the experimental points of Fig. 1. The computed spectral intensity distribution is plotted in Fig. 2a. A number of spectral lines are readily seen. We emphasize that the Fourier

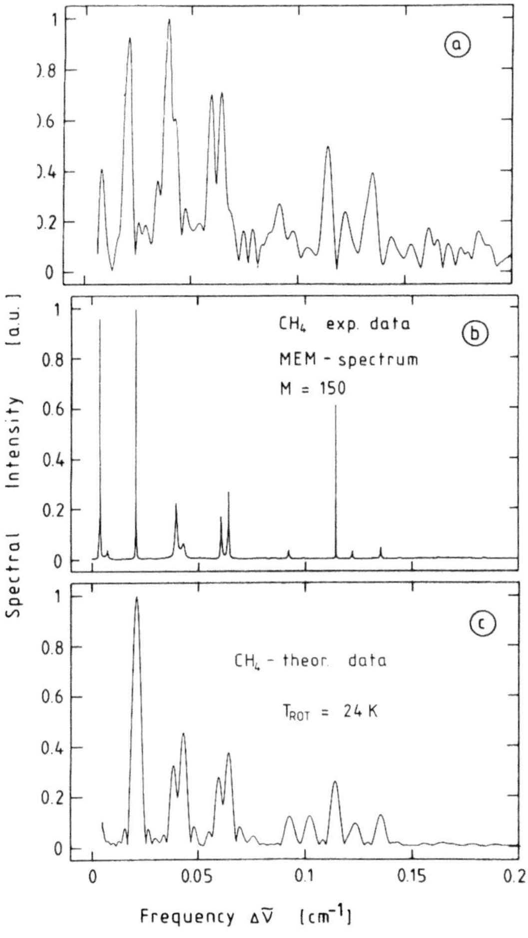

Fig. 2:

High-resolution Raman spectra of CH_4:

a) FT spectrum of experimental data of Fig. 1;
b) MEM spectrum of experimental data of Fig. 1;
c) theoretical FT Raman spectrum

spectrum of Fig. 2a represents new experimental information without use of the available spectroscopic data of CH_4.

For improved frequency resolution we also applied a nonlinear transformation (maximum entropy method, MEM). The result is shown in Fig. 2b. As expected the lines become narrower, but the line strength is different from the FT-spectrum, as noted in the literature /7/.

For comparison, a computer simulation of the difference frequency spectrum is presented in Fig. 2c; i.e. the Fourier transform of the theoretical curve of Fig. 1 implying our knowledge of the CH_4 molecule. The good agreement of the line positions in Figs. 2a and b compared to 2c is noteworthy. Of particular interest is the doublet structure of the two bands around 0.04 cm^{-1} and 0.06 cm^{-1} (see Figs. 2a and b) which verifies the theoretically predicted tensor splitting of the $J = 2$ transitions of the Q-branch. To the best of our knowledge this doublet structure was not measured in previous experimental investigations, and is observed by our high-resolution FT-technique for the first time.

An analysis of our spectroscopic data and a comparison with other results are presented in Table 1. The frequency differences of the

Table 1

Frequency in units of 10^{-3} cm^{-1}	Theory Ref. 6	Theory Ref. 8	Theory Ref. 9	Experiment Spectrum Ref. 6	Experiment FT-Raman this work
$\nu_2-\nu_1$	21.1	20.8	19.8	21.5	20.9 ± 0.2
$\nu_3-\nu_1$	61.0	60.0	57.2	61.5	60.4 ± 1.5
$\nu_4-\nu_1$	63.4	62.3	59.4		63.8 ± 1.5
$\nu_5-\nu_1$	114.0	112.8	107.0	-	113.3 ± 0.3
$\nu_6-\nu_1$	123.0	121.8	115.7	-	122.3 ± 0.5
$\nu_7-\nu_1$	135.2	133.2	126.7	134.5	135.0 ± 0.3
ν_1	2,916,471.9	2,916,472.8		2,916,472	-

observed rotation-vibration transitions with respect to the Q(J=0) frequency ν_1 are listed. There is a good agreement within experimental accuracy with the experimental data of Ref. 6 and the theoretical results of Refs. 6 and 8 while the more recent calculation of Ref. 9 deviates remarkably.

In conclusion it is pointed out that we have demonstrated a novel FT-Raman method with high spectral resolution. Several advantages of the technique should be noted /3,4/. The effect of transit time broadening can be eliminated. Artifacts via the nonresonant part of the third order susceptibility are negligible. A possible dynamic Stark effect during the excitation process does not influence the ns signal transient.

Our approach avoids the stringent stability requirements of Fourier interferometry in the visible. Precise spectroscopic information is provided without narrow-band laser sources.

References:

1 P. Esherik and A. Owyoung, in: Advances in infrared and Raman spectroscopy, Vol. 9, eds. R.J.H. Clark and R.E. Hester (Heyden and Sons, Ltd., London, 1982) p. 130-187, and references cited therein.
2 A. Laubereau and W. Kaiser, Rev. Mod. Phys. 50, 607 (1978); S.A. Akhmanov, N.I. Koroteev, S.A. Magnitskii, V.B. Morozov, A.P. Tarassevich, V.G. Tunkin and I.L. Shumay, in: Ultrafast Phenomena, Vol. IV, eds. D.H. Auston and K.B. Eisenthal (Springer, Berlin, 1984) p. 278-281.
3 H. Graener, A. Laubereau and J.W. Nibler; Optics Lett. 9, 165 (1984).
4 H. Graener and A. Laubereau, Optics Comm. 54, 141 (1985).
5 P. Huber-Wälchli and J.W. Nibler, J. Chem. Phys. 76, 273 (1982).
6 A. Owyoung, C.W. Patterson and R.S. McDowell, Chem. Phys. Lett. 59, 156 (1978).
7 S. Haykin and S. Kesler, in: Nonlinear Methods of Spectral Analysis, eds. S. Haykin (Springer-Verlag, Berlin, Heidelberg, New York, 1979) p. 9-72.
8 J.E. Lolck and A.G. Robiette, Chem. Phys. Lett. 64, 195 (1979).
9 J.E. Lolck, Chem. Phys. Lett. 106, 143 (1984).

Picosecond Transient Raman Spectroscopy Using the High Repetition Rate, Amplified, Synchronously Pumped Dye Laser[1]

T.L. Gustafson, D.A. Chernoff, J.F. Palmer[1]*, and D.M. Roberts*

The Standard Oil Company (Ohio), Corporate Research Center, 4440 Warrensville Center Road, Warrensville Heights, OH 44128, USA

1. Introduction

We are using picosecond transient resonance Raman spectroscopy to study the structures of the first excited singlet states of *trans*-stilbene (tS) and diphenylbutadiene (DPB).[1-4] The importance of *cis-trans* photoisomerization in the diphenylpolyenes as model systems for olefin isomerization is well established and has been reviewed.[5,6] The direct structural and dynamical information obtained by transient Raman spectroscopy is a powerful adjunct to fluorescence and absorption measurements. The observed vibrational frequencies of the excited state provide considerable information about its potential energy surface. This is particularly important in the diphenylpolyenes when considering mechanisms for energy transfer from the optical modes to reaction coordinates. In addition, the intensities in the transient resonance Raman spectra reflect the character of higher electronic states.

Recently several groups have reported the Raman spectra of the first excited singlet state, S_1, of tS and DPB.[1-4,7-12] HAMAGUCHI *et al.*[7-9] and WILBRANDT *et al.*[10,11] used nanosecond lasers with spontaneous Raman detection to obtain the excited singlet state spectra of tS and DPB, respectively. MAEDA and coworkers [12] used nanosecond lasers with coherent Raman detection to obtain the excited singlet state spectra of DPB and other longer chain diphenylpolyenes.

There is presently a disagreement on how to interpret the excited state spectrum of tS. HAMAGUCHI *et al.*[8,9] used two deuterated analogues of tS to assign bands in the excited state at 1567, 1241, and 1148 cm^{-1} to the olefinic C-C double bond stretch, olefinic C-H in-plane bend, and C-phenyl single bond stretch, respectively. These assignments differed from previous work.[1,2,13,14] We agreed with the assignment of the 1567 cm^{-1} as the olefin C-C double bond stretch. But we suggested that the bands at 1241 and 1148 cm^{-1} can be assigned as C-phenyl single bond stretch and C-C-H bend, respectively.[1,2] We proposed that the interpretation of the excited state deuterated spectra is not clear because the normal modes change both form and frequency.[2]

In the present work we addressed the assignment of the excited state spectrum of tS by using tS labelled with ^{13}C in the olefin positions (^{13}C-tS). We anticipated that this would introduce small but perceptible changes in the band positions with no change in the forms of the normal modes. The spectra provided evidence for significant mode mixing in S_1 tS. We also compared the structures of tS and DPB in the excited state.

2. Experimental

We have previously described the use of the high repetition rate, amplified, synchronously pumped dye laser for picosecond transient Raman spectroscopy.[1-4] Briefly, we used a

---

[1]We dedicate this paper to the memory of Joseph Forrest Palmer, who died June 3, 1985, at the age of 23, in a rock climbing accident.

cavity-dumped argon ion laser to pump an amplifier for a synchronously pumped cavity dumped dye laser. We added a second synchronously pumped cavity-dumped dye laser to provide independently tunable pump and probe wavelengths when necessary. A temperature tuned ADA crystal was used for second harmonic generation. The visible probe beam traversed a variable path length and was combined collinearly with the ultraviolet pump beam. We used a single spectrograph with an intensified reticon array detector for observing the Raman signal. We implemented a computer-controlled synchronous detection scheme to suppress baseline drift during data accumulation.

We obtained *trans*-stilbene and diphenylbutadiene from Eastman (scintillation grade) and $\alpha,\alpha',{}^{13}$C-*trans*-stilbene from MSD Isotopes (99.6 atom % ^{13}C). We used these materials without further purification to prepare solutions of 25 mM concentration in *n*-hexane and tetrahydrofuran (Burdick and Jackson) for tS and DPB, respectively.

We aligned the overlap of the pump and probe beams using single wavelength transient absorption kinetics. The absorption kinetics for tS and ^{13}C-tS were indistinguishable, and showed a single exponential decay with a 71 ps time-constant. For DPB the absorption kinetics showed a single exponential decay with an 81 ps time-constant.

3. Results and Discussion

We begin our analysis of the S_1 spectra of tS and ^{13}C-tS by first considering their ground state Raman spectra. In Fig. 1 we show the ground state Raman spectra of tS and ^{13}C-tS in *n*-hexane. Most of the bands in the spectra were solvent bands, but there were several noticeable changes in the 1600 cm^{-1} region. In order to see these changes more clearly, we show the difference spectrum in Fig. 2 (i.e. ^{13}C-tS minus tS, normalized using the 1300 cm^{-1} band in *n*-hexane). The difference spectrum shows an additional feature near 1200 cm^{-1}.

The difference spectrum exhibits two types of changes: band shifts and intensity variations. The derivative type feature near 1200 cm^{-1} represents a shift in the C-phenyl single bond stretch from 1193 cm^{-1} in tS to 1187 cm^{-1} in ^{13}C-tS. The intensity of the

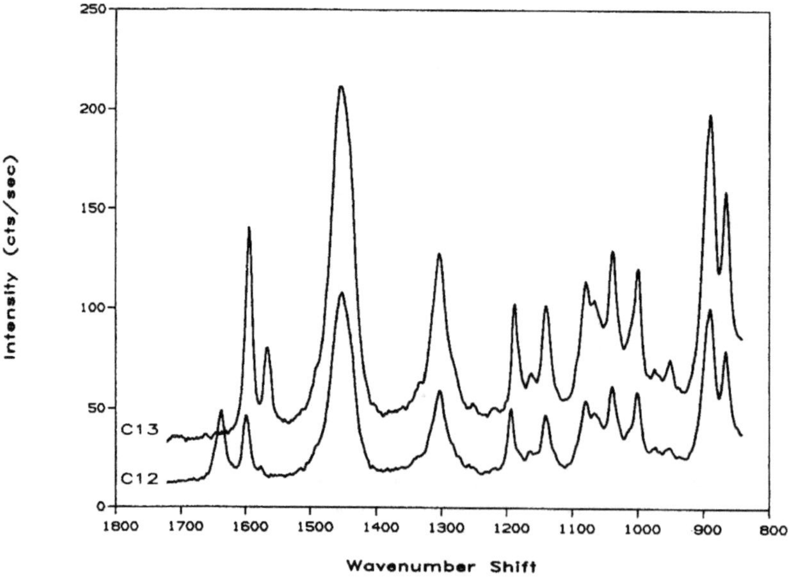

Figure 1. Ground state Raman spectra of *trans*-stilbene and $\alpha,\alpha',{}^{13}$C-*trans*-stilbene in *n*-hexane

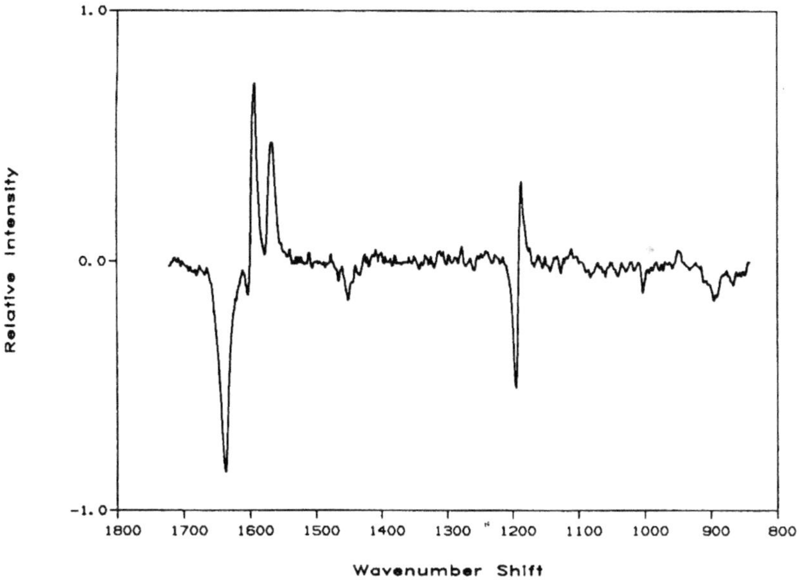

Figure 2. Ground state difference spectrum between *trans*-stilbene and α,α',^{13}C-*trans*-stilbene normalized using the 1300 cm^{-1} band in *n*-hexane

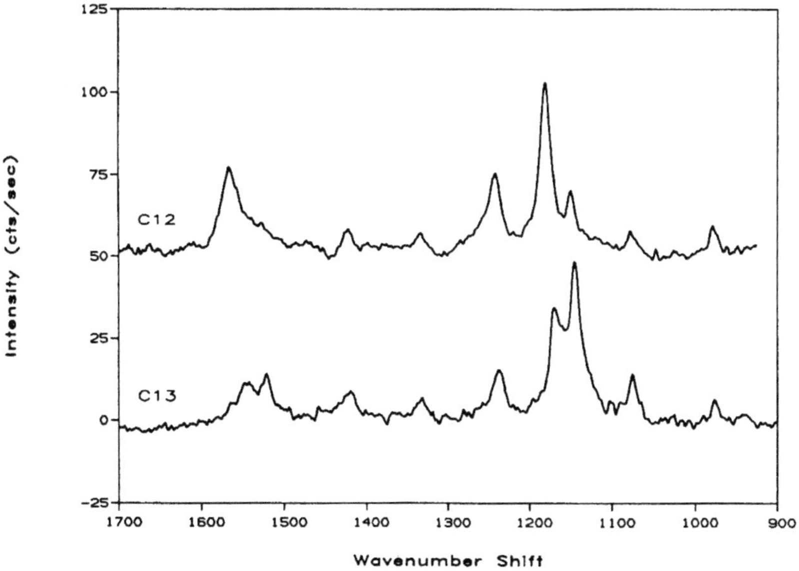

Figure 3. Picosecond transient resonance Raman spectra of *trans*-stilbene and α,α',^{13}C-*trans*-stilbene in *n*-hexane at a delay of 25 ps: 296.3 nm pump and 592.7 nm probe

1187 cm^{-1} band was reduced relative to the 1193 cm^{-1} band. This intensity change is shown in the asymmetry of the derivative feature. The C-C double bond stretching region is not as straightforward to interpret. The band in tS at 1638 cm^{-1} is missing in the ^{13}C-tS spectrum. And a new band appears at 1566 cm^{-1} in ^{13}C-tS. It is unlikely that the 1638 cm^{-1} band and the 1566 cm^{-1} band are directly correlated, since one would not anticipate such a large shift in the olefin C-C double bond stretching frequency with ^{13}C substitution. There is a shift in the position and a change in the intensity of the band at

1597 cm^{-1} in tS. We suggest that the band at 1593 cm^{-1} in ^{13}C-tS arises from a near coincidence between the olefin C-C double bond stretch and the ring C-C double bond stretch. The assignment of the 1566 cm^{-1} band is unclear at this time. Within the experimental noise no other features were distinguishable.

In Fig. 3 we show the S_1 spectra of tS and ^{13}C-tS from 900 to 1700 cm^{-1}. There were several differences between the two spectra in both the band positions and band intensities in the regions near 1200 and 1550 cm^{-1}. In Table 1 we summarize the band positions of both the ground and excited states of tS and ^{13}C-tS in the region of interest. The bands in the excited state of ^{13}C-tS at 1145, 1170, and 1239 cm^{-1} were all shifted lower with respect to comparable bands in tS. These data suggest that each of the three bands has some olefin C-phenyl single bond stretching component in its normal mode. It is particularly significant that the band at 1181 cm^{-1} in tS shows some olefin characteristics. When all ten phenyl hydrogens are deuterated, this band shifts from 1179 to 869 cm^{-1}.[8] A change in the form of the normal modes in the deuterated analogues may account for these changes.

Table 1. Comparison of selected ground state and excited state vibrational frequencies in *trans*-stilbene and α,α',^{13}C-*trans*-stilbene

Ground State		Excited State	
tS	^{13}C-tS	tS	^{13}C-tS
		1150	1145
1193	1187	1181	1170
		1242	1239
	1566		
1597			1546
	1593	1567	
1638			1522

The excited state spectra show that a modest isotopic substitution produced a significant change in the relative intensities of three modes near 1200 cm^{-1}. The intensity of the excited state vibrations is derived from resonance enhancement with a higher electronic state S_n. We have previously used the intensities of the S_1 bands to make several inferences about the structure of S_n.[2] By comparing the normal mode descriptions associated with these three frequencies and by examining the differences caused by isotopic substitution, one may be able to identify a few internal coordinates associated with high Raman intensity. This would provide a basis for a more quantitative comparison of the electronic structure of S_n with S_1.

Building upon the understanding we obtained from our study of tS, we looked at the excited state structure of DPB.[3,4] We found that the relative intensities of the bands varied between the S_1 spectra of DPB and tS, but there appeared to be a one-to-one correspondence in many of the peak positions. Again, the intensities are derived from a higher electronic state and these states may be quite different in DPB and tS. Since the peak positions are indicative of the S_1 excited state structure, the similarity between the DPB and tS spectra suggests that the structure of S_1 DPB in fluid solution is very similar to that of S_1 tS.

References

1. T.L. Gustafson, D.M. Roberts, and D.A. Chernoff, J. Chem. Phys. 79, 1559(1983).
2. T.L. Gustafson, D.M. Roberts, and D.A. Chernoff, J. Chem. Phys. 81, 3438(1984).
3. T.L. Gustafson, D.A. Chernoff, J.F. Palmer, and D.M. Roberts, Ultrafast Phenomena IV, D.H. Auston and K.B. Eisenthal, Ed., (Springer-Verlag, Berlin, 1984), pp. 266.
4. T.L. Gustafson, D.A. Chernoff, J.F. Palmer, and D.M. Roberts, *Ultrashort Pulse Spectroscopy and Applications*, M.J. Soileau, Ed., Proc. SPIE 533, 78(1985).
5. J. Saltiel, J. D'Agostino, E.D. Megarity, L. Metts, K.R. Neuberger, M. Wrighton, and O.C. Zafirou, Org. Photochem. 3, 1(1971).
6. R.M. Hochstrasser, Pure Appl. Chem. 52, 2683(1980).
7. H. Hamaguchi, C. Kato, and M. Tasumi, Chem. Phys. Lett. 100, 3(1983).
8. H. Hamaguchi, T. Urano, and M. Tasumi, Chem. Phys. Lett. 106, 153(1984).
9. H. Hamaguchi, J. Mol. Struct. 126, 125(1985).
10. R. Wilbrandt, N-H. Jensen, and F.W. Langkilde, Chem. Phys. Lett. 111, 123(1984).
11. R. Wilbrandt, W.E.L. Grossman, P.M Killough, J.E. Bennett, and R.E. Hester, Ninth International Conference on Raman Spectroscopy, Tokyo, 1984, (Organizing Committee for the IXth International Conference on Raman Spectroscopy, 1984), pp. 332.
12. T. Kamisuki, H. Kataoka, Y. Adachi, K. Suzuki, A. Kasama, and S. Maeda, Ninth International Conference on Raman Spectroscopy, Tokyo, 1984, (Organizing Committee for the IXth International Conference on Raman Spectroscopy, 1984), pp. 348.
13. R.H. Dyck and D.S. McClure, J. Chem. Phys. 36, 2326(1962).
14. A. Warshel, J. Chem. Phys. 62, 214(1975).

Pulsed Multichannel Raman Technique

A. Deffontaine, M. Delhaye, and M. Bridoux

Laboratoire de Spectrochimie Infrarouge et Raman (LP. 2641 CNRS),
Université des Sciences et Techniques de Lille I - Bâtiment C.5,
F-59655 Villeneuve d'Ascq Cedex, France

INTRODUCTION

The measurements of the rates of formation and decay of transients by optical spectroscopy is essential to the understanding of the mechanisms involved in chemical and biological reactions. The potential of time-resolved vibrational spectroscopy for obtaining structural information about transients intermediates is now well established (1). However, time-resolved Raman spectroscopy, especially in the picosecond time domain, presents many difficulties for the experimentalist. Furthermore, the interpretation of the results is always limited by the performance of the experimental set-up. All the parameters such as time resolution, spectral resolution, signal-to-noise ratio, reproducibility of the experiment, delay time range and discrimination against spurious phenomena have to be considered with much care in each experiment. In this paper we review practical points of prime importance in pulsed spontaneous Raman spectroscopy, and we describe some techniques which we have developed in our laboratory for the detection and characterization of transients with lifetimes in the nanosecond or picosecond time-scale.

SINGLE PULSE RAMAN SPECTROMETRY

A typical time-resolved spontaneous Raman experiment is performed in two steps : first the sample is perturbed by a pulse of electrons or X-rays, a heat pulse or most often by a light pulse. The second step consists of the time-dependent detection of the transients which have been created : free radicals, excited states, unstable isomers, etc. For the studies of photochemically created transients, several approaches may be considered according to whether the laser operates at low-energy and high repetition rate (i.e. 1 nJ, 1 MHz) or at high-energy and low repetition rate (i.e. 1 mJ, 1 Hz).

In our approach, we have, for several reasons, chosen to use a Nd : YAG laser to perform single shots experiments. First, the Nd : YAG laser provides a versatile source of pump and probe pulses because its harmonics (532, 355 and 266 nm) are easily separated, both spatially and temporally, by optical delay lines. Second, the pulse energy is sufficient to pump a large number of molecules into their excited states, in contrast to continuous-wave, synchronously pumped laser pulses. However, as the shot-to-shot repeatability of the pulse width and energy is poor, it is necessary to verify the performance for each shot in order to eliminate non-significative spectra from the signal averaging. It is thus very important to design laser sources with the best stability and highest beam quality (2). As in any pump-probe experiment, the probe should ideally not disturb the system. However, in pulsed Raman spectroscopy the power density inside the sample has to be kept below the threshold for non-linear optical effects such as the stimulated Raman effect. Consequently the beam quality must be good enough to control the generation of non-linear effects, and focusing of the laser beam cannot be as sharp as in conventional Raman spectroscopy. Thus, lens collection of the scattered light must be adapted to the relatively large diameter of the laser beam inside the sample in order to avoid the use of a wide slit, which reduces spectral resolution.

The Raman spectrometer is also of prime importance in pulsed experiments. It is clear that the conventional technique using a scanning monochromator is inadequate for pulsed Raman spectroscopy, because the probability of detection of a rapidly changing or random event is inevitably very weak. On the other hand, in the multichannel technique, the information from all of the spectral elements is simultaneously recorded, thus no spectral event that appears in the spectral range under study is lost. Although multichannel detection of pulsed Raman spectra is not a new technique (3), applications in time-resolved spectroscopy have been developed recently thanks to improvements in the spectrometers and to progresse in the technology of multichannel detectors. The ideal spectrograph should present good luminosity and resolution and superb straylight rejection, while passing a wide band of frequencies. In this respect, we have often used home-built spectrographs with concave holographic gratings. For instance, an unequal arms holographic grating (Jobin-Yvon), with the following characteristics : 1500 gr/mm, front focal length : 658 mm, back focal length : 353 mm, relative aperture : F/3, dispersion : 80 cm^{-1}/mm (at 500 nm) offers a very low stray light level with the minimum number of optical components. However, the optical aberrations of concave holographic gratings are acceptable only in a narrow spectral range. Especially the astigmatism which is strongly dependent on wavelength. severely limits the spatial resolution. For this reason, we preferred in our recent experiments spectrographs with plane gratings and aberration-corrected lenses, which give sharp images and excellent stigmatism over the whole spectral range of interest (cf. poster session). The spectral image is intensified by a gated image intensifier and analyzed by a low-light level T.V. phototube (SIT vidicon). The video image can be stored in a videodisk recorder for further use. A data-handling device permits the selection of a part of the TV picture and its storage after digitization in a computer. A control unit performs the functions of synchronization of the TV camera with the videodisk, trigger of the laser in synchronization with the TV blanking signal and generation of the trigger and sampling signals to the computer.

PULSED RAMAN IMAGERY

The previously described grating presents not only a good stigmatism in the spectral region of interest, but also in the direction parallel to the axis of the slit. This characteristic makes possible a point-to-point correspondence between the scattering track of the laser beam inside the sample and each of the spatial channels in the focal plane of the grating. It is then possible to take advantage of the spatial resolution offered by the multichannel technique (4).

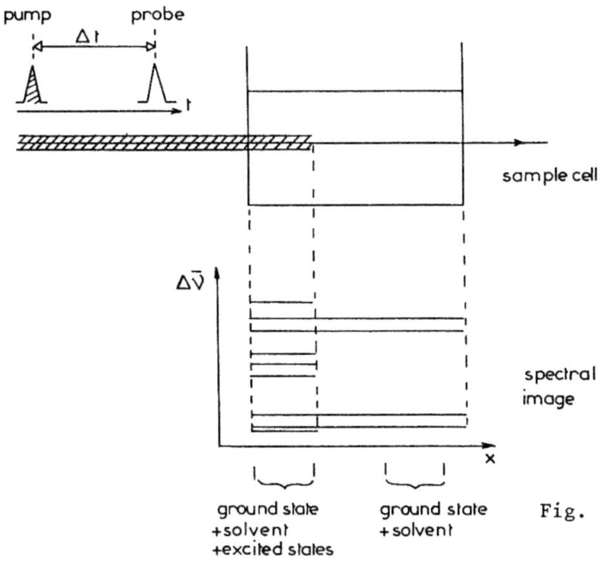

Fig. 1 : Application of Raman imagery to the detection of transients

The effectiveness of our stigmatic multichannel Raman spectrometer is illustrated by the study of the triplet state of all-trans retinal. The same Q-switched Nd : YAG Laser (Quantel NG 24) is used for the pump (third harmonic at 355 nm) and the probe (second harmonic at 532 nm), which are separated and then temporally delayed by an optical delay line. In the sample cell, which is 10 mm long, the concentration of retinal in hexane (10^{-3}M) is such that the photolyzing beam is completely absorbed after a 3 mm travel (Fig. 1). On the spectral image one can observe two different parts corresponding to a perturbed and a non-perturbed sample. Freezing this spectral image and analyzing separately its two parts gives two different spectra from which the Raman spectrum of the triplet state of trans-retinal can be obtained.

Another interesting feature of the Raman imagery technique is connected with the use of a multipass optical delay line, which permits the simultaneous recording of several time-resolved Raman spectra with a single laser pulse. Several multipass delay lines have been designed, which are based on the following principle (Fig. 2). The laser pulse passes N times through the sample, crossing the sample cell at different locations. The scattering tracks are imaged onto the entrance slit of a stigmatic polychromator, resulting in N time-resolved Raman spectra. The delay between the pulses can be adjusted in the 10^{-10} - 10^{-7}s range ; delays of tens of nanoseconds necessitate optical paths of the order of several meters. This imposes very strict requirements on the divergence and spatial distribution of the laser beam. If a shorter time resolution (10^{-12} to 10^{-10}s) is needed, the reflection or transmission stepped-delay echelon technique first developed for absorption spectroscopy (6) can be adapted to Raman spectroscopy. It is also possible to create a train of pulses from a single pulse with a two-mirror system. If the pump pulse arrives later than the first probe pulse, a reference spectrum can be recorded and, hence, background and solvent spectra can easily be subtracted. The intensity of the successive probe pulses is continuously decreasing, but the raw spectra can be normalized. The main advantage of such an experiment is its insensitivity to the fluctuations of the laser : in a single shot experiment, the time duration bandwidth and intensity of the pulse are uniquely and accurately defined.

An example of the multipass delay line is presented in Fig. 2. The pump beam is perpendicular to the probe beam, with two possible geometries. In the case of

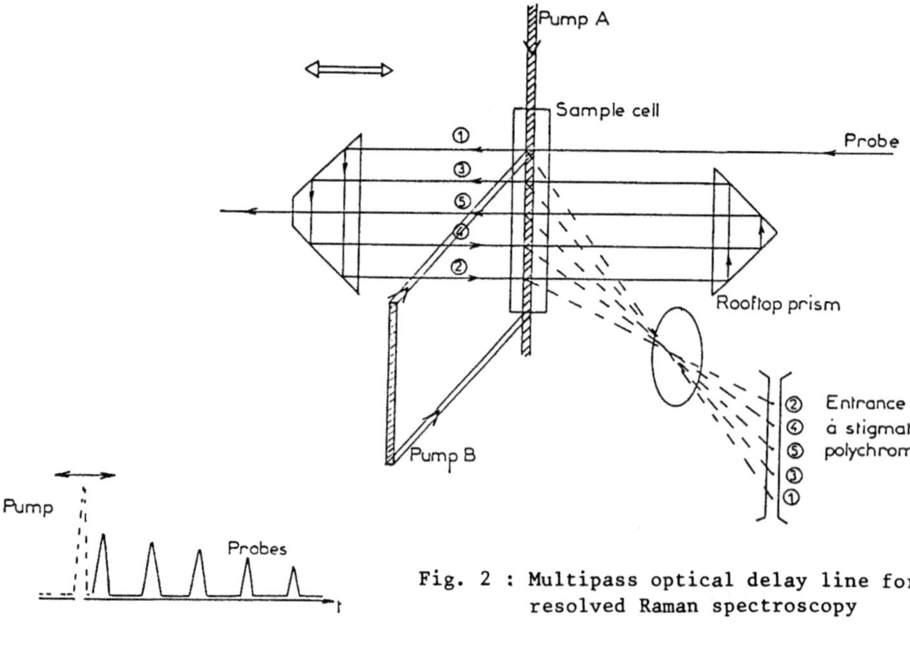

Fig. 2 : Multipass optical delay line for time-resolved Raman spectroscopy

pump A it must be verified that the absorption of the pump is negligible within the sample cell. For the pump B geometry, the laser beam has to be expanded in such a manner that all points of the sample cell are homogeneously illuminated.

PICOSECOND GATING TECHNIQUES

In pulsed Raman spectroscopy the signal-to-noise ratio is greatly increased by gating the detector (3). Nowadays, electronic gating of image intensifiers can be achieved in the nanosecond range, but picosecond gating remains very difficult. In order to discriminate Raman scattering from fluorescence, we have tried both active and passive methods of picosecond gating : excitation of the sample by picosecond laser pulses and synchronous detection of the scattered light could eliminate most of the fluorescence and enhance significantly the signal-to-noise ratio.

One method to obtain time-resolved spectra consists in using a streak camera in the focal plane of a spectrograph. However, coupling a spectrograph and a streak camera is not a trivial matter. It must be remembered that the temporal characteristics of optical pulses are generally modified behind a polychromator. Due to the path difference between monochromatic rays travelling through a grating spectrograph (or monochromator), a time spread is introduced on the pulses (5) : a diffraction grating acts as a light stretcher. The time-broadening can reach much more than 100 picoseconds, depending upon the grating. The applicability of streak cameras in picosecond spectrometry is then limited. It can be noted that a prism spectrograph does not present this drawback. However, prism spectrographs are not the best choice for Raman spectroscopy.

Instead of gating the detector, another solution is to use a light shutter such as a Kerr shutter in front of the spectrometer. Figure 3 shows the experimental set-up which we have used. The fundamental (1.06 μ) and first harmonic (532 nm) of a mode-locked Nd : YAG laser (25 ps) are separated. The green pulse is directed to the sample cell and the scattered light is collected by a lens O_1, which images the track of the laser beam onto the CS_2 cell with a 1:1 magnification. The IR pulse is sent through the 1 mm thick CS_2 cell placed between crossed polarizers (Polaroid HN 22). In the illuminated volume, a birefringence is induced, and the scattered light from the sample is transmitted through the polarizers. The 1.06 μ gating pulse propagates from A to B, so that the Kerr gate opens first at A and later at B. As lens O_1 reverses the image, the 1.06 μ and 532 nm pulses propagate in opposite directions. Uniformity of intensity over time is assured because the trigger pulse and the probe pulse correspond to the same laser shot. The transmitted light is collected by lens O_2 and analyzed by a spectrometer. Last, when it is necessary to discriminate between optical phenomena of different durations, time-resolved picosecond interferometry can be applied (7).

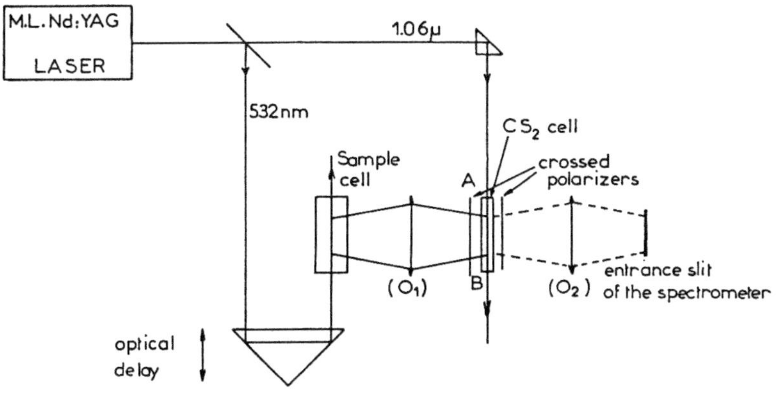

Fig. 3 : Picosecond light gate using Kerr effect

REFERENCES

1. "Time-Resolved Vibrational Spectroscopy" Ed. G.H. Atkinson (John Wiley & Sons, New York) 1983.
2. C.K. Johnson, G.A. Dalickas, S.A. Paynes, A.M. Hochstrasser, Pure & Appl. Chem., 57, 195 (1985).
3. M. Delhaye, Appl. Opt. 7, 2195 (1968).
4. R. Michael-Saade, J.P. Sawerysyn, L.R. Sochet, G. Buntinx, M. Crunelle-Cras, F. Grase, M. Bridoux, "Dynamics of flames and reactive systems", Progress in Astronautics and Aeronautics (Ed. M. Summerfield) 95, 658 (1984).
5. N.H. Schiller, R.R. Alfano, Opt. Comm. 35, 3, 451 (1980).
6. M.R. Topp, P.M. Rentzepis, R.P. Jones, Chem. Phys. Lett., 9, 1, (1971).
7. M. Delhaye, A. Deffontaine, A. Chapput, M. Bridoux, J. Raman Spect. 15, 4, 264 (1984).

Fluorescence Rejection in Raman Spectroscopy

N. Everall, R.W. Jackson, J. Howard, and K. Hutchinson*

Department of Chemistry, University of Durham, South Road, Durham DH1 3LE, U.K.

1. INTRODUCTION

The masking of Raman transitions by intense fluorescence has led to many attempts to reduce this fluorescence contribution. Amongst the methods available, those which are based on the temporal discrimination of the Raman and fluorescence photons appear to have the most general applicability [1-4]. We have investigated two different experimental approaches involving different detectors. The first utilises a gated diode array (DAD), which is switched on and off synchronously with the laser pulses, and the second a photomultiplier tube (PMT). In the latter case, it is the output of the PMT which is electronically gated to select the appropriate time-regime.

2. EXPERIMENTAL

A block diagram of the instrumental configuration for the DAD experiments is shown in Fig. 1. Picosecond (∿15 ps) pulses were provided by a Spectra Physics sync. pumped, mode-locked, cavity-dumped laser system. The gated DAD (Spectroscopy Instruments) consisted of an IRY700 detector head, ST110 controller, MD100 fast pulser and LSI11/23 computer. Since the synchronous electrical output signal from the cavity dumper driver was not of TTL level, it was necessary to insert a signal generator, with TTL output, (Phillips PM5712) between the driver and the fast pulser. Both the fast pulser and the signal generator contain variable delays which enable synchronisation of the detector gate with the photons from the sample to be achieved. A 5 ns gate was used for all measurements.

The PMT (RCA 31034-05) system is similar to that previously reported [1,3]. In order to evaluate the techniques, the 992 cm^{-1} Raman band of benzene was studied in solutions doped with various concentrations of a fluorescent dye (rubrene). The decay-time for a degassed solution is 15.9 ns while it is 12.5 ns otherwise.

3. DIODE ARRAY SYSTEM

In order to synchronise the detector gate with the arrival of the photons, the variable delay on the fast pulser was altered while observing the detector signal. Figure 2 shows the intensity observed (5×10^{-3} M solution, not degassed) at 992 cm^{-1} (Raman plus that part of the fluorescence contained within the 5 ns gate) as a function of the delay. Negative times indicate that the gate is positioned prior to the arrival of the photon pulse. Clearly, instead of a curve originating at zero and with a single maximum at positive time, several curves (maxima separated by ∿15 ns intervals) are observed. This has not previously been reported, and we have shown it to be due to multiple opening of the detector gate for a single trigger pulse. This problem has very recently been overcome by the manufacturer, but

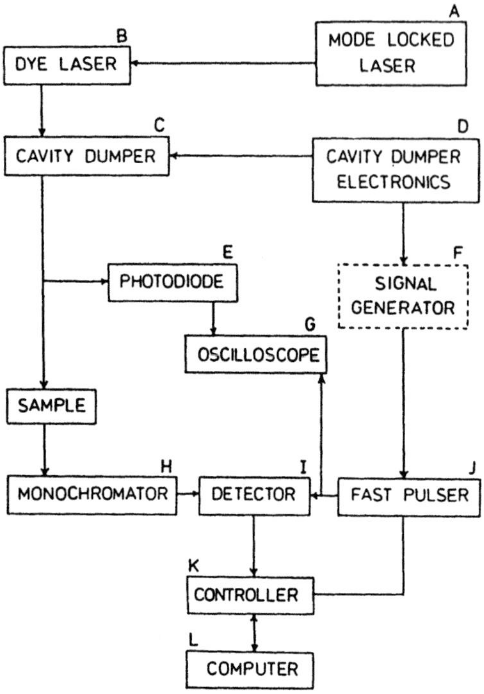

Fig. 1. Experimental arrangement for the measurements. (A) Spectra Physics model 342 Mode-Locking argon ion laser, (B) 375 Dye laser, (C) 344S Cavity Dumper, (D) 454 Cavity Dumper driver, (E) 403B Photodiode, (F) Phillips PM5712 Pulse Generator, (G) Tektronix 2215A oscilloscope, (H) Spex Ramalog 4 Raman spectrometer, (I) Spectroscopy Instruments model IRY700 Diode Array Detector, (J) MD100 Fast Pulser, (K) ST110 controller, (L) LSI 11/23 computer

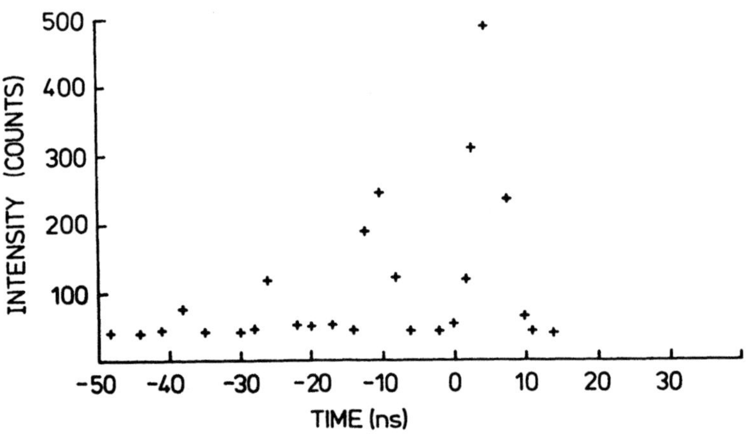

Fig. 2. Fluorescence intensity as a function of the gate position "t" for a wavenumber shift of 993 cm^{-1}. The sample was 5×10^{-3} M solution of rubrene in benzene. Negative time values indicate that the gate is positioned prior to the arrival of the photon pulse.

it is probably still a feature of gated DAD systems from other suppliers. The multiple gate opening is clearly a disadvantage in many applications. For instance, it seriously increased the detected fluorescence contribution in our measurements by including contributions from the tail of the fluorescence decay in with the counts detected with the first, and desired, gate opening.

4. FLUORESCENCE REJECTION USING THE DIODE ARRAY

The response function of the diode array is trapezoidal (3.5 ns top and 9 ns base) and the question arises as to the best position, relative to the Raman photons, at which to begin to open the gate.

We define the gate position "t" in Fig. 3. A computer programme has been written which models the total instrument response to a fluorescing sample and an example of the results is shown in Fig. 4.

In order to assess the improvement in the fluorescence rejection (Raman intensity/fluorescence intensity = S(R)/S(F)) and, more importantly, in the signal (S(R)) to noise (N) ratios for various gate positions (t) we have calculated and measured these parameters in the form:

$$\frac{\text{values with the detector gate operating (5 ns)}}{\text{values with the detector gate left open}}$$

This procedure has the merits that the values are independent of laser performance (power, wavelength etc.) and of parameters such as detector efficiency etc. The calculated and measured values are shown in Table 1. For the gated spectrum the gate was operated at 4 kHz, which is the maximum value currently feasible. From Table 1 it can be seen that there is generally good agreement between the calculated and observed values, and that for small values of "t" there is a substantial benefit in overcoming the problem of multiple gate opening. The

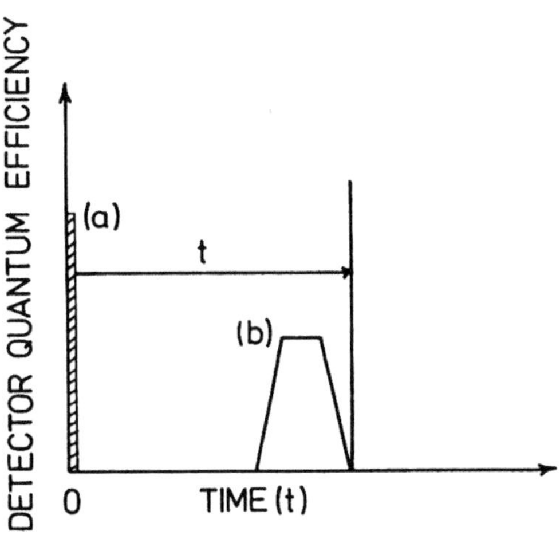

Fig. 3. Definition of gate position "t".
(a) Raman photons,
(b) detector gate

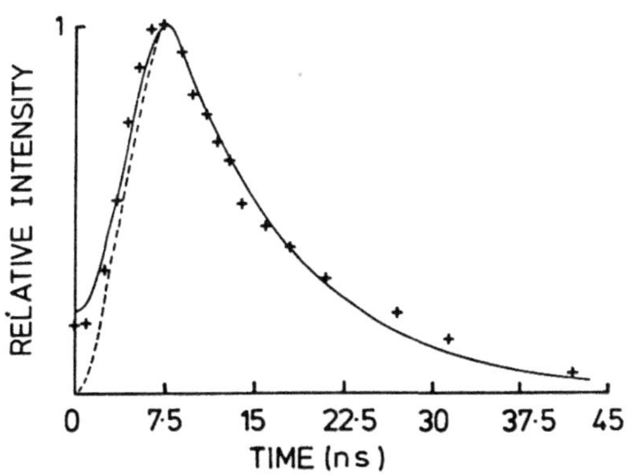

Fig. 4. Cross correlation of an exponential function ($K = 0.95\ ns^{-1}$) with a trapezoidal impulse response. The broken line is the result obtained when detector ringing is ignored, and the solid line when this effect is included. + measured values

Table 1. Improvements in S(R)/S(F) and S(R)/N ratios in gated (5 ns) compared with ungated operation of the diode array detector

t[ns]	IMPROVEMENT IN S(R)/S(F)			IMPROVEMENT IN S(R)/N		
	CALC		OBS	CALC		OBS
	single opening	multiple opening	multiple opening	single opening	multiple opening	multiple opening
1.0	18.4	2.9	5.0	1.0	0.4	0.5
2.5	8.5	4.8	5.3	1.0	0.7	0.7
3.5	5.3	3.9	4.3	0.9	0.7	0.7
5.5	3.0	2.6	2.7	0.6	0.6	0.6
7.5	1.3	1.3	1.3	0.3	0.3	0.3

important observation is, however, that despite the improved rejection the signal-to-noise ratio is not improved and can actually be degraded! Since clearly S(R)/N improvement is the most significant factor, this result appears disappointing! Our work does, however, show how much-improved performance can be achieved. We have shown that:

$$\frac{S(R)/N\ \text{gated}}{S(R)/N\ \text{ungated}} \propto \left[\frac{I_m}{I_T}\right]^{\frac{1}{2}} \qquad (1)$$

where I_m is the intensity of a cavity-dumped pulse and I_T is the total intensity between the start of two adjacent cavity-dumped pulses. In theory $I_m = I_T$, however, in practice there is leakage out of the cavity of the ca 82MHz component in the cavity. At 4 kHz typically $I_m/I_T=0.005$ and so while only 1/200 of the laser power is being used in the gated measurement, it is all being used in the ungated case! If I_m/I_T could be made equal to unity then S(R)/N would improve by a factor of $\sim$14. We are working on this with the manufacturers. Another alternative is simply to increase the repetition rate of the DAD system, which would also increase I_m/I_T and this too we are working on. An additional improvement (factor of 3.5) would be obtained by reducing the rise and fall times of the gate.

5. PHOTOMULTIPLIER TUBE DETECTOR

The laser system was the same as that described above, and the method of gating is similar to that described previously [1,3]. For the PMT the instrument response is well described by an isosceles triangle of 1.75 ns FWHM. An example of the data is shown in Fig. 5. We conclude that: for maximum S(R)/N improvement (a) the gate should always be set so as to accept the first photons to reach the detector. This differs from the procedure used in the literature [1,3] of setting the gate to accept the FWHH of the instrument response function. (b) improvements in S(R)/N of ca 3 and in S(R)/S(F) of ca 14 can be obtained (1×10^{-3} M solution of rubrene in benzene, τ_f = 12.5 ns). (c) the optimum gate-width is not simply that which will accept 67% of the Raman photons, as has been suggested [2,4], but is in fact a function of the width of the instrument response and the fluorescence lifetime. Thus if the FWHM = 1.0 ns the optimum Raman fraction collected is $\gtrsim$70% if τ_f < 5 ns and $\lesssim$67% if τ_f > 5 ns. The fraction required increases monotonically with decreasing lifetime at a fixed width of the instrument response function. (d) when the noise did not follow Poisson statistics, the observed S(R)/N improvement was at least a factor of two greater than calculated.

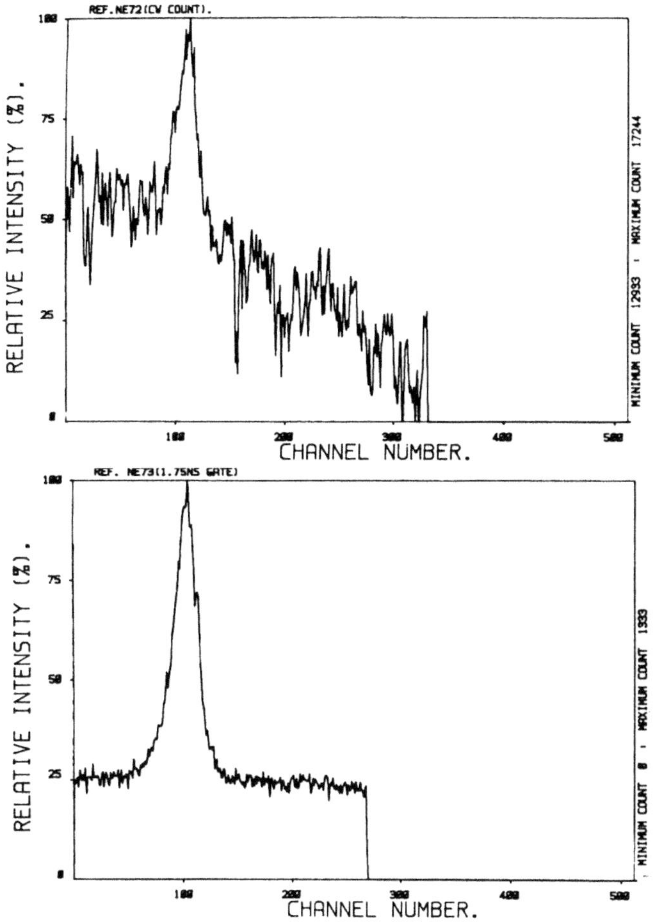

Fig. 5. Raman spectra of the 992 cm^{-1} band of benzene from a 4 x 10^{-4}M solution of rubrene in benzene (τ=15.9 ns) (a) ungated (b) 1.75 ns gate

6. CONCLUSION

Successful discrimination was achieved with both detector systems, however further improvements in the performance of components is required if the DAD system is to prove useful. Our calculations indicate a limiting value of S(R)/N of ca 36 corresponding to a gate width of ca 24 ps using our laser system and a PMT.

7. REFERENCES

1. R.P. Duyne, D.L. Jeanmaire and D.F. Shriver, Anal. Chem. 46(2), 213 (1974).
2. J.M. Harris, R.W. Chrisman, F.E. Lytle and R.S. Tobias, Anal. Chem. 48(13), 1937 (1976).
3. S. Burgess and I.W. Shepherd, J. Phys. E. Sci. Inst. 10, 617 (1977).
4. T.L. Gustafson and F.E. Lytle, Anal. Chem. 4, 634 (1982).

Our thanks go to the SERC (UK), Ministry of Defence (extra mural research contract D/ER1A/9/4/2051/026) and British Petroleum Ltd. for supporting this work.

Measurement of the Third-Order Electronic Susceptibility of GaP by Picosecond CARS Spectroscopy

J. Kuhl

MPI für Festkörperforschung, D-7000 Stuttgart 80, F.R.G.

B.K. Rhee and W.E. Bron

Department of Physics, Indiana University, Bloomington, IN 47405, USA

In principle, investigations of different nonlinear effects such as third harmonic generation, frequency mixing, beam self-focusing and intensity-induced transmission, reflectivity or polarization changes permit the determination of the nonlinear tensor components $\chi^{(3)}_{ijkl}$ defined by

$$P_i^{(3)}(\omega_4) = \chi^{(3)}_{ijkl}(-\omega_4,\omega_1,\omega_2,\omega_3)\cdot E_j(\omega_1)\cdot E_k(\omega_2)\cdot E_l(\omega_3) \quad (1)$$

Here $P^{(3)}$ is the 3rd order nonlinear polarization at ω_4 due to the nonlinear interaction of the fields $E(\omega_m)$ $m = 1,2,3$. Whereas in general, electronic nonlinearities as well as nuclear rearrangements give rise to $P^{(3)}$ the lattice contributions are only observable at resonance. The interference between both effects at a Raman resonance can be used to measure $\chi^{(3)}$ in terms of known Raman cross-sections in a coherent anti-Stokes Raman scattering (CARS) experiment [1]. So far these studies have been performed in the frequency domain, although it is evident that each technique has its corresponding alternative approach in the time domain [2]. The optical modes of the lattice vibration are characterized by relaxation times of the order 1-20 ps whereas the nonresonant electronic processes seem to be temporally instantaneous (response time $< 10^{-13}$s). Thus the present technology for the generation of ultrashort optical pulses allows to discriminate the resonant and nonresonant terms owing to their different time-dependence. Recently we reported on the first application of CARS spectroscopy with picosecond optical pulses to determine all the relevant components of $\chi^{(3)}$ for a crystal [3].

In a CARS experiment 3 input beams with frequencies ω_1, $\omega_s = \omega_1 - \omega_o$ (ω_o = frequency of the lattice mode), $\omega_p = \omega_1$ and wave vectors $\vec{k}_1$, $\vec{k}_s$ and $\vec{k}_p$ generate a fourth wave at $\omega_c = 2\omega_1 - \omega_s$. First two photons $\hbar\omega_1$ and $\hbar\omega_s$ are mixed to create a phonon $\hbar\omega_o$. In a second step this phonon absorbes a photon and a scattered photon $\hbar\omega_c$ is excited. Effective excitation of the CARS signal necessitates perfect phase-matching between the interacting photons $\Delta\vec{k} = \vec{k}_c - (\vec{k}_1 - \vec{k}_s + \vec{k}_p) = 0$. If the experiment is performed with ps-light pulses for excitation and probing the decay of the coherent vibrational amplitude can be observed by recording the CARS-signal as a function of the delay between the probe pulse and the two excitation pulses. In addition to the Raman term, which is purely imaginary at resonance and the nonresonant electronic term due to virtual excitation of higher electronic levels $P^{(3)}$ may contain resonant electronic contributions resulting from two-photon absorption. The latter process yields a complex contribution to $\chi^{(3)}$ whereas the part due to the nonresonant electronic polarization is real.

The time-integrated signal at ω_c measured in the ps pulse experiment at a given delay Δt is proportional to the time-integral of $[P^{(3)}]^2$ induced in the sample.

$$I(\Delta t) \sim S(\Delta k) \int_{-\infty}^{+\infty} dt \left| \vec{E}_p(t-\Delta t) \left[\cdot N \cdot R_A \cdot Q(t) + 3\, \chi^{(3)} \cdot \vec{E}_1(t) \cdot \vec{E}_s^{*}(t) \right] \right|^2 \quad (2)$$

with

$$S(\Delta k) = \sin^2(\Delta k \cdot L/2)/(\Delta k \cdot L/2)^2 \quad (3)$$

Here R_A is the Raman tensor, N the number of primitive cells per cm^3 and L the effective interaction length. The time evolution of the coherent amplitude Q of the lattice vibration is described by the differential equation

$$\ddot{Q} + \Gamma \cdot \dot{Q} + \omega_0^2 \cdot Q = \mu^{-1} \cdot R_A \cdot \vec{E}_1(t) \cdot \vec{E}_s^{*}(t) \quad (4)$$

where $\Gamma^{-1} = T_2$ is the phonon dephasing time and μ the reduced lattice mass.

In the experiment the optical fields $E_1(\omega_1)$, $E_s(\omega_s)$ and $E_p(\omega_p = \omega_1)$ are generated by two synchronously mode-locked Rhodamine 6G dye lasers pumped by the same acousto-optically mode-locked Ar^+ laser. These lasers provide two synchronized pulse trains with a pulse duration of 2.7 ps and an average power of about 40 mW at 76.5 MHz repetition rate. The residual small jitter of 1-2ps between the two pulse trains derived from cross correlation measurements sets a lower limit of about 1.3 ps for the time resolution of this CARS spectrometer. Further details of the laser system and the experimental set-up have been described in ref.[4].

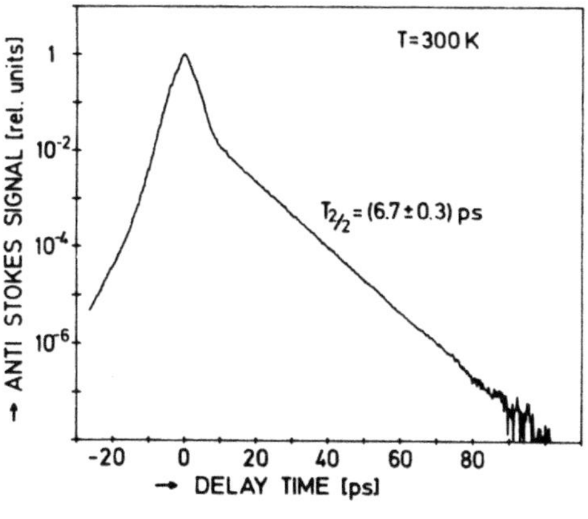

Fig.1: Time-resolved CARS signal at the LO phonon resonance of GaP

Figure 1 shows a semilog plot of the time-resolved CARS signal at the LO phonon resonance of GaP. The signal-to-noise ratio is representative for a measurement time of 20 minutes for the 500 data points. The rapidly changing signal near $\Delta t = 0$ is primarily due to the nonresonant electronic term of $\chi^{(3)}$. The full shape of this signal is observed if $\omega_1 - \omega_s$ is detuned against ω_0 by a few cm^{-1} (see Fig. 2 in ref.[4]).

The exponentially decaying signal beyond 20 ps is a measure of the temporal evolution of the coherent amplitude of the excited Raman mode. The slope of this curve yields a dephasing time of $T_2/2 = 6.7 \pm 0.3$ ps. The relative contribution of the electronic and Raman process to $I(\Delta t)$ can be specified by a quantity Y defined as the ratio of the peak intensity at $\Delta t = 0$ to the intensity of the phonon component extrapolated back from the exponential decay to $\Delta t = 0$. If the laser pulse profile is known, a detailed analysis of the shape of $I(\Delta t)$ permits the decurate determination of $\chi^{(3)}$. Auto- and cross-correlation functions, as well as the purely electronic four-wave mixing signal recorded for $\omega_l - \omega_s \neq \omega_o$ showed that our optical pulses could be well described by the slightly asymmetric biexponential function

$$f(t) = \begin{cases} \exp(\delta\gamma t) & \text{for } t < 0 \\ \exp(-\gamma t) & \text{for } t > 0 \end{cases} \qquad (5)$$

with $\delta = 1.3$.

The experimentally found broadening of the cross correlation due to the jitter has been considered by convoluting the profile of the s-laser with a gaussian disribution (characterized by a parameter z) of its center position relative to the center position of the l-pulse. A slight mismatch of the temporal overlap between the two pulse trains has been regarded by a parameter τ describing the temporal shift between the two maxima. At this point $Q(t)$ and $I(\Delta t)$ are specified in terms of the parameters $\gamma, z, \delta, \tau, T_2, Y$ and the Raman tensor R_A which can be taken from the literature [5]. The optimum values of the parameters and thus $\chi^{(3)}$ are determined for each measured trace of $I(\Delta t)$ by a least square fit. The analysis has to take into account that R_A and $\chi^{(3)}$ are effective quantities, and that their magnitudes vary with the polarization of the fields E_l, E_s and E_p relative to the crystal axes. We have used the 7 different configurations shown in Fig. 2 to measure the real and imaginary part of the 3 active independent components of $\chi^{(3)}$.

EFFECTIVE $\chi^{(3)}$ FOR DIFFERENT CRYSTAL ORIENTATIONS AND POLARIZATIONS OF THE LASER FIELDS

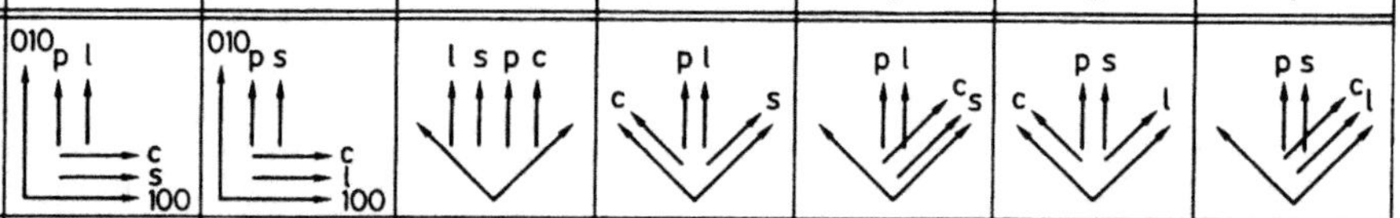

CONFIG	1	2	3	4	5	6	7
$\chi^{(3)}_{eff}$	χ_{1221}	χ_{1122}	$\frac{1}{2}(\chi_{1111} + 2\chi_{1122} + \chi_{1221})$	χ_{1122}	$\frac{1}{2}(\chi_{1111} + \chi_{1221})$	$\frac{1}{2}(\chi_{1221} + \chi_{1122})$	$\frac{1}{2}(\chi_{1111} + \chi_{1122})$

Fig.2

Three typical recordings for configurations 1, 2 and 3 of Fig. 2 are depicted in Fig. 3. The analysis of more than 32 different runs yields the following values for the real and imaginary parts of $\chi^{(3)}$ (in units of 10^{-10} esu).

$\chi'_{1221} = 2.1 \pm 0.15$ $\qquad$ $\chi''_{1221} = -0.48 \pm 0.24$

$\chi'_{1122} = 1.8 \pm 0.44$ $\qquad$ $\chi''_{1122} = -0.7 \pm 0.48$

$\chi'_{1111} = 2.1 \pm 0.7$ $\qquad$ $\chi''_{1111} = -0.73 \pm 0.81$

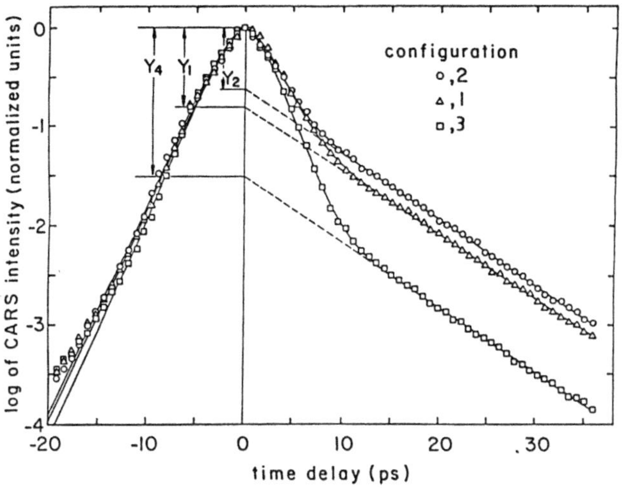

Fig. 3: Experimental data and fitted I(Δt) (solid lines) for three polarization configurations defined in Fig.2.

These results definitely show that in our excitation regime $\chi^{(3)}$ of GaP is complex owing to the nonneglectible contributions from two-photon absorption (In our experiment $2\hbar\omega_1 = 4.30$ eV is larger than the band gap of GaP ($E_g = 2.26$eV)). This is supported by the finding that we have no Kleinman symmetry ($\chi'_{1122} \neq \chi'_{1221}$). The relatively large uncertainty of the imaginary part χ'' is explained by the fact that χ'' for values $< 10^{-10}$ esu only slightly influences the curvature of I(Δt) in the region $0 < \Delta t < 10$ ps. Unfortunately it is impossible to compare our results with values obtained by different techniques or theoretical calculations, since no $\chi^{(3)}$ data for GaP have been published so far.

In conclusion we have analyzed the time-resolved CARS spectroscopy with ps light pulses as a powerful tool to measure 3rd order nonlinear electronic susceptibilities. Its potentials have been demonstrated by the first determination of all the relevant complex tensor components of $\chi^{(3)}$ for GaP. Although this method is strongly related to the traditional measurements in the frequency domain the time-domain technique has distinct advantages for certain conditions.

Whereas the frequency domain technique can be applied if the Raman term is comparable to or stronger than the electronic contribution, the time-domain approach is suitable for the opposite situation. The determination of $\chi^{(3)}$ from spectral domain experiments is facilitated if the phonon dephasing time is short, so that the Raman line width can be easily resolved. On the other hand, accurate measurements in the time-regime necessitate that $T_2/2$ is larger than the temporal resolution of the spectrometer. In addition, it can be shown that in the time-domain experiment contributions from two-step two-wave mixing [$(\chi^{(2)})^2$] to the third-order nonlinear polarization can be neglected in a first approximation [3].

References

1. M.D. Levenson, N. Bloembergen, Phys. Rev. B.10, 4447 (1974); C. Flytzanis in "Quantum Electronics" ed. by H. Rubin and C.L. Tang, (Academic Press New York, 1978) Vol. I

2. W. Zinth, A. Laubereau, W. Kaiser, Opt. Commun. 26, 457 (1978); J. Etchepare, G. Grillon, I. Thomazeau, A. Migus, A. Antonetti, Journ. Opt. Soc. Am. B 2, 649 (1984)

3. B.K. Rhee, W. E. Bron, J. Kuhl, Phys. Rev. B 30, 7358 (1984)

4. J. Kuhl, D. von der Linde in "Picosecond Phenomena III", ed. by K. B. Eisenthal, R. M. Hochstrasser, W. Kaiser and A. Laubereau (Springer-Verlag Berlin, 1982)

5. J.M. Calleja, H. Vogt, M. Cardona, Philos. Mag. A 45, 239 (1982)

Polarization Effects of Time-Resolved Coherent Raman Scattering in Liquids

N. Kohles and A. Laubereau

Physikalisches Institut, Universität Bayreuth, D-8580 Bayreuth, F.R.G.

Ultrafast excitation and subsequent probing of molecular vibrations in liquids and gases and of lattice vibrations in solids have received increasing interest in recent years /1,2/. Using coherent Raman scattering of delayed probing pulses after stimulated Raman excitation, the dephasing of molecular vibrations or of lattice modes was directly studied. Most of the previous work has been performed under special polarization conditions with parallel polarization for the excitation and the probing process. The present paper is devoted to more general studies in the liquid case. Theoretical and experimental data will be presented, demonstrating that the observed signal transients and the obtained information are vastly different, depending on the specific polarization conditions.

The theoretical treatment of coherent Raman probing is well known for parallel polarization of the exciting laser and Stokes pulses (field amplitudes $E_L || E_S$) and of the anti-Stokes scattering of the delayed probing pulses ($E_A || E_p$). In experimental studies the angle between the polarization planes of the excitation and probing process has been often chosen to be 90° for practical reasons. This polarization geometry is depicted in Fig. 1 as case 1a. Fig. 1 also includes the situation with all four field vectors parallel (case 1b). These cases will be called "parallel polarization". Physically different is the situation with perpendicular polarization planes of the two excitation pulses, $E_L \perp E_S$, and correspondingly also of the two probe field components, $E_p \perp E_A$. Such conditions will be termed in the following "crossed polarization" (cases 2a and 2b in Fig. 1). For general pump and probe experiments finally the question arises if parallel pump polarization may be combined with perpendicular probe polarization and vice versa ("mixed polarization", see Fig. 1). For these situations our theory predicts a vanishing scattering signal, i.e. the probe scattering does not couple to the excitation process.

We have performed a detailed theoretical study for time-resolved CARS under the polarization conditions of Fig. 1, taking into account the isotropic and anisotropic parts of the Raman polarizability tensor with coefficients, a and γ, respectively /3/. The fast orientational motion is treated classically /2/. Our calculations show that three factors contribute to the coherent probe scattering signal:

(i) coherent scattering off the isotropic part of the (resonant) material excitation via the isotropic scattering coefficient a; this contribution is of purely vibrational origin and relaxes with the vibrational dephasing time T_2;
(ii) coherent scattering via the anisotropic coefficient γ from a second component of the (resonant) material excitation, which has rotation-vibration character and decays with the anisotropic time-constant τ_{an};

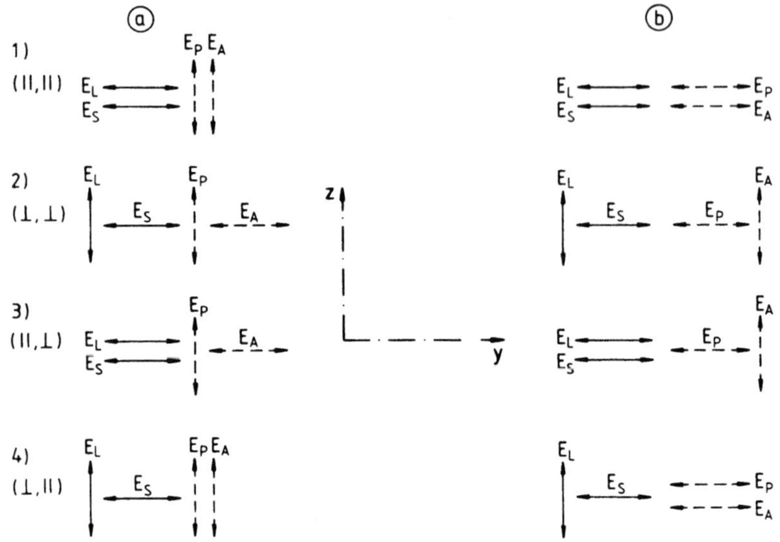

Fig. 1: Polarization geometries of coherent Raman scattering; E_L,E_S: excitation pulses; E_P, E_A: incident and scattered probe field

(iii) four wave mixing via the nonresonant part χ_{NR} of the third order susceptibility; this interaction occurs for temporal overlap between excitation and probing pulses and represents a nonresonant (electronic) component of the material excitation with almost instantaneous response to the incident radiation.

For parallel polarization the first contribution dominates in general, while for crossed polarization interaction (i) is absent and mechanisms (ii) and (iii) have to be considered. Correspondingly a notably different time-dependence is expected for the observed scattering signal.

Our theoretical results have been verified by an experimental investigation. Our experimental system is described as follows. Single bandwidth-limited pulses of 5 ps are generated by a Nd:glass laser system and a subsequent KDP frequency doubler. Passing a first beam splitter, a small fraction of the green pulse is removed and serves as probing pulse with well-defined time delay. By the help of a second beam splitter, 50 % of the remainder is focussed into a Raman generator and produces a Stokes shifted pulse by stimulated Raman scattering. CCl_4 and benzene are used as generator substances. The Stokes pulse and the rest of the green pulse are simultaneously directed into the sample cell (1-5 cm length) and excite the vibrational transition of interest by transient stimulated Raman amplification with small scattering efficiency. The coherent anti-Stokes scattering of the probe pulse is detected in phase-matching direction by a photomultiplier and a small grating spectrometer. Using various polarizers and high quality optical components, parallel (case 1a, Fig. 1) and crossed polarization (case 2a, Fig. 1) can be adjusted. For part of our investigations more general polarization conditions are established by means of a $\lambda/4$-plate inserted in front of the sample cell.

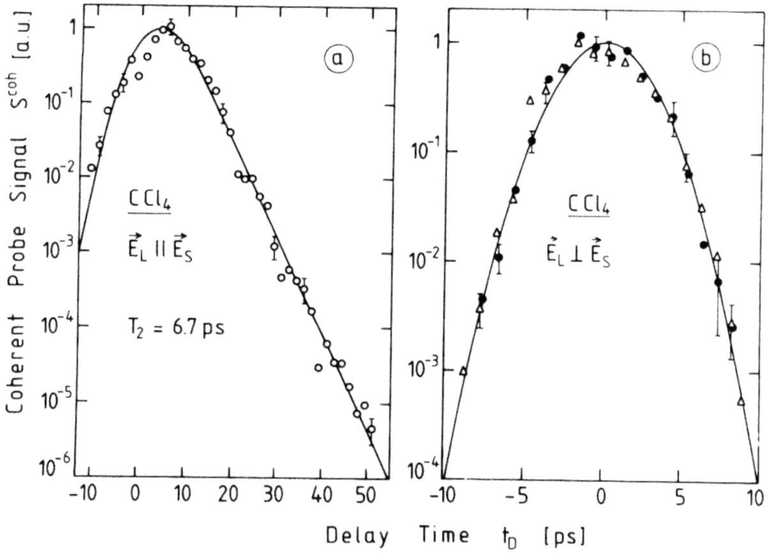

Fig. 2: Coherent anti-Stokes scattering of CCl_4 versus delay time:
a) parallel polarization;
b) crossed polarization; open triangles refer to a C_6H_{12} sample, where the excitation process is far off resonance. Theoretical curves; the different time-dependence of Figs. a and b should be noted.

An experimental example is presented in Fig. 2. The totally symmetric tetrahedron vibration ν_1 of neat CCl_4 at 459 cm^{-1} is investigated. The data for parallel pump and probe polarization are depicted in Fig. 2a. The rapid increase and subsequent exponential decay of the scattering signal $S_{\parallel}(t_D)$ over nearly six orders of magnitude represent the time evolution of the isotropic material excitation. The vibrational dephasing time $T_2 = 6.7 \pm 0.6$ ps is directly obtained from the signal transient, in good agreement with earlier results.

The solid line is calculated from the theory summarized above. The isotopic multiplicity of the vibrational transition has been omitted in the computation for the sake of simplicity. Because of the small depolarization factor corresponding to $\gamma^2/a^2 \approx 0.035$ the effect of the orientational motion on the material excitation (mechanism (ii)) is negligible. The nonresonant contribution via χ_{NR} is also small, so that the data of Fig. 2a represent purely vibrational dephasing, as noted previously /2/.

Of particular interest are the scattering data of Fig. 2b for crossed polarization. We note a rapid increase and also decay of the scattering signal $S_{\perp}(t_D)$ (full points). Because of the smallness of γ and the short rotational time-constant the anisotropic resonant material excitation is negligible, and the contribution via the nonresonant susceptibility χ_{NR} dominates. For the Gaussian shape of our light pulses, the calculated signal transient (solid curve) is also Gaussian, in good agreement with the experimental results (full points). It is interesting to note that we have measured the same time-dependence replacing CCl_4 in the sample cell by C_6H_{12}, where the excitation process is far off resonance /4/ (open triangles in Fig. 2b).

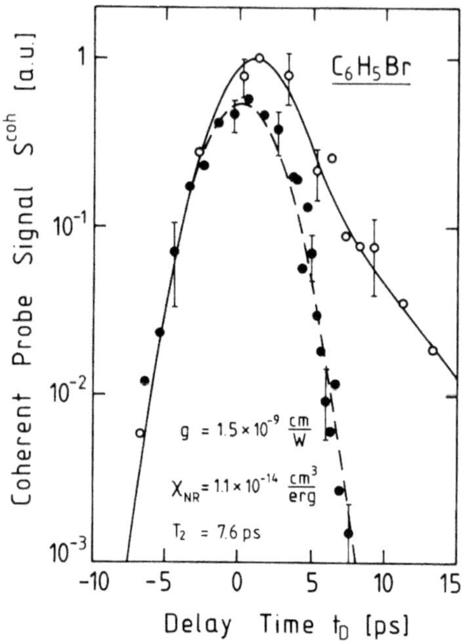

Fig. 3:

Coherent anti-Stokes scattering of C_6H_5Br versus delay-time for crossed polarization (full points) and more general polarization conditions (open circles). Theoretical curves

A comparison of the data of Figs. a and b for $S_\parallel$ and $S_\perp$, respectively, gives information on the relative magnitude of the nonresonant and resonant contributions to the scattering process. The measured ratio $S^{max}_\perp / S^{max}_\parallel \approx 10^{-3}$ agrees with theoretical estimates using reported values of the stimulated gain factor /5/ and the nonresonant third-order susceptibility /6/.

More accurate data on the nonresonant and resonant excitations may be obtained when the ratio of the two signal transients $S_\perp(t_D)$ and $S_\parallel(t_D)$ is measured with minor changes of the experimental system. An example is shown in Fig. 3. The symmetric ring breathing mode of bromobenzene at 1000 cm^{-1} is investigated. As generator substance we use benzene (992 cm^{-1}). The observed scattering $S_\perp(t_D)$ for crossed polarization is represented by full points. We note a rapid rise and decay similar to Fig. 2b in good agreement with the calculated curve (broken line). Also for the small depolarization factor of the ring mode of this molecule (small γ), the nonresonant contribution via χ_{NR} dominates around t_D=0. Now, a low-order $\lambda/4$ quartz plate is placed in front of the sample cell, keeping the other experimental parameters constant. The specific plate generates circular polarization for the pump and probe fields, and a well-defined elliptical polarization for the Stokes pulse. The experimental situation combines parallel and crossed polarization conditions (cases 1b and 2a of Fig. 1). The measured scattering signal is indicated by the open circles in Fig. 3. We note a first rapid increase and decay of the signal curve, which receives large contributions from the nonresonant interaction around t_D=0. For larger values of t_D, the resonant isotropic excitation survives, which decreases according to the vibrational dephasing time T_2.

The solid line in the Fig. is calculated and represents a superposition of $S_\perp$ and $S_\parallel$ plus additional interference terms. From the data of Fig. 3 and the steady gain factor $g = 1.5 \times 10^{-9}$ cm/W of bromobenzene we determine $\chi_{NR\,yzzy} = 1.1 \times 10^{-14}$ cm³/erg in satisfactory agreement with published data /6,7/.

In conclusion, it is pointed out that we have investigated coherent Raman scattering in liquids under generalized polarization conditions. Basic features of our theoretical treatment have been verified experimentally. The observed signal transients depend drastically on the chosen polarization geometry, because of the varying role of different components of the involved material excitation.

References:

1 See for example "Ultrafast Phenomena IV", editors D.H. Auston and K.B. Eisenthal, Springer Verlag (Berlin 1984)
2 A. Laubereau and W. Kaiser, Rev. Mod. Phys. 50, 3607 (1978)
3 N. Kohles and A. Laubereau, to be published
4 W. Zinth, H.J. Polland, A. Laubereau, and W. Kaiser, Appl. Phys. B26, 77 (1981)
5 H. Görner, M. Maier, and W. Kaiser, J. Raman Spectr. 2, 363 (1974)
6 M.D. Levenson and N. Bloembergen, J. Chem. Phys. 60, 1323 (1974)
7 S. Saikan and G. Marowsky, Opt. Commun. 26, 466 (1978)

Part II

Vibrational Dynamics of Molecules

Ultrafast Intramolecular Vibrational Redistribution and Intermolecular Energy Dissipation of Polyatomic Molecules in Solution

W. Kaiser and A. Seilmeier

Physik Department E 11, Technische Universität München, D-8000 München, F.R.G.

In the liquid phase, interaction processes show very short time constants. For instance, the intramolecular redistribution of vibrational energy in the electronic ground state of large molecules is found to be smaller than 2 ps and the transfer of energy from a hot molecule to the solvent is measured to be of the order of tenths of picoseconds. In addition, there is strong evidence of a very rapid reverse process, where vibrationally excited solvent molecules give their energy to the solute molecules which act as molecular energy sensors/1/.

In our investigations, energy is supplied to the molecules i) by infrared photons of approximately 3000 cm^{-1} exciting NH - or CH - stretching modes and ii) by visible photons of 19,000 cm^{-1} making transitions to the first electronic state with subsequent rapid internal conversion to the vibrational manifold of the electronic ground state. The momentary state of the molecules i.e. the redistribution of vibrational energy is monitored measuring changes of absorption (or fluorescence) at the long wavelength tail of the electronic absorption of the probe molecule. In most molecules the edge of absorption represents optical transitions from the thermally populated vibrational manifold to the first excited electronic S_1 state. Transient changes of the vibrational population give rise to a variation of the edge absorption (the excitation spectrum of the fluorescence). In fact, the shape of the absorption edge measured several picoseconds after excitation gives valuable information on the distribution of excitation energy in the molecule. Our data indicate that the supplied energy is rapidly distributed over the vibrational manifold, as if the molecule has acquired an enhanced internal temperature. This transient internal temperature can be compared with the value calculated under the assumption that all vibrational modes participate in the redistribution of the excitation energy. For a number of molecules investigated so far there is good agreement between the experimentally determined and theoretically estimated internal temperature.

In Fig. 1, the room-temperature absorption edge of anthracene (1) is compared with the transient absorption (2) taken 7 ps after excitation with infrared photons of 3050 cm^{-1} /2/. The small shoulder at the 300K curve is due to three vibrational hot bands around 1400 cm^{-1}. After IR excitation of the CH - modes at 3050 cm^{-1}, the energy is rapidly redistributed and the hot band absorption at 1400 cm^{-1} increases drastically. An increase of internal temperature of 170K is estimated from the experimental data. The same value is calculated when the energy of 3050 cm^{-1} is distributed over all vibrational modes of anthracene. In Fig.2, the build up and decay of the excess fluorescence is presented for a probing frequency of 24,760 cm^{-1} (Note the 00-transition of anthracene is at 26,380 cm^{-1}). The figure clearly shows the fast

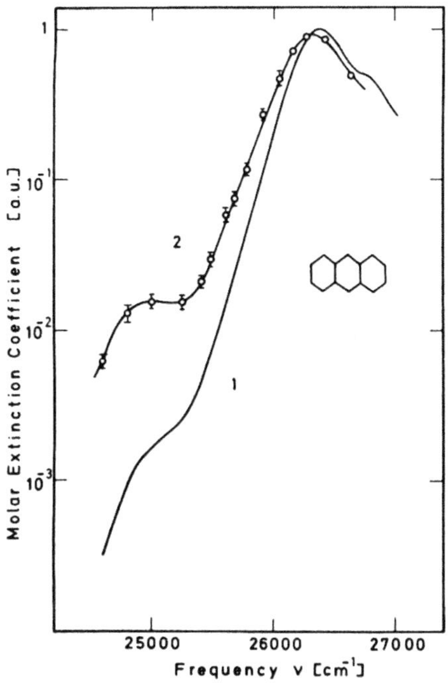

Fig.1
Standard absorption edge of anthracene at room temperature (1). Transient absorption taken 7 ps after IR excitation by photons of 3050 cm^{-1} (2). Note the enhancement of the hot bands of frequencies of 1400 cm^{-1}.

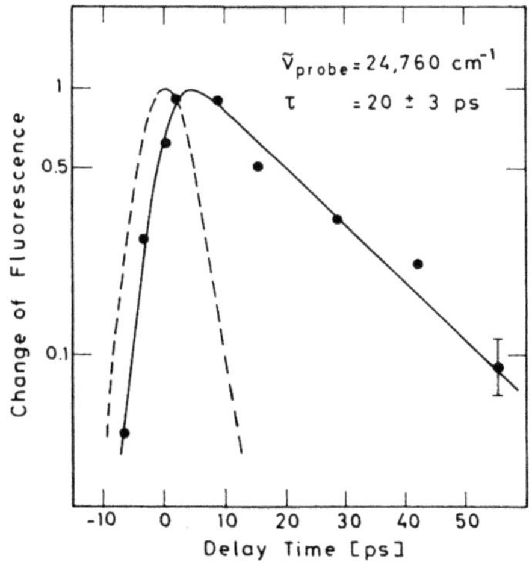

Fig.2
Temporal evolution of the excess fluorescence of anthracene after IR excitation. The signal rises rapidly during the excitation and decays by intermolecular interaction. The broken curve gives the correlation curve of excitation and probe pulses.

intramolecular redistribution of population within the time resolution of the system (2 ps),and a relatively slower decay with 20ps which corresponds to the intermolecular energy-transfer to the molecular surrounding.

Next, the large dye molecule Coumarin 7 is discussed (see insert in Fig.3)/3/. The curve on the r.h.s. of Fig.3 represents the long wavelength absorption tail of the molecule at room temperature. The exponential slope allows to estimate the molecular temperature of 300K. The points on the l.h.s. of the figure correspond to fluorescence data taken 8.5 ps after excitation of the Coumarin molecules via NH - stretching modes with infrared pulses of 3400 cm^{-1}. The line through the experimental points suggest a transient temperature of 400K. The same internal temperature is calculated for a redistribution of an energy of 3400 cm^{-1} over the many vibrational modes of the molecule.

According to the preceding arguments, very high temperatures are anticipated when medium size molecules with a limited number of vibrational degrees of freedom are excited with energies of ten thousands of cm^{-1}. This notion was confirmed by exciting azulene to the first excited electronic state with photons of 19,000 cm^{-1} and measuring the long wavelength absorption after 10 ps/4/. In

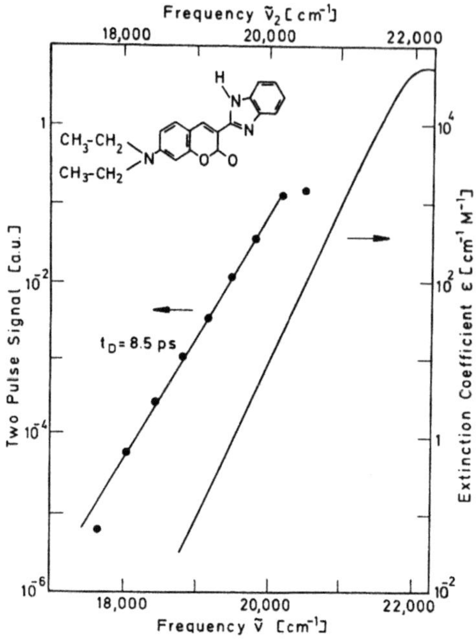

Fig.3
Absorption edge of Coumarin 7 at 300K (r.h.s.). The slope of fluorescence excitation data (l.h.s.) measured 8.5 ps after excitation by infrared pulses with 3400 cm^{-1} suggests a transient internal temperature of 400K as a result of intramolecular vibrational redistribution.

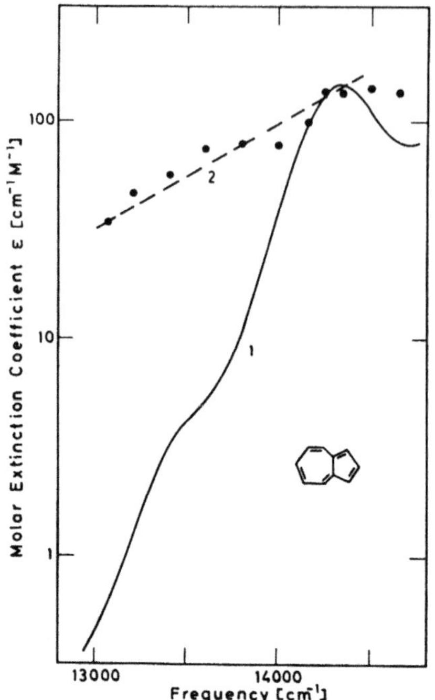

Fig.4
Absorption edge of azulene at 300K (1) and after excitation by visible pulses of 19000 cm^{-1} (2). The broken line corresponds to an internal temperature of 1200K.

Fig.4, the room-temperature absorption edge (1) is depicted together with the data of the excited molecules (2). The drastic change of absorption points to a large temperature increase of the azulene molecules. In fact, the broken line through the experimental points corresponds to a Boltzmann slope of 1200K. The same internal temperature results from an estimate of the specific heat of azulene and from the supplied energy of 19,000cm^{-1}. Of special interest is the fact that the dissipation of the vibrational excess energy to the surrounding solvent molecules proceeds with a time-constant of several tenths of picoseconds. Obviously, molecules of very high internal temperatures can exist in the liquid phase for times of several 10^{-11} s.

References

1 A. Seilmeier, P. O. J. Scherer, W. Kaiser: Chem. Phys. Lett. 105, 140 (1984)
2 N. H. Gottfried, A. Seilmeier, W. Kaiser: Chem. Phys. Lett. 111, 326 (1984)
3 F. Wondrazek, A. Seilmeier, W. Kaiser: Chem. Phys. Lett. 104, 121 (1984)
4 W. Wild, N. H. Gottfried, A. Seilmeier, W. Kaiser: to be published 1985

Time-Resolved Observation of Vibrationally Highly Excited Polyatomic Molecules

H. Hippler

Institut für Physikalische Chemie, Universität Göttingen, Tammannstraße 6, D-3400 Göttingen, F.R.G.

1. Introduction

Vibrationally highly excited molecular states play an important role in a vast class of chemical reactions. For this reason, a spectroscopic detection of these states is of great interest. Various spectroscopic techniques can be used for a time-resolved observation of vibrationally excited state populations. Fluorescence of multiphoton ionization is particularly sensitive. Whenever selected individual states can be separated, state-resolved absorption, fluorescence excitation, or stimulated emission are useful. There exist, however, situations where the densities of states is so large that one cannot realize a resolution of single states, or where neither fluorescence nor multiphoton ionization signals can be detected, Such situations frequently occur with vibrationally highly excited polyatomic molecules in the electronic ground state. In this case, "hot UV-absorption spectroscopy" as developed in our laboratory [1-4] can be used for timeresolved observation of excited state populations. This technique allowed us to study the dynamics of intramolecular processes, such as isomerization [5,6] and bond fission [7,8] under isolated molecule conditions, as well as of collisional intermolecular energy-transfer [9-13] .

In our experiments UV laser absorption produces molecules in an electronically excited state. After internal conversion, an almost microcanonical ensemble of vibrationally highly excited molecules in the ground electronic state is prepared. Under collision-free conditions, these molecules can now chemically react to products. Using "hot UV absorption spectroscopy" the temporal evolution of the population of these highly excited molecules can be directly observed. This allowed us to study specific rate-constants for unimolecular reactions. Specific rate-constants for unimolecular isomerization of cycloheptatrienes [5,6] and for bond fission of toluene [7] and other aromatics [8] have been investigated in our laboratory. In the presence of a heat bath of sufficient high pressure, collisional deactivation will compete with the unimolecular reaction and may even quench the reaction path completely. Also in this case "hot UV absorption spectroscopy" allows us to directly monitor the population of these highly excited molecules during collisional deactivation. We have performed in our laboratory investigations of collisional deactivation of the following vibrationally highly excited molecules: toluene [9], substituted cycloheptatrienes [10] , azulene [11] , and the two triatomic molecules CS_2 [12] and SO_2 [13] where the vibrationally excited molecules are prepared by efficient mixing of electronically excited states with the ground electronic state.

"Hot UV absorption spectroscopy" allows for a direct detection of populations of highly excited states. Furthermore, even the average internal vibrational energy of the excited molecules can be determined, more or less independent of the energy distributions. In addition, "hot UV absorption spectroscopy", apart from indicating the presence of excited states, can serve as a spectroscopic thermometer. The response of this thermometer is very fast, as demonstrated by Kaiser et al. [15]. In order to prove this statement, key experiments with vibrationally highly excited azulene will be reported.

2. Experimental and Results

Details of the experimental set-up are given elsewhere [4, 6, 9, 11, 12] and hence will only briefly be illustrated. The molecules were excited with radiation at 248 nm, 308 nm, 337 nm, and 351 nm from an excimer laser (Lambda Physik EMG 100) with the appropriate gas mixtures. Transient UV absorptions were measured directly using a c.w. Xe-Hg high-pressure arc lamp (Hanovia Type 901B), a quartz monochromator (Zeiss M4 QIII), and a photomultiplier (RCA 1P28). A fast transient recorder (Tektronix R7912) allowed to average up to 16000 shots if necessary.

Fig. 1 shows three spectra of azulene at 300 K, 800 K, and 1500 K, which we have measured in a shock tube. With increasing temperature the structure disappears, the value of the maximum decreases while the width increases. In addition to the spectra of the canonical ensembles from thermal experiments, the triangles mark the absorption coefficients of an almost microcanonical ensemble of azulene molecules produced via UV absorption from a nitrogen laser at 337 nm followed by internal conversion. The vibrational energy of about 30000 cm^{-1} would in a thermal system correspond to a temperature of about 1500 K. The similarity of the two spectra of the two different ensembles is obvious, and the reason for that will be illustrated in the following.

The energy distributions of a canonical and of an almost microcanonical ensemble of azulene are very different even when the average internal vibrational energies are identical. At 1500 K the energy distribution of the canonical ensemble is very broad. While an almost microcanonical ensemble with the same average internal energy, as obtained by UV absorption at 337 nm and 300 K followed by internal

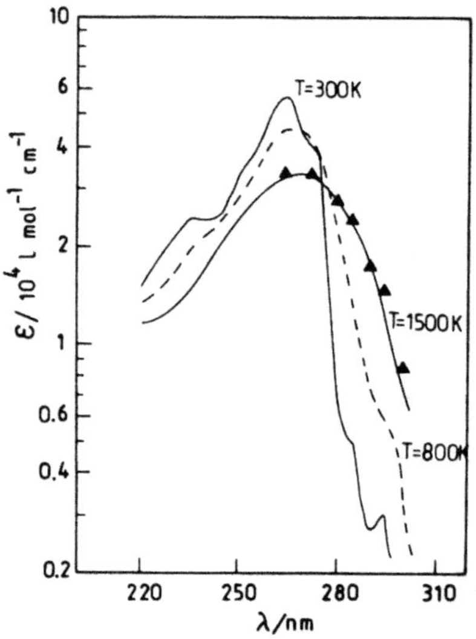

Fig. 1

Absorption coefficient of azulene (see text)

conversion , yields a much more narrow energy distribution. Its average energy is determined by the wavelength of the laser and the initial thermal energy at 300 K. The width of the distribution is essentially given by the width of the initial thermal distribution. Although the two distributions are very different, their UV absorption spectra are very similar. UV absorption cross-sections are essentially determined by the highest populated levels of those oscillators who contribute to the transition moment. Therefore, for similar absorption spectra of two different ensembles, the population of the lowest levels of the molecular oscillators should be similar.

In Fig. 2 the relative population of one oscillator in azulene with an average internal energy of 30600 cm^{-1} is shown. The upper curve (a) stands for a canonical ensemble and the lower curve (b) for a microcanonical ensemble. Details of the calculation are given in ref. [11]. At low excitation energies, where the population of the oscillator is still high, the two curves are identical. The reason for this can be visualized by the following argument: In canonical ensembles an oscillator feels the heat bath of the same oscillator in all the other molecules, while in a microcanonical ensemble all the other oscillators of the same molecule act as a heat bath. Since only highly populated levels can contribute to the absorption cross-section, the UV absorption spectrum of a microcanonical ensemble must be very similar to the UV absorption spectrum of a canonical ensemble, provided the average internal energies are identical. The functional form of energy distribution will only be of minor influence on the UV spectrum.

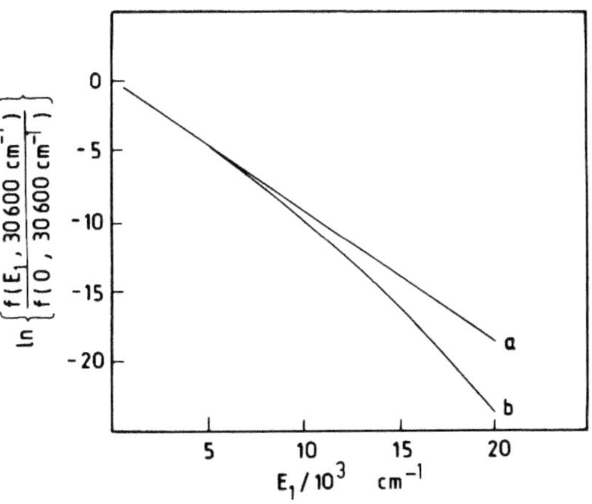

Fig. 2

Energy distribution $f(E_1, \langle E \rangle)$ in a single classical oscillator (energy E_1) of azulene at total average energy $\langle E \rangle = 30600\ cm^{-1}$ (see text)

The absorption cross-section depends on temperature as shown in Fig. 1. Statistical thermodynamics relates temperature with the average internal energy of the molecules. Therefore, the UV absorption cross-section becomes a detector for the internal energy of the excited molecules. At 290 nm, for instance, the cross-section increases with increasing internal energy, while it decreases at the maximum. In between, at about 275 nm the cross-section is insensitive to vibrational excitation of the molecules. For collisional deactivation experiments, one chooses a wavelength where exists a simple correlation between absorption cross-section and average internal energy.

Figure 3 shows the experimental calibration of the absorption cross-section af azulene at 290 nm. The full circles are experimental results from shock tube studies. The corresponding temperatures are indicated by the upper scale. The full triangles are results from photochemical

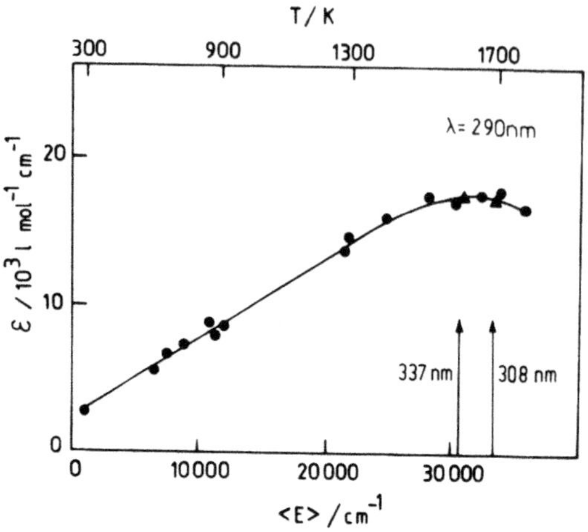

Fig. 3

Absorption coefficients of thermally and laser-excited azulene (see text)

excitation with a nitrogen laser and a xenon chloride laser. Over a wide energy-range there is a linear relation between absorption cross-section and excitation energy. This means for collisional deactivation experiments that absorption time profiles can be directly translated into energy time profiles via this calibration curve.

Figure 4 represents a typical experimental result. The absorption has been translated into average internal energies. 7 mtorr of azulene in the presence of 590 mtorr of Ar were exposed to a pulse of a nitrogen laser. The increase of the signal indicates the production of vibrationally excited azulene molecules due to absorption followed by internal conversion. The decay reflects the deactivation of azulene in collisions with argon. The energy-loss curve is nearly a linear function of time. This means that the average amount of energy transferred per collision does not depend on the excitation energy. Only at low energies the average amount of energy transferred per collision decreases with decreasing the degree of excitation.

More details about the average amount of energy transferred per collision are given in Fig. 5. It presents our results for collisional deactivation of vibrationally highly excited azulene molecules. The full lines are for the series of n-alkanes and the dashed lines are for n-perfluoroalkanes. All in common are the almost energy-independent average amounts of energy transferred per collision. Only below about 10000 cm^{-1} is there a substantial decrease with decreasing excitation energy. This energy-dependence of $\langle \Delta E \rangle$ values is very

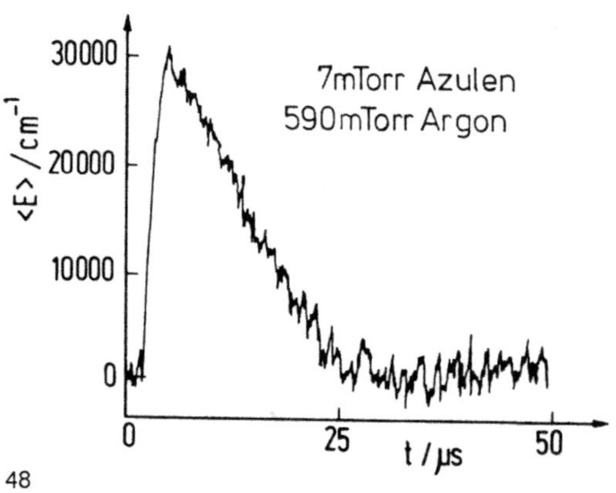

Fig. 4

Absorption signal of vibrationally highly excited azulene after laser excitation at 337 nm. Signal decay due to collisional deactivation (see text)

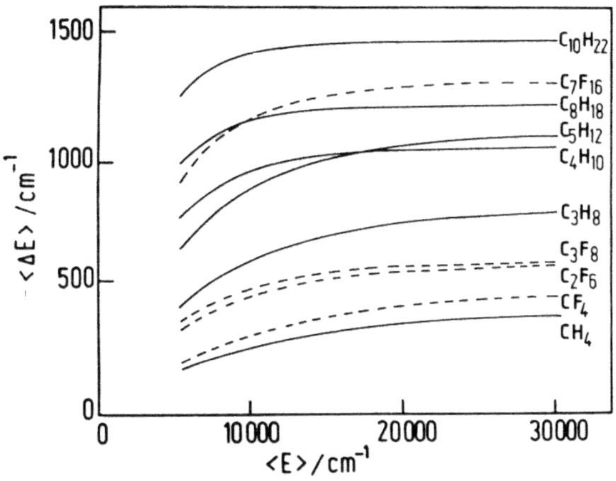

Fig. 5

Average energies transferred per collision $\langle \Delta E \rangle$ of excited azulene as a function of the energy $\langle E \rangle$ in various bath gases (see text)

similar to our results on collisional deactivation of vibrationally highly excited toluene [9] and cycloheptatrienes [10] . Even these energy-independent $\langle \Delta E \rangle$ values are almost identical for the same cooling collider, and hardly depend on the nature of the excited molecules.

This comparison is shown in Fig. 6. The energy-independent $\langle \Delta E \rangle$ values are plotted against the number of atoms in the cooling collider for the series of the n-alkanes. For the same collision partner but different excited molecules there are the same $\langle \Delta E \rangle$ values within the experimental error. A more complete comparison of our results with the literature is given elsewhere [1,11] . For much smaller highly excited molecules like CS_2 and SO_2, however, the situation is very different.

As an example, some of our results on collisional deactivation of CS_2 [12] are illustrated in Fig. 7. Average amounts of energy transferred per collision as a function of the internal energy of the CS_2 molecule are shown. A slope of 45 degrees, as in the case of Ar reflects a quadratic energy-dependence. In general the energy-dependences for small molecules are much stronger then for large molecules. However, at high excitation energies, e.g. 30000 cm^{-1} , the average amounts of energy transferred per collision is of the same order of magnitude as for the

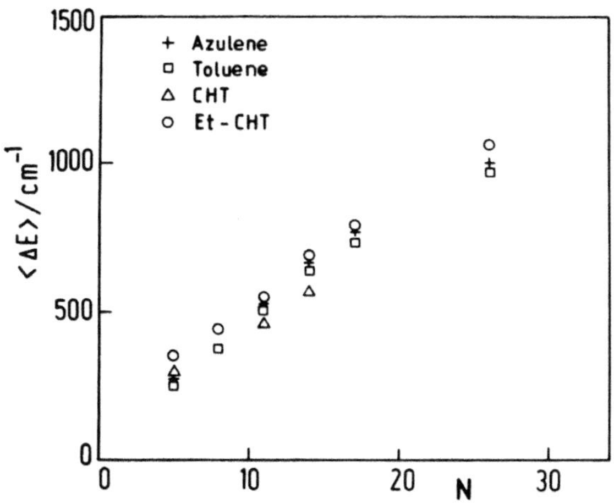

Fig. 6

Correlation of average energies transferred per collision $\langle \Delta E \rangle$ from different excited molecules with the number of atoms in the cooling colliders (n-alkanes). (see text)

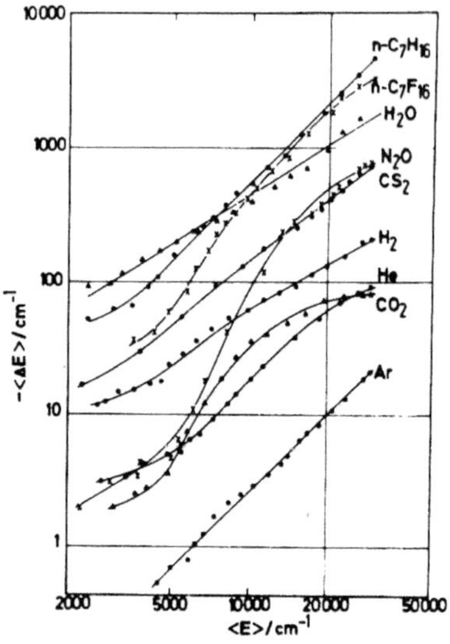

Fig. 7

Average energies transferred per collision $\langle \Delta E \rangle$ of excited CS_2 as a function of the energy $\langle E \rangle$ in various bath gases. (see text)

large molecules, and compares well with indirect results from thermal recombination of triatomic molecules [13,15].

We have directly measured collisional deactivation in the gas phase. From these results a prediction of collisional deactivation of vibrationally excited molecules in the liquid phase should be possible. However, the density in the liquid phase is much higher than in our gas phase experiments. Therefore the actual time for collisional deactivation in the liquid phase will be in the picosecond domain . With picosecond spectroscopy these phenomena become experimentally accessible. Experiments on the collisional deactivation of cis-stilbene with an initial vibrational energy of 33000 cm^{-1} in the gas phase and in the liquid phase are under way in our laboratory [16].

In Fig. 8 the predicted energy-loss profile of azulene in liquid CCl_4 is illustrated. The initial internal energy is 17500 cm^{-1} and the time-scale is well in the reach of the available picosecond techniques. The process can be observed either by "hot UV absorption spectroscopy" or by adding fast molecular thermometers, indicating the raise of temperature of the heat bath during collisional deactivation. The curve in Fig. 8 has been calculated by assuming that collisional energy-transfer in the liquid phase is similar to collisional energy-transfer in the gas phase. Therefore, the important quantity which enters the prediction is the binary collisional frequency of the highly excited molecules with the solvent molecules. In earlier experiments, we have shown that collisional deactivation of vibrationally highly excited I_2 [17] and Br_2 [18] scales with the inverse of the binary diffusion coefficient. This is true not only for highly compressed gases but also for compressed liquids up to 7 kbar. Also, Chatelet et al. [19] have shown that collisional deactivation of vibrationally excited hydrogen in highly compressed fluids scales with the inverse of the diffusion coefficient. Therefore, we determined our collisional frequency for deactivation in the liquid phase relative to the collisional frequency for deactivation in the gas phase by using the corresponding diffusion coefficients. The diffusion coefficients have been determined from the viscosity of CCl_4 and the Lennard-Jones parameters using the low-pressure extension of the Stokes-Einstein relation [20]. A comparison of the prediction with liquid phase experiments is of considerable interest. Discrepancies may have their origin in altered

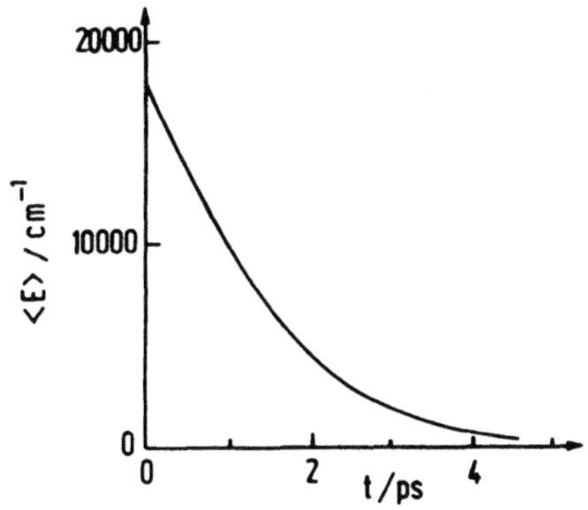

Fig. 8

Predicted energy-loss profile of vibrationally highly excited azulene in liquid CCl_4. (see text)

collision frequencies in the liquid phase solvation shell surrounding the excited molecules. The reason for that may be a local heating of the solvation shell because of too slow energy-transport from the solvation shell into the free liquid. Also, energy-transfer may be altered by the collective natur of liquid-phase collisions. Finally it should be noticed that in azulene-CCl_4 mixtures the energy-loss may be so fast that the internal conversion in azulene from $S_1 \longrightarrow S_o$ provides a bottle-neck.

3. Conclusion

To conclude, "hot UV absorption spectroscopy" is a powerful and simple method to determine the internal energy of vibrationally highly excited molecules. One can use this technique for direct collisional deactivation experiments. Under collision-free conditions, it is possible to directly observe intramolecular dynamics as isomerization or bond fission.

4. Acknowledgements

I want to thank all the people who contributed to the work. Especially Prof. Troe who was involved in all of it.

5. References

1. H. Hippler, Ber. Bunsenges. Physik. Chem. 89, 303, (1985)
2. D.C. Astholz, L. Brouwer, and J. Troe, Ber. Bunsenges. Physik. Chem. 85, 559, (1981); D.C. Astholz, A.E. Croce, and J. Troe, J. Phys. Chem. 86, 696 (1982)
3. H.Hippler, J. Troe, and H.J. Wendelken, J. Chem. Phys. 78, 5351, (1983);J.E. Dove, H. Hippler, H.J. Plach, and J. Troe. J. Chem. Phys. 81, 1209 (1984)
4. L. Brouwer, H. Hippler, L. Lindemann, and J. Troe, J. Phys. Chem. in press (1985)
5. H. Hippler, K. Luther, J. Troe, and R. Walsh, J. Chem. Phys. 68, 323, (1978); H. Hippler, K. Luther, and J. Troe, Faraday Discuss. Chem. Soc. 67, 173, (1979)
6. H. Hippler, K. Luther, J. Troe, and H.J. Wendelken, J. Chem. Phys. 79, 239, (1983)
7. H. Hippler, V. Schubert, J. Troe, and H.J. Wendelken, Chem. Phys. Lett. 84, 253, (1981)
8. H. Hippler, L. Lindemann, and J. Troe, to be published

9. H. Hippler, J. Troe, and H.J. Wendelken, J. Chem. Phys. 78, 6709, (1983); M. Heymann, H. Hippler, and J. Troe, J. Chem. Phys. 80, 1853, (1984)
10. H. Hippler, J. Troe, and H.J. Wendelken, J. Chem. Phys. 78, 6718 (1983)
11. H. Hippler, L. Lindemann, and J. Troe, J. Phys. Chem. in press (1985)
12. J.E. Dove, H. Hippler, and J. Troe, J. Chem. Phys. 82, 1907, (1985)
13. H. Hippler, D. Nahr, and J. Troe, to be published
14. W. Kaiser, Ber. Bunsenges. Physik. Chem. 89, 213, (1985); A. Seilmeier, P.O.J. Scherer, and W. Kaiser, Chem. Phys. Lett. 105, 140, (1984)
15. C. Cobos, H. Hippler, and J. Troe, J. Phys. Chem. 89, 1178 (1985)
16. J. Schroeder, J. Troe, and F. Voß, to be published
17. B. Otto, J. Schroeder, and J. Troe, J. Chem. Phys. 87, 2054 (1983)
18. H. Hippler, V. Schubert, and J. Troe, J. Chem. Phys. 81, 3931, (1985)
19. M. Chatelet, J. Kiefer, and B. Oksengorn, Chem. Phys. 79, 413, (1983)
20. H. Hippler, V. Schubert, and J. Troe, Ber. Bunsenges. Physik. Chem. in press (1985)

Time-Resolved Raman Spectroscopy of Highly Excited Vibrational States of Polyatomic Molecules

V.S. Letokhov and E.A. Ryabov

Institute of Spectroscopy, USSR Academy of Sciences,
SU-142092 Moscow Region, Troitzk, USSR

1. Introduction

The vibrational excitation of a polyatomic molecule can drastically change the character of intramolecular motion. As a result of mode interaction at certain level of excitation, the intramolecular motion changes qualitatively from regular to stochastic (see e.g. [1,2]).

The stochastic motion is consistent with the statistical distribution of vibrational energy over all the molecular modes. Therefore the stochastization effect can be studied by measuring the dynamics of energy-partitioning among different modes, one of them being under resonant excitation. Such measurements, as well as the studies of intermolecular energy distribution can be made by spontaneous Raman spectroscopy (RS) technique. Using this method it was shown in our earlier works [3,4] that at multiple-photon excitation (MPE) of SF_6 and CF_3I by CO_2 laser radiation the modes not coupled directly with the radiation were nevertheless excited during the laser pulse ($\sim 10^{-8}$ s) provided that the total vibrational energy of a molecule exceeds some specific value (of 3900 ± 500 cm^{-1} for SF_6 and of 6000 ± 500 cm^{-1} for CF_3I). It was assumed that these energy values correspond to the transition to stochastic motion in SF_6 and CF_3I. The experiments carried out in [5] with a time-resolution of $\sim 10^{-9}$ s confirmed the basic conclusions drawn for SF_6 in ref. [4].

The present talk treats the results of RS experiments on the energy distribution formed at MPE of CF_2HCl. This molecule differs greatly from SF_6 and CF_3I in its spectral properties. The energy being the same, it has less density of vibrational states. Besides, it has a high-frequency ν_1 vibration corresponding to C-H stretching mode. So, it was supposed that these differences can affect the type of inter- and intramolecular distribution in these molecules.

2. Method and Experimental Setup

There are simple relations between the spectrum - integrated RS signal and the average vibrational energy ε_i stored in the ν_i mode being probed

$$I_i^{AS} = A \cdot \bar{\varepsilon}_i / h\nu_i \qquad (1)$$

for the anti-Stokes component and

$$I_i^{S} = B \cdot (1 + \bar{\varepsilon}_i / h\nu_i) \qquad (2)$$

for the Stokes component. By calibrating A or B constants at known gas temperature one can measure then from (1) or (2) the absolute values of $\bar{\varepsilon}_i$ for an arbitrary energy-distribution function. Using short pump and probe pulses the RS signal can be measured on the time-resolved basis.

The experiments were performed using the setup similar to that described in ref. [4]. The molecules were excited by TEA CO_2 laser radiation with pulse width τ_{IR} = 25 ÷ 30 ns. The second harmonic of Nd:YAG laser with pulse duration τ_{vis} = 10 ns was used for RS probing. The delay between these two pulses τ_d could be continuously changed with accuracy $\Delta\tau \lesssim 10$ ns.

The spectrum - integrated signal was detected by a photon-counting system with a photomultiplier tube. A multichannel system was applied to detect the RS spectrum.

3. Results

The CF_2HCl molecule has C_s-symmetry and nine normal modes with their frequencies being [6,7]: ν_1 - 3023 cm^{-1}, ν_2 - 1311 cm^{-1}, ν_3 - 1100 cm^{-1}, ν_4 - 809 cm^{-1}, ν_5 - 595 cm^{-1}, ν_6 - 422 cm^{-1}, ν_7 - 1347 cm^{-1}, ν_8 - 1118 cm^{-1}, ν_9 - 365 cm^{-1}. Raman transitions are allowed for all these modes. The ν_3 and ν_8 modes, corresponding to symmetric and antisymmetric stretching of C-F bond [6], can be pumped by CO_2 laser radiation. MPE and MPD of CF_2HCl have been investigated in a number of papers and in [8,9] in particular.

All our measurements in the case of CF_2HCl were carried out at the 9R30 line of CO_2 laser (ω = 1084.6 cm^{-1}). The energy fluence Φ varied over the range 0.7-3.5 J/cm^2, when the yield of MPD is less than 0.01 [9].

The spectral measurements showed that the intermolecular distribution at MPE of CF_2HCl is highly nonequilibrium. Figure 1 presents two RS spectra of the ν_5 mode in the absence of excitation (a) and with Φ = 1.3 J/cm^2 (b). The spectrum (b) was obtained at $P(CF_2HCl)\cdot\mathcal{T}(CO_2)$ = 1 μs·Torr and $\mathcal{T}_d$ = 0. Even though collisions are significant under these conditions of excitation, the energy distribution of molecules is still nonequilibrium. As in the case of SF_6 [4], there are at least two maxima in the spectrum. One of them, with a large red shift, can be attributed to highly excited, "hot", molecules, and the second one - to less excited, "cold", molecules. As the $\mathcal{T}_d$ delay increases, an equilibrium distribution with single maximum can be observed.

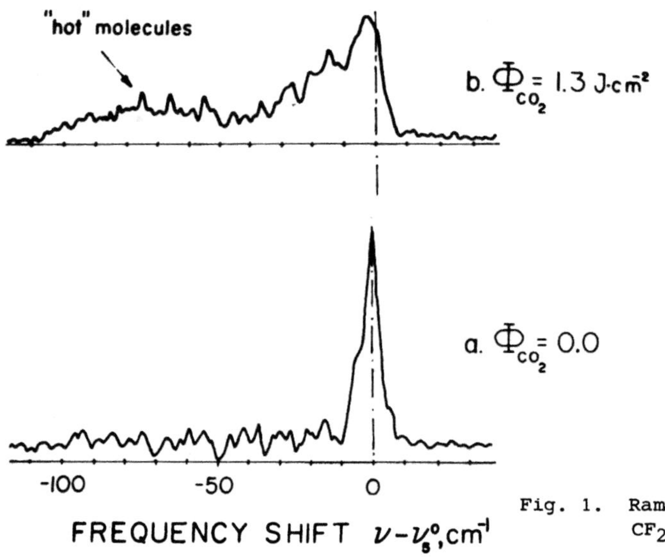

Fig. 1. Raman spectra of the ν_5 mode of CF_2HCl

The energy stored in different modes of CF_2HCl molecules as a result of their excitation by CO_2 laser was measured in the anti-Stokes region using the above-described procedure. One of the most essential results of these measurements is that at MPE of CF_2HCl the RS signal was observed at collisionless conditions <u>in all the modes</u> including the C-H bond.

Figures 2 and 3 show the energy $\bar{\varepsilon}_i$ in the mode being probed as a fuction of CO_2 laser energy fluence Φ for one of the nonresonant - ν_4 and the resonant - ν_3, ν_8 modes. There are "instantaneous" ($\mathcal{T}_d$ = 5 ÷ 10 ns) values of $\bar{\varepsilon}_i^s$ (curves b) and those measured with $\mathcal{T}_d$ = 4 μs (curves a) when the vibrational equilibrium is established

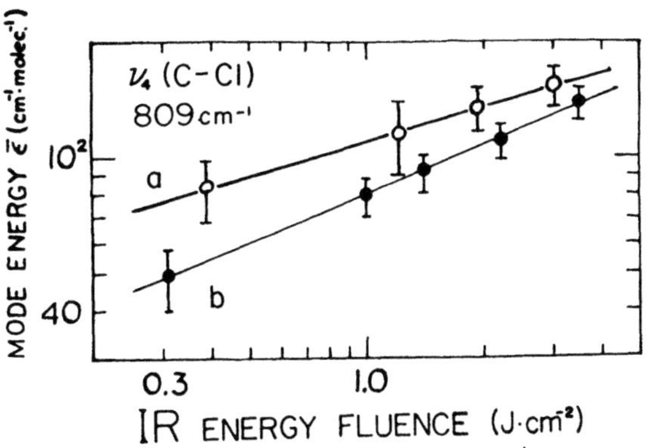

Fig. 2. The fluence dependence of ν_4 mode energy. $P(CF_2HCl)$ = 1.0 Torr

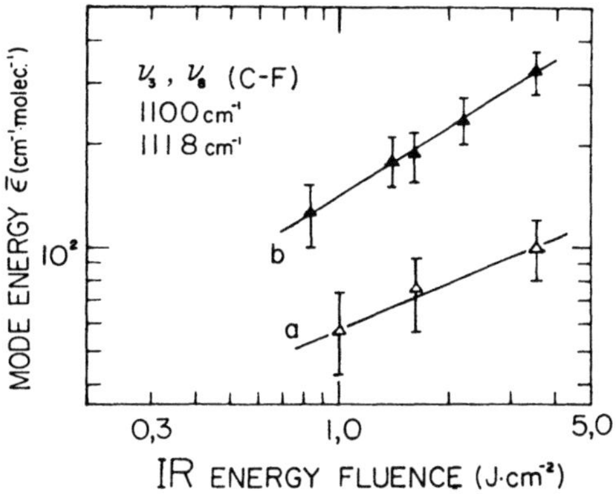

Fig. 3. The fluence dependence of ν_3, ν_8 mode energy. $P(CF_2HCl)$ = 1.0 Torr

(see below). For all the modes $\bar{\varepsilon}_i$ grows with increasing Φ. It was also found that, as equilibrium is established, the energy in the low-frequency ν_4, ν_5, ν_7, ν_9 modes increases (see e.g. Fig. 2) and in the high-frequency ν_1, ν_2, ν_3, ν_8 it drops (see e.g. Fig. 3).

As stated above, an equilibrium vibrational distribution takes place when $P\cdot\tau_d$ = 4 μs·Torr. This is confirmed by experiments. The vibrational temperatures $T_i(\Phi)$ were calculated from the values of $\bar{\varepsilon}_i(\Phi)$ measured with $P\cdot\tau_d$ = 4 μs·Torr. At fixed values of Φ the temperatures in all the modes were found to be equal. This result allows us to state that with $P\cdot\tau_d$ = 4 μs·Torr equilibrium distribution of absorbed energy takes place only for the vibrational degrees of freedom and the "heating" of the rotational and translational degrees of freedom is still small. This follows from comparing the "instantaneous" value of average vibrational energy of molecules $\bar{\varepsilon}^s = \sum_i \bar{\varepsilon}_i^s$, that is taken right after the CO_2 laser pulse, and its value $\bar{\varepsilon}^f$ at τ_d = 4 μs. The values of $\bar{\varepsilon}^s$ and $\bar{\varepsilon}^f$ within the accuracy of measurements are coincident. With Φ = 1.6 J/cm^2 for example, $\bar{\varepsilon}^s$ = 1090±120 cm^{-1} and $\bar{\varepsilon}^f$ = 1050±20 cm^{-1}.

As it is seen from Fig. 2 and 3, the establishment of vibrational equilibrium cauced by collisions with delay τ_d increase results in energy redistribution in different molecular modes. The dynamics of this redistribution was measured for the resonant ν_3 and ν_8 modes. It was found that this process can be described well by exponential law with the time-constant $P\cdot\tau$ = 1.4±0.4 μs·Torr.

4. Discussion of Results

Thus, the experimental results show that, as the CO_2 laser radiation resonantly acts on the ν_3 and ν_8 modes of the CF_2HCl molecule, all the modes of the molecule become excited during a laser pulse. The collisions bring about energy redistribution in the modes. The question arises - what type of energy distribution is formed at laser excitation of these molecules?

The analysis of the experimental results (for details see [8]) showed that they consist well with the following type of inter- and intramolecular energy distribution - see Fig. 4. The two groups (ensembles) of CF_2HCl molecules are formed as a result of CO_2 laser excitation. The first one consists of vibrationally excited molecules with their fraction q_1 and the average energy $\bar{\varepsilon}^1$. The latter being $\bar{\varepsilon}^1 \geqslant$ 4000 ÷ 5000 cm^{-1}, a statistical energy distribution over all the modes takes place. Then there is another, lower ensemble of molecules, their fraction being $(1 - q_1)$. The energy of nonresonant modes of this ensemble corresponds to initial room temperature T_0. The energy (or temperature) of resonant modes is higher, as a result of laser excitation. However, the average level of molecular excitation in these modes doesn't exceed 1 ÷ 2 quanta.

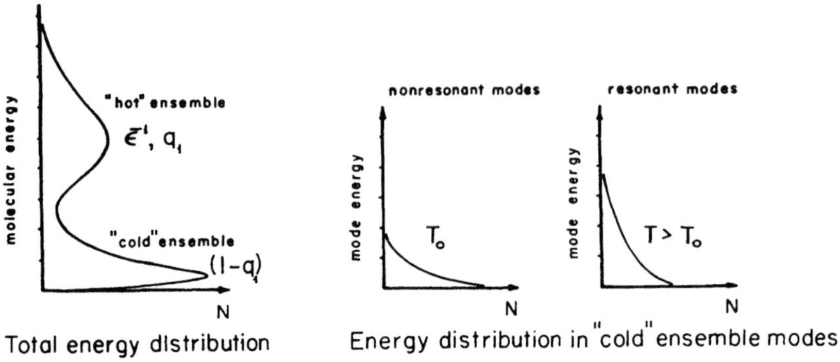

Fig. 4. Model of energy distribution in CF_2HCl molecules

The total fraction q of excited CF_2HCl molecules was estimated. For example, at $\Phi = 1.6\ J/cm^2$ $q \simeq 0.3$ of all molecules absorb laser radiation. Approximately a half of them ($q_1 \simeq 0.15$) forms the "hot" ensemble. The second half remains in low-lying states of resonant modes.

Collisions lead to the relaxation of initial nonequilibrium intermolecular relaxation to the equilibrium one. The molecular energy is redistributing among all the modes in accordance with their heat capacity. Since the latter depends on the oscillator's frequency and temperature, the high-frequency modes will be "switched off" and the energy will be transferred to the low-frequency modes, as it takes place in the experiment - see Figs. 2,3.

The results presented show that a possibility of mode-selective excitation of CF_2HCl (and probably of some other molecules) by MPprocess is very limited, at least on the time-scale $\sim 10^{-8}$ s. Absorption of as few as 4 - 5 quanta results in excitation of all molecular modes, and only if 1 - 2 quanta are absorbed does the energy remain localized in resonant modes.

One of the important conclusion drawn from analysis of experimental results is that a statistically equilibrium energy distribution over all the modes can be observed at a comparatively low excitaion level of CF_2HCl, 4000 - 5000 cm^{-1}. Such a distribution may, in principle, arise in two cases. It may reflect the property of the molecule itself and result from the method of excitation in use.

The available results enable us to single out at least two main potential mechanisms of statistical equilibrium distribution of vibrational energy over different modes of molecules under MP excitation. The first mechanism is vibrational motion stochastization at a certain energy level. In the case of CF_2HCl this is consistent with 4000 - 5000 cm^{-1}. The second mechanism may be associated with excitation of a great set of different combination vibrations as a result of multiple-photon absorption of several laser quanta. Further studies are required to choose between them.

References

1. R. A. Marcus, Laser Chemistry, 2, 203 (1983)
2. V. N. Bagratashvili, V. S. Letokhov, A. A. Makarov, E. A. Ryabov, Laser Chemistry, 1, 211 (1983); 4, 171 (1984)
3. V. N. Bagratashvili, Yu. G. Vainer, V. S. Doljikov, S. F. Kol'yakov, A. A. Makarov, L. P. Malyavkin, E. A. Ryabov, E. G. Sil'kis, V. D. Titov, Sov. Phys. JETP Lett., 30, 471 (1979)
4. V. N. Bagratashvili, Yu. G. Vainer, V. S. Doljikov, S. F. Kol'yakov, V. S. Letokhov, A. A. Makarov, L. P. Malyavkin, E. A. Ryabov, E. G. Silkis, V. D. Titiov, Sov. Phys. JETP, 53, 512 (1981)

5. E. Mazur, I. Burak, N. Bloembergen, Chem. Phys. Lett., 105, 258 (1984)
6. J. G. McLaughlin, M. Poliakoff, J. J. Turner, J. Mol. Struct., 82, 51 (1982)
7. E. K. Plyler, W. G. Benedict, J. Res. Nat. Bur. Stand., 47, 202 (1951)
8. V. S. Doljikov, Yu. S. Doljikov, V. S. Letokhov, A. A. Makarov, A. L. Malinovsky, E. A. Ryabov, Chem. Phys., (in press)

Transient Vibrational Excitation of Dye Molecules in a Solid Polymer Matrix

A. Seilmeier and W. Kaiser

Physik-Department E 11, Technische Universität München, D-8000 München, F.R.G.

The vibrational relaxation in liquids and in liquid solutions has been studied extensively in recent years. In this paper we want to report on time-resolved investigations of vibrationally excited dye molecules in a solid plastic film. Such systems have found increased attention for the application in mass storage systems and for collectors of sun energy in front of photoelectric cells.

In our investigations the molecules are vibrationally excited by resonant absorption of an infrared picosecond light pulse at frequency ν_1. The instantaneous population is monitored by a second picosecond light pulse in the visible which promotes the excited molecules to the fluorescing S_1-state. The frequency of the probe pulse ν_2 is chosen to be smaller than the 0-0 transition frequency ν_{00}. Transitions to the S_1-state are only possible in molecules with occupied vibrational states above $\nu_{00}-\nu_2$.

In our experiments we used a modelocked Nd:glass laser system,which produces single pulses of approximately 6 ps duration. The pulses are divided into two parts. One pulse pumps a parametric generator-amplifier system - consisting of two $LiNbO_3$ crystals - which generates infrared pulses tunable between 2800 cm^{-1} and 3500 cm^{-1}. These pulses serve as pump pulses for the vibrational excitation. The second part is frequency-doubled in a KDP crystal. This green pulse at $\tilde{\nu}_2$=18,940 cm^{-1} is properly delayed in order to probe the momentary excitation.

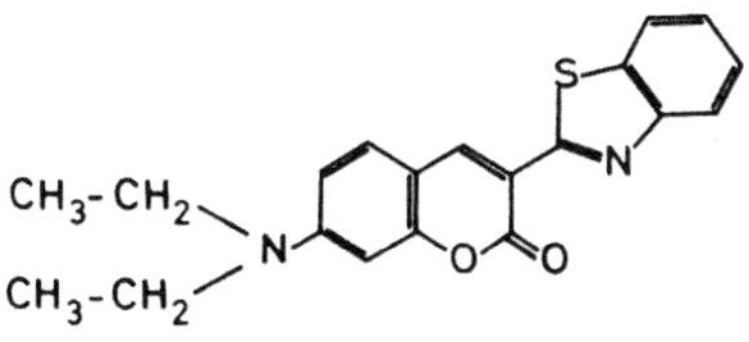

Coumarin 6

In the following data are reported on Coumarin 6 incorporated in a polyester film composed of terephthalic acid and ethylene glycol (Hostaphan) . The thickness of the film was approximately 15 μm, the dye concentration 10-2 M. The electronic 0-0 transition frequency of Coumarin 6 is located approximately at 20,800 cm^{-1}. Thus the probe pulse at $\tilde{\nu}_2$=18,940 cm^{-1} monitors the population of vibrational states around 2000 cm^{-1}.

The infrared absorption of the sample is shown in Fig.1. The absorbance A = -log T of the film is plotted as a function of frequency in Fig.1a. The spectrum is dominated by the absorption of the plastic film. At 2975 cm^{-1} approximately 50% of the infrared energy is absorbed by CH - modes of the polyester matrix. Figure 1b shows the difference spectrum between a doped and an undoped polyester film. The infrared spectrum of the dye Coumarin 6 is clearly seen /1/. The absorbance of the dye at 2970 cm^{-1} is smaller by approximately a factor of 20 due to the small concentration of 10^{-2} M.

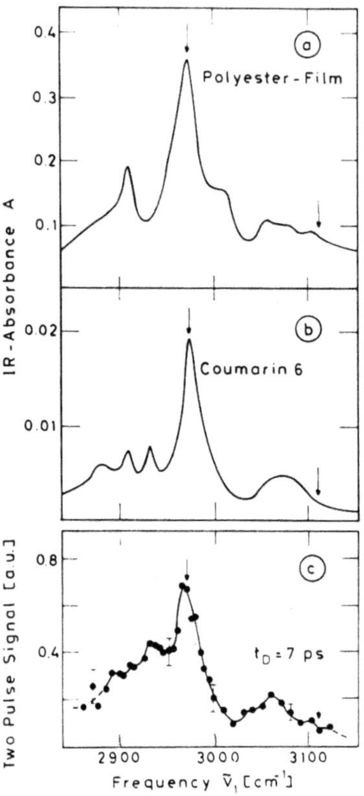

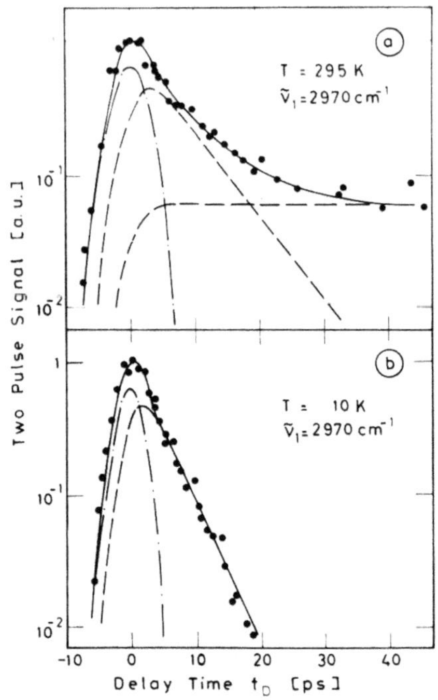

Fig.1
Infrared spectra of a polyester film containing Coumarin 6
a) Infrared spectrum of the polyester film
b) Difference spectrum between a doped and an undoped film
c) Fluorescence signal as a function of the infrared frequency measured at a delay time of 7 ps

Fig.2
Fluorescence signal as a function of the delay time at a) room-temperature, and b) 10K. The dash-dotted line represents the cross correlation of the excitation and the probe pulse. The solid line is the sum of the dash-dotted and the broken curves.

Firstly, time-resolved investigations at room-temperature are discussed . In Fig.2a the fluorescence signal is presented as a function of the delay time between the exciting infrared pulse at $\tilde{\nu}_1$=2970 cm^{-1} and the probing pulse at $\tilde{\nu}_2$=18,940 cm^{-1}. Three different dynamic events contribute to the fluorescence signal. The solid line through the experimental points is the sum of the dash-dotted and the two broken curves. For short delay times the curve follows the cross correlation curve (dash-dotted line) which was measured simultaneously. Additional frequency-resolved measurements indicate that this signal is generated either by resonant two-photon transitions or by two-step transitions via infrared intermediate states with large Franck-Condon factors /2/. These transitions have been investigated in liquid solutions in more detail /1/.

The situation is different at delay times of approximately 10 ps. A second component with a relaxation time of ~ 7 ps dominates the signal in Fig.2a. More information about this excitation process is obtained from frequency-resolved experiments. Figure 1c shows the fluorescence signal as a function of the infrared excitation frequency ν_1 at a delay time of 7 ps. The spectrum follows closely the infrared absorption spectrum of the dye molecules in Fig.1b. The CH, CH_2, and CH_3 - stretching modes are clearly resolved. The corresponding second fluorescence component is generated via resonant absorption of the CH - stretching modes of the dye molecules. The absorbed energy is rapidly redistributed via intramolecular processes, resulting in a distribution which is quite well described by an increased internal temperature /3/. The excess population due to the temperature-rise of the dye molecules is probed in our experiment. The fluorescence signal in Fig.1c, i.e. the number of excited molecules, is proportional to the infrared absorption. The time-constant of 7 ps in Fig.2a describes the cooling of the dye molecules via intermolecular processes.

After a delay time of approximately 30 ps a third component of the fluorescence signal appears, which persists for more than several hundred picoseconds (see broken curve in Fig.2a). This signal is generated by infrared heating of the polyester matrix which absorbs ~ 50% of the pump pulse. Thermal energy is transferred from the matrix to the dye molecules /4/. The rapid heating of the polyester film is detected via population changes in the dye molecules. We estimate a temperature rise of 15K from the specific heat of polyester, the interaction volume and the pulse energy. This value is consistent with the magnitude of the signal in Fig.2a. Macroscopic thermal conduction to the surrounding restores the starting temperature within milliseconds.

Figure 2b shows experimental data of a time - resolved experiment where the film was cooled to 10K. Two components (dash-dotted and broken curve) contribute to the fluorescence signal. At early delay times the signal is generated by transitions to the S_1-state via intermediate states with favorable matrix elements. The signal follows the dash-dotted cross correlation curve (see Fig.2a) . For longer delay times we observe a second fluorescence component with a decay time of 4±1.5 ps. This time-constant represents the intermolecular energy-transfer from the directly excited dye molecules to the plastic matrix. The third component is not observed in the low-temperature experiment. We estimate a temperature increase to 60K due to the infrared absorption of the polymer matrix. The population of the probed vibrational levels of the dye at this temperature is too small to lead to a measureable fluorescence excitation.

In our investigations we observe a shorter time-constant at low temperatures while a longer time-constant is generally expected for the energy dissipation at reduced temperatures. We recall that the depopulation of vibrational states is observed in our experiments and not directly the energy dissipation. It is possible to estimate the time-constant for energy dissipation from the measured lifetime, the relation between occupation and temperature for a Boltzmann absorption edge, and from the temperature-dependent specific heat. Time-constants for the energy dissipation of ~10 ps and ~20 ps result for the room-temperature and the low-temperature experiment.

Of special interest is a comparison of Coumarin 6 in liquid solution and in the plastic matrix: i) The long-lived component due to excitation of the surrounding is absent in liquid solutions which are not absorbing in this wavelength region /1/. ii) The time - constant for the energy-transfer from the dye molecules to the surrounding is the same in the solution and the plastic film. We conclude that the dye molecules are not bonded to the polyester matrix.

References

1 J. P. Maier, A. Seilmeier, W. Kaiser: Chem. Phys. Lett. 70 , 591 (1980)

2 A. Seilmeier, J. P. Maier, F. Wondrazek, W. Kaiser: to be published 1985

3 F. Wondrazek, A.Seilmeier, W. Kaiser: Chem. Phys. Lett. 104 , 121 (1984); N. H. Gottfried, A. Seilmeier, W. Kaiser: Chem. Phys. Lett. 111 , 326 (1984)

4 A. Seilmeier, P. O. J. Scherer, W. Kaiser: Chem. Phys. Lett. 105 , 140 (1984)

Ultrafast Vibronic Dynamics of Dye Molecules Studied by the Induced Grating Method

P. Troeger, C.-H. Liu, and A. Laubereau

Physikalisches Institut, Universität Bayreuth, D-8580 Bayreuth, F.R.G.

In recent years numerous papers have investigated the induced grating method for the study of the ultrafast dynamics of dye molecules /1/. A quantitative understanding of the signal transients observed at short times, where the so-called coherence peak occurs, and of the polarization dependence of this effect has been lacking. In this article we present new theoretical and experimental data on the picosecond time scale for the elastic (Rayleigh) light scattering of delayed probing pulses from induced population gratings. The scattering mechanism for the so-called coherence peak is interpreted as a resonant two-step, one-photon process, which depends on the vibronic relaxation of the terminating levels in the excited electronic state. The contributions of other nonlinear processes, i.e. single-step multiphoton interactions, are found to be negligible.

We have performed detailed model calculations which follow the treatment of Ref. 2 and give quantitative information on the probe scattering signal under various polarization conditions. The dye molecules are treated by a generalized 3-level model (see Fig. 1 a). Linearly polarized light with frequency ω_L promotes dye molecules from the ground state (1) to excited vibrational levels (2) in the upper electronic state S_1. The transient population of levels (2) subsequently decays by radiationless processes, which are described by a vibronic relaxation time τ_v, to the bottom (3) of the S_1 state. A frequency difference of several hundred cm^{-1} between levels (2) and (3) is considered so that the levels are clearly separated as compared with the homogeneous linewidth of the electronic transition /3/. Intersystem crossing to triplet states is negligible on the picosecond time scale considered. Molecular rotation is treated classically on account of the large moments of inertia of dye molecules (small rotational quanta) and the large damping in liquid solution at room temperature. Inhomogeneous broadening is included in the calculation introducing a distribution of 3-level systems.

The beam geometry is depicted schematically in Fig. 1 b. The angles between the four light beams are small ($\leq 2^o$) and are largely exaggerated in the Fig. A novel 3-dimensional wave vector geometry is chosen which allows perfect phase-matching between the k-vectors of the two pump pulses, k_{L1} and k_{L2}, the incident probe pulse k_{pr} and the observed scattering signal. Two grating vectors k_1, k_2 (broken lines in Fig. 1) have to be considered for the material excitation:

$$\vec{k}_1 = \vec{k}_{L2} - \vec{k}_{L1} \text{ and } \vec{k}_2 = \vec{k}_{pr} - \vec{k}_{L1}$$

Light scattering from the two population gratings k_1, k_2, represents two interaction channels. The second channel via k_2 gives rise to a signal enhancement ("coherence peak") for coincident pump and probe pulses, i.e. short delay times. A calculated example for the

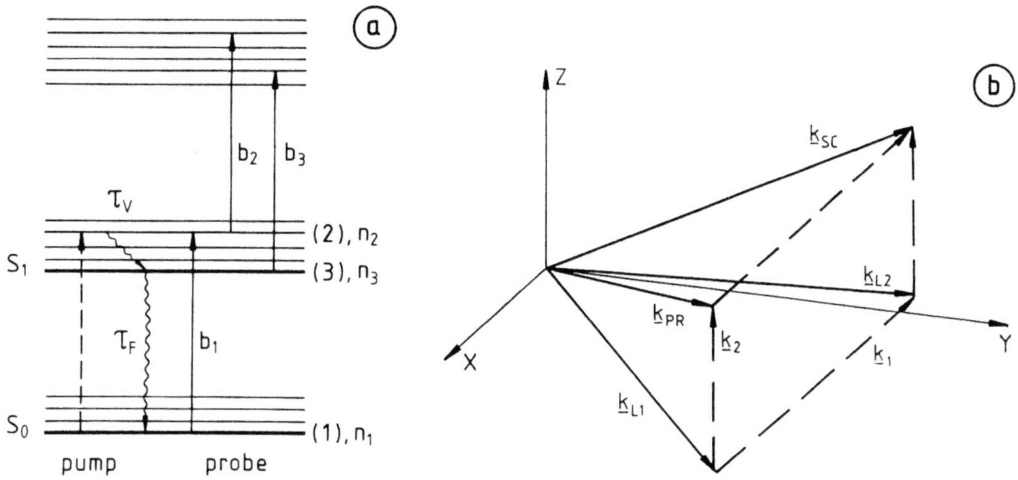

Fig. 1: a) Energy levels of dye molecules (schematically);
b) three-dimensional wave vector geometry with perfect phasematching

effect of vibronic relaxation on the coherence peak is depicted in Fig. 2.

A typical situation for dye molecules is considered with τ_F = 3 ns, τ_R = 200 ps and values of the vibronic relaxation time in the range τ_V = 0 to 1.6 ps. For the absorption coefficients b_2, b_3 of excited state absorption from levels (2) and (3) we assume $b_2 = b_3 = 0.75\ b_1$, where b_1 represents the ground state absorption. The time-integrated scattering signal $S_{\parallel}(t_D)$ is plotted versus delay time t_D of the

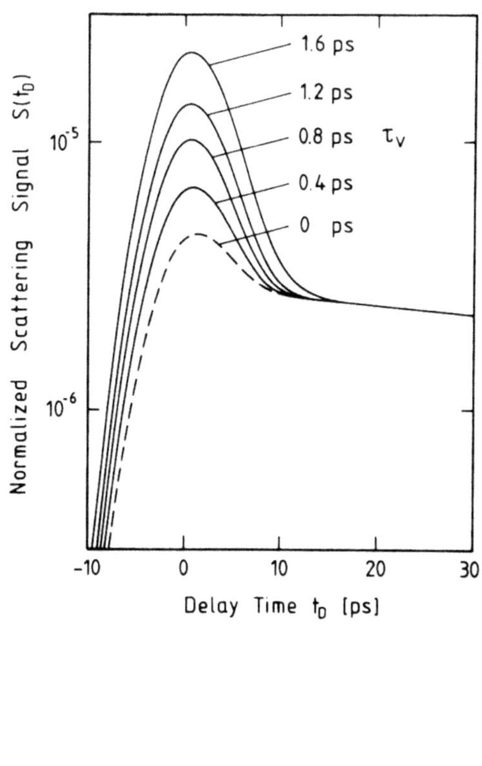

Fig. 2:
Calculated probe scattering signal $S_{\parallel}(t_D)$ versus delay time t_D between excitation and probing pulses

probe pulse E_{pr}, for parallel polarization of the input pulses. The signal rises to a maximum at $t_D \approx 0$ and then decays rapidly. For larger t_D the signal decays much more slowly according to the relaxation times τ_F and τ_R. The signal overshoot at $t_D = 0$ represents the coherence peak noted in previous investigations /4,5/. Fig. 2 shows that the peak height depends on τ_V. Two factors contribute to the enhanced signal: the transient population of the excited vibronic levels (2) and the second component of the material excitation (k_2) representing pump-probe coupling. The latter point accounts solely for the coherence peak if $\tau_V = 0$ (broken curve).

We have performed measurements of the coherence peak for the determination of the vibronic lifetime. Our experimental system is described as follows. Single pulses of 5 ps duration are generated by a modelocked Nd: glass laser system followed by a KDP frequency doubler. By the help of a first beam splitter part of the pulse is used for the probing process. The remainder of the pulse is split into two equal parts which serve as excitation pulses. The three beams are directed into the sample according to the three-dimensional wave vector geometry discussed above. Perfect phase-matching is facilitated using short sample cells of 10 µm to 1000 µm. The elastic (Bragg) scattering of the probe pulse from the induced grating in the sample is measured in a small solid angle of acceptance by the help of interference filters and a photomultiplier. The polarization of the input pulses and of the scattered radiation is adjusted by rotable polarizers.

Some experimental results for rhodamine 6G (Rh) in ethanol and phenoxazone 9 (Ph) in dioxane are presented in Figs. 3 a-d. Concentration is 5×10^{-4} M/l and 3×10^{-3} M/l for Rh and Ph, respectively,

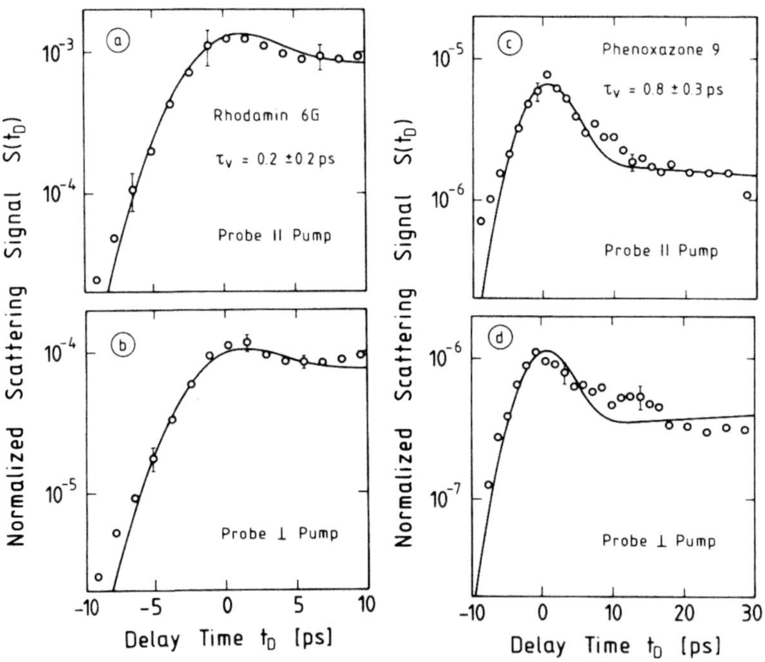

Fig. 3: Measured scattering signal $S(t_D)$ for parallel (a,c) and perpendicular polarization (b,d) vs. delay time for rhodamine 6G (a,b) and phenoxazone 9 (c,d); experimental points; theoretical curves

corresponding to a sample transmission of 20 % at 527 nm. The vibronic levels populated by the absorption process are situated approximately 700 cm^{-1} (Rh) and 1000 cm^{-1} (Ph) above the bottom of the S_1 level. Linear polarization of the pump pulses is adjusted perpendicular to the plane of incidence. Probe polarization is parallel (∥) or perpendicular (⊥) with respect to the pump polarization. The scattering signal $S_\parallel(t_D)$ and $S_\perp(t_D)$ are plotted versus delay time between pump and probe pulses for a short time interval $t_D \leq 30$ ps. The signal rises to a maximum at $t_D = 0$. For larger delay, there is a certain decrease with a subsequent slow time behaviour depending on the polarization conditions. For parallel polarization the signal decreases according to the rotational relaxation time and fluorescence lifetime (Figs. a and c). For perpendicular polarization our data indicate a signal increase to a broad maximum at several hundred picosecond delay (not shown in Figs. b and d).

The solid curves in the Figs. are calculated using τ_V as a fitting parameter. For the excited state absorption we take $b_2 \approx b_3$ and the data of Ref. 2 for the ratio b_3/b_1. For phenoxazone 9 there is a minor contribution of the excited state absorption to the signal enhancement around $t_D = 0$, while this effect is negligible for rhodamine. The results for the vibronic relaxation time are $\tau_V = 0.2 \pm 0.2$ ps and $\tau_V = 0.8 \pm 0.2$ ps for Rh and Ph, respectively. The intensity ratio for the scattering signal with parallel and perpendicular polarization should be noted. Our data indicate the ratio $S_\parallel(0)/S_\perp(0) = 9 \pm 1$. This result is in good agreement with our theoretical expectations, but differs somewhat from results of Refs. 6.

It is interesting to compare our results on τ_V with other data. We note agreement within experimental accuracy with the values obtained in Ref. 2 from measurements of induced dichroism. Our result on Rh is also in accordance with measurements of nonlinear absorption changes /7/. Very recently population lifetimes of a few hundred fs were reported for several dye solutions at room temperature, e.g. a value of 190 fs for rhodamine 640 /8/. A special deconvolution technique and pulses of 70 fs were used. These investigations consistently indicate vibronic time constants in the subpicosecond range. The findings, however are at variance with Ref. 9 reporting the vibrational relaxation of rhodamine 640 and several other dyes to be faster than 20 fs.

An example of a different polarization geometry is depicted in Fig. 4. The induced dichroism of uranine is investigated by a novel approach applying the induced grating technique. The polarization situation is illustrated in Fig. 4 a. The field vector of the probe pulse is oriented at 45° with respect to the polarization of the two pump pulses generating the population grating. The scattering signal is detected behind a blocking polarizer. The signal decay, which is followed in Fig. 4 b over several hundred picoseconds, gives direct information on molecular rotation. The time constant $\tau_R = 200$ ps is directly obtained from the signal slope. It is important that the technique used for the data of Fig. 4 does not require the high quality optical components necessary for conventional dichroism studies. The large null-effect radiation which occurs in previous transient dichroism studies is avoided in our investigation; this radiation may perturb the experimental data /10/ and produce undesirable background signals.

In conclusion, we have shown that the coherence peak observed in transient grating studies depends on the vibronic relaxation time. A special wave vector geometry for perfect phase-matching and adjustable polarizations has been developed. Our four-beam technique can be used

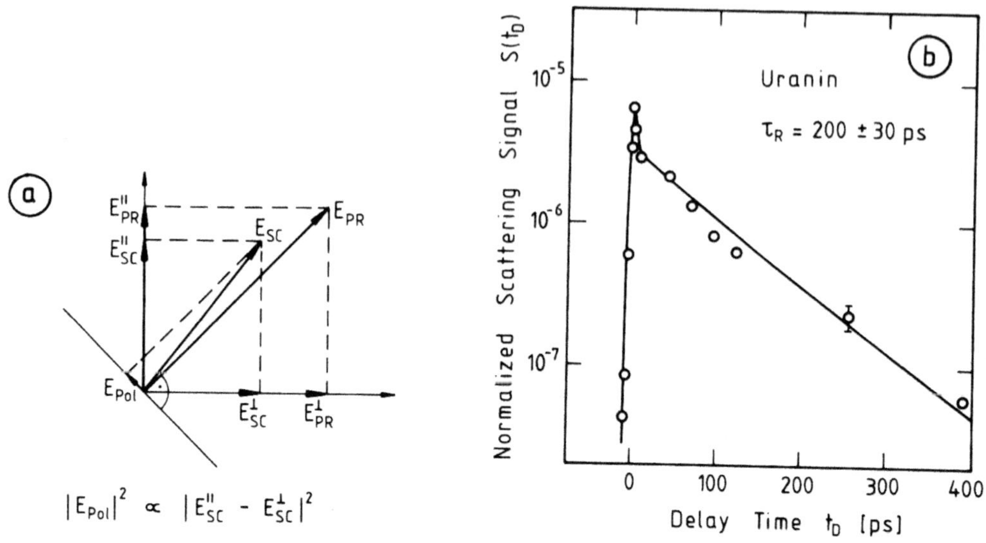

Fig. 4: Induced dichroism studied by the transient grating technique
a) polarization geometry
b) measured scattering signal vs. delay time t_D of the probe pulse (E_{pr}); experimental points; theoretical curve

for the study of a variety of dynamic processes of molecules in the excited electronic or ground state. An important advantage compared to nonlinear absorption or induced dichroism studies is that the scattering method allows us to work at low excitation level and avoids disturbing background signals.

References

1 See, for example "Ultrafast Phenomena IV", eds. D.H. Auston and K.B. Eisenthal (Springer Berlin, Heidelberg, 1984).
2 D. Reiser and A. Laubereau, Appl. Phys. B 27, 115 (1982); Ber. Bunsenges. Phys. Chem. 86, 1106 (1982).
3 H. Somma, Y. Taira and T. Yajima, in "Picosecond Phenomena", p.224.
4 C.V. Shank and E.P. Ippen, Appl. Phys. Lett. 26, 62 (1975).
5 H.E. Lessing, A. von Jena and M. Reichert, Chem. Phys. Lett. 36, 517 (1975).
6 A. von Jena and H.E. Lessing, Opt. Quant. Electr. 11, 419 (1979); C.R. Gochanour and M.D. Fayer, J. Phys. Chem. 85, 1989 (1981).
7 A. Penzkofer, W. Falkenstein and W. Kaiser, Chem. Phys. Lett. 44, 82 (1976).
8 A.M. Weiner and E.P. Ippen, Chem. Phys. Lett. 114, 456 (1985).
9 A.J. Taylor, D.J. Erskine and C.L. Tang, in "Ultrafast Phenomena IV", p. 137, eds. D.H. Auston and K.B. Eisenthal (Springer Berlin, Heidelberg, 1984).
10 D. Waldeck, A.J. Cross, Jr., D.B. McDonald and G.R. Fleming, J. Chem. Phys. 74, 3381 (1981).

Vibrational Fluorescence from Matrix-Isolated CO and NO Pumped with a Color Center Laser

J.P. Galaup, J.Y. Harbec, J.J. Zondy, R. Charneau, and H. Dubost

Laboratoire de Photophysique Moléculaire du C. N. R. S.*, Bât. 213, Université de Paris-Sud, F-91405 Orsay-Cédex, France

We have studied laser-induced vibrational fluorescence of matrix-isolated CO and NO in the 1.2 to 6 μm region. Samples of $^{13}C^{18}O$ or $^{14}N^{16}O$ in nitrogen or argon at the desired concentration ratio are deposited onto a gold coated copper mirror held at liquid He temperatures. Guest molecules are optically pumped in their v = 2 state by the light of a modulated CW KCl:Li color center laser delivering 5 to 20 mW. For time-resolved studies, a 300 μs to 15 ms excitation square pulse is tailored with an acousto-optic modulator. In the course of these experiments, several interesting phenomena have been observed, including stimulated emission, strong anharmonic V-V pumping, intermolecular vibrational to electronic (V-E) transfer and possibly IR-induced photodissociation. These results are described in detail in several forthcoming papers [1,2].

1. Stimulated Emission

Overtone pumping of CO or NO molecules in Ar or N_2 crystals leads to the occurrence of stimulated emission between the directly populated state v = 2 and the initially empty v = 1 state. In the case of N_2 matrix, direct evidence for stimulated emission is provided by the observation of intense, irregular and short spikes superimposed on the background fluorescence signal corresponding to the v = 2 → v = 1 transition. The occurrence of spiking is critically dependent on the optical adjustment. Sometimes, regular oscillations with a period in the microsecond range such as those shown in Fig. 1 are observed. The period of such oscillations is three orders of magnitude shorter than the spontaneous fluorescence lifetime of several tens of ms. A similar phenomenon is well known to occur

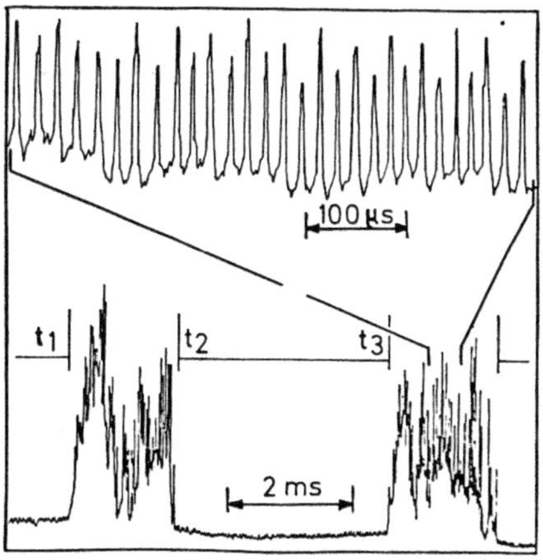

Figure 1

Time-resolved fundamental emission of 0.05 % $^{13}C^{18}O$ in a nitrogen matrix at T = 2.7 K under optical pumping in the first overtone. Single-shot signals of the overall fluorescence are recorded using a transient digitizer. The laser pump is on from time t_1 to t_2 then off from t_2 to t_3 and the cycle is repeated again.

* *Laboratoire associé à l'Université de Paris-Sud.*

in solid-state laser systems with long-lived species. Recently, laser action has been observed from the KBr:CN$^-$ system under overtone pumping [3]. Although no cavity has been intentionally designed in our experiment, the intensity of the spikes is about 100 times larger than the fluorescence background intensity. Nonlinear equations describing the time-evolution of the populations and of the stimulated photon density predict oscillations ranging from ns to μs time-scale depending on the optical path. In our case, the latter can be lengthened by reflection onto the gold mirror substrate and by total internal reflection at the surface of the sample. Spiking emission is no more observed at temperatures greater than 7-9 K because the gain is very sensitive to the decrease in population-inversion induced by the endothermic transfer process :

$$M(v=2) + M(v=0) + \Delta E_{02} \rightarrow M(v=1) + M(v=1)$$

More than the technical difficulties associated with the use of liquid He, the severe restriction on the tunability of such potential solid-state vibrational laser systems makes them not very attractive in practice. However, the vibrational lines exhibit a strong temperature-broadening. Operation at higher temperatures could be achieved by optical pumping through higher overtones. Indeed direct pumping of a vibrational level with $v > 2$ will result in stimulated emission on the $v \rightarrow v-1$ transition. The population inversion between v and v-1 would tend to be destroyed by the process :

$$M(v) + M(0) \rightarrow M(v-1) + M(1)$$

which has an activation energy $\Delta E = (v-1)\ \Delta E_{02}$.

2. Strong Anharmonic V-V Pumping

In the case of CO in solid Ar, the anharmonic V-V pumping is surprisingly efficient. With a sample containing 1 % $^{13}C^{18}O$ at $T \leqslant 4.7$ K, IR emission has been observed from high-lying vibrational levels up to v = 30. The time evolution of the intensity of the IR-fluorescence coming from levels v = 2 to v = 28 has been recorded. For the sake of clarity, only the populations of states v = 2 to v = 9 have been drawn in Fig. 2. A few remarks about the time-evolution have to be pointed out :

. The population of v = 2 grows during 150 μs, decreases markedly during the rest of the pumping time then drops down in 500 μs after the pump is turned off.
. The upper levels are strongly populated during the pumping time, a maximum beeing reached for v = 8. In addition, successive population-inversions establish from v = 4 to v = 8.
. The populations N_4 to N_6 benefit from a sudden increase during a few milliseconds after the pump is turned off. Correspondingly, N_8 and later N_7 exhibit a slight decrease.

Numerical simulations using the set of differential equations describing the evolution of vibrational populations have been performed. From previous studies on the Ar:CO system [4], it has been proved that V-V pumping results from the migration-activated fusion process :

$$CO(v=n) + CO(v=1) \rightarrow CO(v=n+1) + CO(v=0) + \Delta E$$

The corresponding concentration-dependent transfer rates have been measured for $v \leqslant 6$. The time evolution of Fig. 2 cannot be reproduced by a simulation taking into account only V-V transfer. Contrary to what is experimentally observed, the pumped v = 2 level keeps growing as long as the pump pulse is on,and then undergoes radiative decay during several tens of ms. Most of the pumped molecules remain trapped in the v = 2 state and the transfer to $v > 2$ states is inefficient.

The situation is quite different if one allows v = 2 to relax towards v = 1 through stimulated emission. It is then necessary to couple the previous set of equations with another one describing the population of the resident photons at

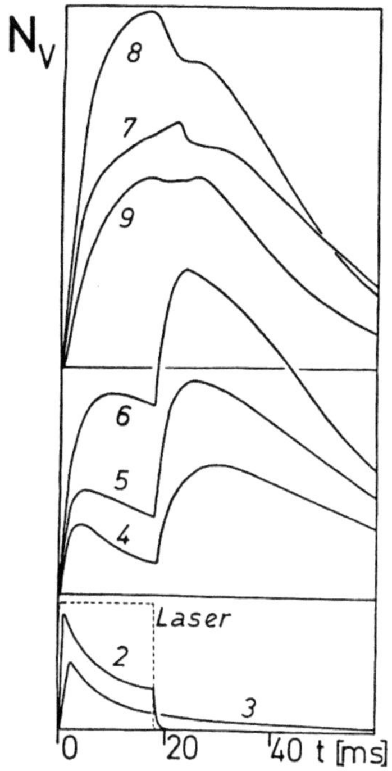

Figure 2

Experimental time-evolution of the vibrational populations of $^{13}C^{18}O$ in solid argon. (c = 1 %, T = 4.7 K). The relative populations have been extracted from the intensity of the $\Delta v = 1$ fluorescence lines through the relation :

$$N_v/N_2 = \frac{I_{v\to v-1}}{I_{2\to 1}} \times \frac{A_{2\to 1}}{A_{v\to v-1}} ,$$

where $I_{v\to v-1}$ is the fluorescence intensity of the v→v-1 transition and $A_{v\to v-1}$ the corresponding EINSTEIN coefficient.

the frequency $\nu_{2\to 1}$, assuming a loss-rate corresponding to their transit time throughout the sample. N_1 grows then at the same rate as N_2 and V-V transfers to the upper states are strongly enhanced, leading to the observed population inversions. The time evolution of N_2 is also comparable to that experimentally observed. This result shows that even though the spiking behaviour is not observed in the case of Ar matrix, stimulated emission still occurs from v = 2 to v = 1. The bumpy features on N_4 to N_6 result from the radiative relaxation,which is slower from N_8 to N_4,combined with the population inversions. A series of inversions of weaker magnitude is still observed at 10 K. The evolution of N_2 still remains the same,which is the evidence that reverse temperature-activated transfer does not appreciably compete with the stimulated emission. In conclusion, despite the weakness of the transition dipole moment, overtone pumping-induced stimulated emission is a very efficient method for the production of highly vibrationally excited molecules.

3. Intermolecular Vibrational to Electronic-Transfer

Since in the CO molecule the energy of the vibrational levels near v = 28 matches that of the lowest electronic triplet state $A^3\Pi$, a search for an UV luminescence has been attempted. However,the detection of the phosphorescence CAMERON bands of CO remained unsuccessful, in spite of the fact that a detection apparatus very sensitive in the 200 nm region was used. Actually an UV-visible fluorescence consisting of an intense complex system in the 650-800 nm region and of a weaker single vibrational progression extending from 320 to 640 nm has been recorded. The single progression is unambiguously identified with the $C^3\Delta_u \to X^3\Sigma_g^-$ transition system of molecular $^{16}O_2$ in solid Ar. The nine most intense lines can be satisfactorily assigned to the $\Delta v = 0$ and 1 sequences of the $b^1\Sigma_g^+ \to X^3\Sigma_g^-$ band of $^{18}O_2$ with v' = 0 to 5. Several weak lines are identified with the same sequences of $^{16}O_2$. Excitation spectra and time-evolution have been recorded for the most intense lines of each system, at 421.4 nm and at 763.0 nm respectively. For both

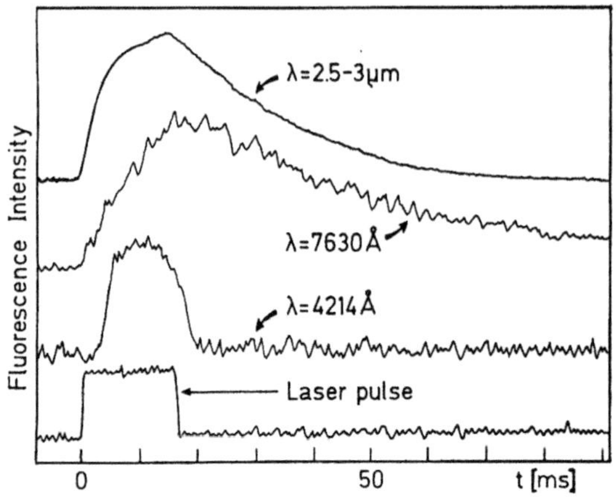

Figure 3

Time-evolution of the $^{13}C^{18}O$ infrared fluorescence, of the $^{18}O_2$ red and of the $^{16}O_2$ violet emission bands.

lines excitation is into the CO overtone absorption. Time-resolved signals are shown in Fig. 3. The onset of the violet fluorescence at λ = 421.4 nm is delayed with respect to the IR radiation by $\sim$ 3 ms. On the contrary, the red fluorescence at 763.0 nm is not delayed but starts growing at a rate similar to that of the IR fluorescence coming from CO vibrational levels between v = 20 and v = 25.

These observations demonstrate that O_2 is electronically excited by energy-transfer from the vibrational levels of CO. Two independant intermolecular V-E processes are supposed to occur, one giving rise to the complex red system :

$$CO(X^1\Sigma^+, 20<v<25) + O_2(X^3\Sigma_g^-) \rightarrow CO(X^1\Sigma^+, v'<v) + O_2(b^1\Sigma_g^+, v<6) \quad (1)$$

and the other responsible for the UV-visible single progression :

$$CO(X^1\Sigma^+, v>25) + O_2(X^3\Sigma_g^-) \rightarrow CO(X^1\Sigma^+, v'<v) + O_2(C^3\Delta_u, v=0) \quad (2)$$

The temporal behaviour of the intensity of the red and violet fluorescence lines suggests that process (1) occurs from CO vibrational levels lower than those involved in process (2). The strong intensity of the b→X band of $^{18}O_2$ suggests that the molecular oxygen present in the sample is isotopically enriched. $^{18}O_2$ molecules could originate either from impurities in the enriched CO gas or from the in-situ photodissociation of $^{13}C^{18}O$. The former hypothesis does not explain why the C→X band of $^{18}O_2$ is missing. In addition,it does not account for the large changes in the intensity ratio between the b→X band of $^{18}O_2$ and the C→X band of $^{16}O_2$ which are observed from one experiment to another. The latter assumption accounts more satisfactorily for the experimental facts. Indeed the dissociation yield of $^{13}C^{18}O$ is expected to be critically dependent upon the laser intensity. The output power of the color center laser could easily fluctuate by a factor of two from day to day. Moreover, if $^{18}O_2$ molecules originate from the dissociation of $^{13}C^{18}O$, they must be formed in regions where the CO density is high. The distance between vibrationally excited $^{13}C^{18}O$ and $^{18}O_2$ molecules may be short enough to make process (1) predominant.

References

[1] J.P. Galaup, J.Y. Harbec, R. Charneau and H. Dubost : Chem. Phys. Lett., in the press

[2] J.P. Galaup, J.Y. Harbec, R. Charneau and H. Dubost : submitted to Chem. Phys. Lett.

[3] T.R. Gosnell, A.J. Sievers and C.R. Pollock : Optics Letters, 10, 125 (1985)

[4] H. Dubost and R. Charneau : Chem. Phys. 41, 329 (1979).

Vibrational Energy Decay of Surface Adsorbates

E.J. Heilweil, M.P. Casassa, R.R. Cavanagh, and J.C. Stephenson

National Bureau of Standards, Center for Chemical Physics, Gaithersburg, MD 20899, USA

1. Introduction

A more complete understanding of the involvement of vibrational energy during chemical reactions of ground electronic state molecular systems may be obtained from direct measurements of vibrational population lifetimes (T_1). In particular, knowledge of the magnitude of energy relaxation rates of adsorbate vibrational modes at surfaces and at reaction temperatures should yield insight into the mechanisms and macroscopic kinetics of surface chemistry and catalysis.

While there has been some experimental [1] and theoretical [2,3] work to determine the magnitude of surface vibrational decay times (especially for molecules adsorbed on metallic substrates [2,3]), direct measurements of surface vibrational T_1 have only recently been reported [4-6]. By applying established picosecond laser techniques developed for liquid phase T_1 determinations [7] to systems with high surface area (large effective adsorbate concentration), T_1 lifetimes have been obtained for several species chemisorbed on colloidal silica particles which are pressed into disks or dispersed in liquids [8]. The picosecond method used to obtain decay times and results for high-frequency adsorbate stretching modes are summarized here. Additional measurements of the T_1 temperature-dependence for OH(v=1) in bulk silica [9] and of relaxation times for model OH-containing molecules in dilute solution are also reviewed.

2. Experimental [10]

The picosecond laser system and technique used for this investigation has been previously described in detail [5]. The experimental system is based on a conventional actively/passively modelocked Nd^{+3}:YAG laser. Pulses of 35 mJ at 1.06 microns are used to pump a three-stage $LiNbO_3$ Optical Parametric Amplifier (OPA) chain. This arrangement, which is similar to that used by LAUBEREAU and KAISER [7] and others [11], generates tunable (2700-7200 cm^{-1}) IR pulses of about 18 ps duration, 10-20 cm^{-1} FWHM bandwidth and ~300 μJ of energy at a 10 Hz repetition rate.

In the present case, the OPA output (ω_{IR}) is tuned to an infrared $v=0\rightarrow1$ transition of an adsorbate vibrational mode. The major portion of the OPA pulse is focused into the sample to saturate this transition. When a significant fraction of population is transferred to the v=1 state, and provided that $v=1\rightarrow2$ transitions are not resonant with ω_{IR}, then the sample will become more transparent (bleached) at ω_{IR}. To measure the sample transmission recovery, a second weaker probe pulse (about 1% of the pumping pulse energy) is used to interrogate the transmission of the excited volume. The transmission of this probing pulse through the sample is then monitored as a function of delay time (t_D) between the two pulses. For two-level systems, the transient bleaching recovery is a direct measure of the v=1 T_1 lifetime.

The physical and chemical characteristics of the samples used in this investigation deserve particular attention. In order to perform time-resolved IR transmission measurements, fairly large "effective" adsorbate concentrations are required to obtain significant pulse absorptions. By studying small SiO_2 particles of ca. 100 Å diameter, large surface areas can be confined to small sample volumes. As an example, there are ~4 OH groups bound to 100 $Å^2$ of SiO_2 surface area. For typical pressed disks or solvent dispersions, the effective OH concentration is ~0.1 molar and samples of ~50% transmission can easily be achieved. By proper choice of solvent for refractive index-matching, low light scattering levels are also obtained.

Commercial grades of fumed silica (Aerosil 200, 200 m^2/g or Cabosil Corp. EH-5, 400 m^2/g) were used. Removal of adsorbed water and substitution of -OD, $-NH_2$, $-OCH_3$ and -BOH species for surface -OH groups was accomplished by chemical reaction methods in a heated pyrex vessel attached to a glass vacuum system [12]. Dispersions were prepared by mixing the silica powder into solvents with stirring and then transferring the sample to conventional 1 mm pathlength IR absorption cells. Temperature-dependent measurements for OH(v=1) embedded in a fused silica flat were conducted by heating the sample in a propane-oxygen flame or cooled in a liquid-N_2 cell [9]. Studies of model molecules containing -SiOH (silanols), -COH (alcohols) and their deuterated (OD) analogs were performed at room temperature in CCl_4 solution.

3. Results and Discussion

IR transmission spectra of the samples studied may be found in numerous publications [8,12]. Dried silica disks, for example, exhibit a strong, sharp absorption arising from isolated surface OH groups at 3749 cm^{-1} (~8 cm^{-1} FWHM) accompanied by a broad weaker shoulder to lower wavenumbers which has been attributed to hydrogen-bonded OH groups. SiOD and BOH modified samples produce spectra that closely resemble those of SiOH (ν_{OD} = 2750 cm^{-1}, ν_{BOH} = 3705 cm^{-1}), while $-NH_2$ and $-OCH_3$ exchanged adsorbates gives rise to additional absorption bands (ν_{NH} = 3420, 3510 cm^{-1} ν_{CH} ≅ 2870, 2960, 2990 cm^{-1}). Spectra of silica dispersed in solvents, however, exhibit a broadening (about 30-60 cm^{-1} FWHM) and red-shift (depending on solvent) of the adsorbate spectral features.

Representative decay data for the above adsorbate vibrational modes may be found in earlier papers [4-6,8,9]. T_1 decay times obtained for OH(v=1) relaxation in a variety of solvents and for the stretching modes of the other adsorbate species are summarized in Table I.

Our initial results for OH(v=1) on SiO_2 in vacuum and dispersed in CCl_4 at room-temperature yielded T_1 lifetimes of 204 and 159 ps, respectively [4]. These lifetimes correspond to roughly 2 x 10^4 vibrational periods and cannot be extracted from the broad and asymmetric linewidths alone. Further studies of silica in various solvents indicate that the OH(v=1) T_1 lifetime decreases as the highest solvent fundamental vibrational mode frequencies approach the 3700 cm^{-1} OH frequency. This is particularly evident for CH-stretch (~3000 cm^{-1}) containing molecules such as CH_2Cl_2 and C_6H_6 where T_1 = 102 ± 20 ps and 87 ± 30 ps, respectively. In the presence of adsorbed H_2O, which possesses a broad, partially overlapping OH absorption with SiOH, T_1 reduces even further to ~56 ps. In this case, direct V-V transfer from the surface OH to adsorbed water may participate in the deactivation process.

Insight into vibrational energy-transfer at surfaces may be gained by studying other species bound to the silica substrate. In this way the potential accepting modes of the bulk material (e.g., vibrational modes or phonons) remain unchanged while the adsorbate local modes and adsorbate-substrate coupling are modified. The simplest substitution for the OH group is isotopic exchange to OD. In this case T_1 is reduced from 204 ps for SiOH to 149 ps for SiOD (as pressed disks in vacuum) [14]. Decay of the lower energy OD quantum into the substrate was expected to require fewer quantum transitions than for OH (i.e., $\Delta V \geq 3$ vs. $\Delta V \geq 4$) and result in a faster decay rate. However, the observed isotope effect was smaller than anticipated.

TABLE I: T_1 vibrational decay times at room-temperature for the various adsorbate/silica systems studied.

System	T_1 (ps)	$k(10^9\ s^{-1})$	$k'(10^9\ s^{-1})^a$	Notes
SiOH/vacuum	204±20	4.9	-	Pressed SiO_2 disk
SiOH/CCl_4	159±16	6.3	1.4	Dry SiO_2 dispersion
SiOH/CF_2Br_2	140+30	7.1	2.2	"
SiOH/CH_2Cl_2	102±20	9.8	4.9	"
SiOH/C_6H_6	87±30	11.0	6.1	"
SiOH/H_2O/CCl_4	56±10	18.0	13.0	SiOH T_1, ~5H_2O/100Å^2 physisorbed
SiOD/vacuum	149±15	-	-	OD T_1 with 67% OH replaced by OD
BOH/vacuum BOH/CCl_4	~70	~14.3	-	Early decay time; long time component present
$SiNH_2$/vacuum $SiNH_2$/CCl_4	$\lesssim$20	$\gtrsim$50	-	Pulsewidth limited signal, ΔT/T ~5% for both stretches
$SiOCH_3$/vacuum	-	-	-	No pulse saturation observed for any CH-stretching mode

$^a k' = k - k_{vacuum}$

The hydroxyl T_1 lifetime for BOH modified silica (ν_{OH} ~3700 cm^{-1}) in vacuum or CCl_4 is found to be ~70 ps. Since the boron mass and coordination to the surface oxygens is different from silicon, changes in the local site-bending and stretching frequencies must occur. This surface modification is found to substantially alter the relaxation lifetime of the surface OH species.

Finally, we briefly address the T_1 decays of silica $-NH_2$ and $-OCH_3$ adsorbates. Both NH-stretching modes of $-NH_2$ were found to saturate to the same extent (ΔT/T ~ 5%) although time-resolved measurements produced an instrumental pulsewidth limited response. While precise T_1 values could not be deduced from these results, the two modes apparently exhange population rapidly and have comparable lifetimes (≤ 20 ps). Attempts to saturate the CH-stretching transitions of $-OCH_3$ were unsuccessful, however. This may arise because of complex adsorbate vibrational level coupling (Fermi resonances)[13] or low molecular absorption cross-sections. The above results for adsorbates on SiO_2 suggest that increasing the complexity of adsorbate vibrational level structure leads to more rapid vibrational energy relaxation at the silica surface.

The mechanisms responsible for adsorbate vibrational energy-transfer may be discerned by measuring T_1 as a function of temperature. Data for OH impurities in fused silica over the 100-1500 K temperature-range [9] and more recently for OH and OD on the silica surface [14] have thus been obtained. The results are interpreted in terms of multiphonon relaxation theory, which suggests (from the order of the dependence) that the OH quantum decays into four quanta of ~1000 cm^{-1} each. This result suggests that the Si-OH stretching motion (~970 cm^{-1}) may play a major role in accepting the initial excitation.

If only vibrational motions near the adsorbate are responsible for the observed energy-transfer rates, then one may consider the above results in terms of specific functional groups attached to a large molecule. Measurements of T_1 for small molecules in solution which contain such groups (e.g., OH, NH_2) are expected to exhibit similar lifetimes as for the surface systems. Studies of molecules containing SiOH and SiOD groups dissolved in CCl_4 at room-temperature have therefore been performed. Preliminary experiments on $(C_2H_5)_3SiOH$ and $(C_2H_5)_3SiOD$ gave T_1 lifetimes of about 170 ps and 220 ps, respectively. These values are comparable to the results for silica-OH in CCl_4 and are surprisingly large when one considers the nearly resonant CH-stretches as intramolecular accepting modes. Measurements on other SiOD-containing molecules also gave $T_1 \approx$ 240 ps. Attempts to study analogous alcohol systems (e.g., $(C_2H_5)_3COH$, C_6H_5OH, CH_3OH) produced varied results. While a few systems exhibited signals too small to be measured, CH_3OD and C_6H_5OD gave $T_1 \leq 40$ ps. It is suggested that modes neighboring the OH (with frequencies which change upon carbon substitution) are again responsible for this lifetime decrease. Future experiments will explore the relation between local molecular structure and T_1.

4. CONCLUSIONS

Direct time-resolved picosecond infrared saturation and ground state recovery measurements of the vibrational population decay for -OH, -OD, -BOH, $-NH_2$, and $-OCH_3$ adsorbate modes on the surface of room-temperature silica have been reviewed. T_1 lifetimes for these systems in vacuum and in contact with several organic solvents range from several picoseconds for $-NH_2$ to 200 ps for -OH at the silica-vacuum interface. Solvent interaction increases the rate of vibrational energy relaxation,and seems to depend on the solvent internal vibrational level structure. These results, in conjunction with the temperature-dependence and model molecule experiments, support the idea that the magnitude of the T_1 lifetime depends critically on the adsorbate molecular level structure and local mode frequencies.

This research was supported in part by the Air Force Office of Scientific Research.

5. REFERENCES

1. B. N. J. Persson and R. Ryberg, Phys. Rev. Lett. 54, 2119 (1985); J. C. Tully, T. J. Chabal, K. Raghavachari, J. M. Bowman, R. R. Lucchese, Phys. Rev. B. 31, 1184 (1985).
2. J. W. Gadzuk and A. C. Luntz, Surf. Sci. 144, 429 (1984); J. C. Arizasu, D. L. Mills, K. G. Lloyd and J. C. Hemminger, Phys. Rev. B 30, 507 (1984).
3. B. Hellsing and M. Persson, Physica Scripta 29, 360 (1984); B. N. J. Persson, J. Phys. C 17, 4741 (1984).
4. E. J. Heilweil, M. P. Casassa, R. R. Cavanagh, and J. C. Stephenson, J. Chem. Phys. 81, 2856 (1984).
5. E. J. Heilweil, M. P. Casassa, R. R. Cavanagh, and J. C. Stephenson, J. Chem. Phys. 82, 5216 (1985)
6. M. P. Casassa, E. J. Heilweil, J. C. Stephenson, and R. R. Cavanagh, J. Vac. Sci. A 3, 1655 (1985).
7. A. Laubereau and W. Kaiser, Rev. Mod. Phys., 50, 607 (1978); "Ultrafast Phenomena IV," D. H. Auston and K. B. Eisenthal, eds. (Springer Verlag, NY 1984).
8. E. J. Heilweil, M. P. Casassa, R. R. Cavanagh, and J. C. Stephenson, in SPIE Proceedings, Ultrashort Pulse Spectroscopy and Applications, Vol. 533, pp. 15, 1985; E. J. Heilweil, M. P. Casassa, R. R. Cavanagh, and J. C. Stephenson, J. Vac. Sci. Tech. B, (in press).
9. E. J. Heilweil, M. P. Casassa, R. R. Cavanagh, and J. C. Stephenson, Chem. Phys. Lett., 117, 185 (1985).
10. Certain commercial equipment, instruments, or materials are identified in this paper in order to adequately specify the experimental procedure. In no case does such identification imply recommendation or endorsement by the

National Bureau of Standards, nor does it imply that the materials or equipment identified are necessarily the best available for the purpose.
11. J. Chesnoy and D. Ricard, Chem. Phys. 67, 347 (1982); A Seilmeier, P. D. J. Schere and W. Kaiser, Chem. Phys. Lett. 105, 140 (1984).
12. -OD prepared by exposure to D_2O; $-NH_2$ by ammination after chlorination: J. B. Peri, J. Phys. Chem. 70, 2937 (1966); $-OCH_3$ by methylation: R. S. McDonald, J. Phys. Chem. 62, 1168 (1958); and BOH by exposure to BCl_3 and water: M. L. Hair and W. Hertl, J. Phys. Chem. 74, 91 (1970).
13. A. Fendt, S. F. Fischer, and W. Kaiser, Chem. Phys., 57, 55 (1981).
14. M. P. Casassa, E. J. Heilweil, J. C. Stephenson, and R. R. Cavanagh, submitted to Phys. Rev. Lett.

Vibrational Energy Relaxation of Hydrogen Chloride in Liquid and Solid Xenon

J. Chesnoy

Laboratoire d'Optique Quantique du C.N.R.S., Ecole Polytechnique, F-91128 Palaiseau Cedex, France

Introduction

Vibrational relaxation and spectroscopy of impurity molecules trapped in low-temperature matrices have been intensively studied [1]. In the liquid phase, vibrational relaxation attracts a great deal of attention as well [2], but very few studies make a link between these two phases. This paper presents the vibrational population relaxation for hydrogen chloride (HCl) diluted in liquid and solid xenon (Xe), on both sides of the fusion point studied by laser-induced fluorescence (LIF).

The v=1 relaxation time is measured in the liquid and in the solid versus concentration and temperature. On solidification of the xenon, we observe the continuity of the relaxation times for the HCl isolated in the solution. In the solid phase, we measured the relaxation times for the first three excited vibrational states versus temperature up to the fusion point. We can understand our results in terms of binary approaches to vibrational relaxation currently applied to the liquid phase.

The aim of this work was to compare the relaxation and spectroscopic behaviour at the phase-transition and to attempt an explanation of the results in a common framework, as studies in liquid and crystalline phase are usually carried out from disconnected points of view.

Experiment and results

The LIF technique employs a passively Q switched neodymium YAG laser (25 ns duration) pumping a lithium niobate angle-tuned parametric oscillator (OPO) delivering 2mJ around 3.5μm to excite the v=1 level, and 10 mJ around 1.8μm to populated the v=2 level. Frequency doubling by a second lithium niobate crystal, of the output at 2.4μm, produces 2mJ around 1.2μm, permitting the excitation of the v=3 level. The fluorescence signal is collected via a monochromator and dielectric filters onto a photovoltaïc InSb detector. Exponential decay curves are followed over a dynamic measurement range of about 20. Relaxation as well as spectroscopic measurement can be achieved on the same sample. Vibrational absorption in the fundamental band allows the concentration of HCl in the Xe solvent to be monitored.

1 Spectroscopy

Fig.1 a presents the fundamental absorption cross-section of HCl diluted in Xe in the liquid and the solid on both sides of the fusion point,while the fluorescence spectra are displayed in fig. 1 b. The spectra have two shoulders that are very similar to the gas P and R branches without the fine structure. But a central "Q" branch is clearly apparent in the spectrum of the liquid,as was already observed, while it is absent in the solid. The induction of a forbidden Q branch for molecules in solvents has been extensively studied and interpreted as due to the mixing of free rotational states $|J,m\rangle$ with the neighbouring states $|J\pm1,m\rangle$. The absence of a central branch in the crystal manisfests the symmetry of inversion for the rotational barrier undergone by the HCl molecule at the center of its site in Xe. As developed in Ref.3, it implies that the rotational states are primarily sensi-

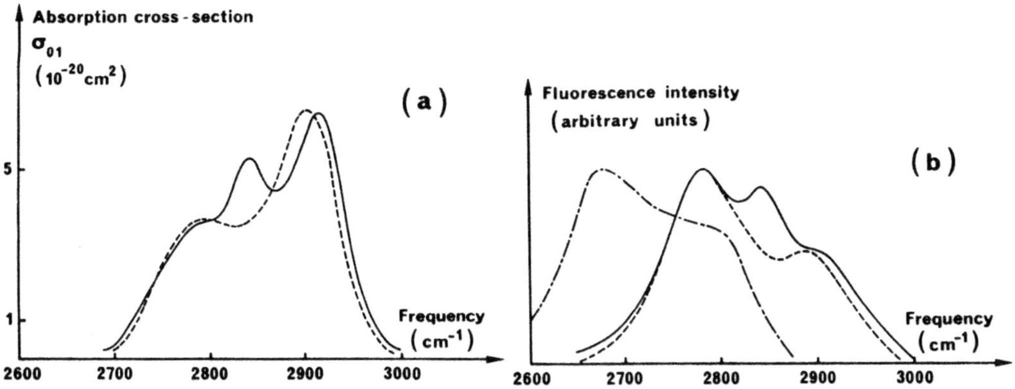

Fig.1 : Vibrational spectra of HCl diluted in Xe close to the fusion point, in the liquid at 163 K (full lines) and in the solid at 158 K (dotted lines). The upper part of the figure displays the absorption spectra from v=0 to v=1 whilst, in the lower part, the fluorescence spectra from v=1 to v=0 are presented as well as the fluorescence spectrum from v=2 to v=1 (dash-dotted line) in the solid. The relative magnitudes of the fluorescence intensities are arbitrary

tive to the mean barrier and not to the thermal motion,which average on a time-scale shorter than the rotational period. Notice that absorption and fluorescence spectra are symmetric with respect to each other by mirror reflection. The fluorescence spectrum for v=2 could be recorded only in the solid (dash-dotted line). The fluorescence spectra for v=2 and v=1 are similar but shifted by the anharmonicity, conserved from the gas phase, so that a spectral window can easily be found to observe the fluorescence decay from a single vibrational level.

2 Liquid Phase Relaxation

The relaxation of HCl/Xe solutions in the liquid at 163 K, just above the fusion point, is presented in Fig.2 a. The relaxation rate (inverse relaxation time) $1/\tau_{10}$ is plotted against the mole fraction x of HCl in Xe. A linear concentration dependence is observed, due to the relaxation constant ks of HCl by itself in Xe. The non-zero initial value of $1/\tau_{10}$ gives the relaxation time t_{10}^{∞} of HCl isolated in liquid xenon. ρ_0 being the Xe density (molecule / cm^3) :

$$\frac{1}{\tau_{10}} = \frac{1}{\tau_{10}^{\infty}} + ks\ \rho_0\ x \qquad (1)$$

with $\tau_{10}^{\infty} = 11\mu s$ and ks $\simeq 8.4 \pm 0.5 \times 10^{-14}\ s^{-1}\ cm^3$. An important observation concerns the very large difference bewteen the relaxation efficiencies of HCl and Xe for the deexcitation of HCl : HCl is about 20 000 times more efficient than Xe.

3 The v=1 Relaxation in the Matrix

Fig.2 b shows the variation of the relaxation rate of HCl as a function of its mole fraction in solidified Xe. The variation, less pronounced that in the liquid, comes from a different origin. As is often observed in low-temperature matrices [4], the variation of τ_{10} with the concentration comes reasonably from a Förster transfer to impurities, and experiments on HCl/rare gases in low-temperature matrices have shown that the dimer HCl-HCl could play the role of an energy trap.

We obtain an important result that the inverse relaxation time extrapolated to the fusion point gives a value around 650 ms^{-1} that is not far from the value $1/\tau_{10}^{\infty}$ measured in the liquid ($\sim$ 900 ms^{-1}). This gives a direct observation of the continuity of the relaxation time across the matrix/liquid transition. We will now be concerned only with dilute solutions of HCl in Xe where the guest molecule can be

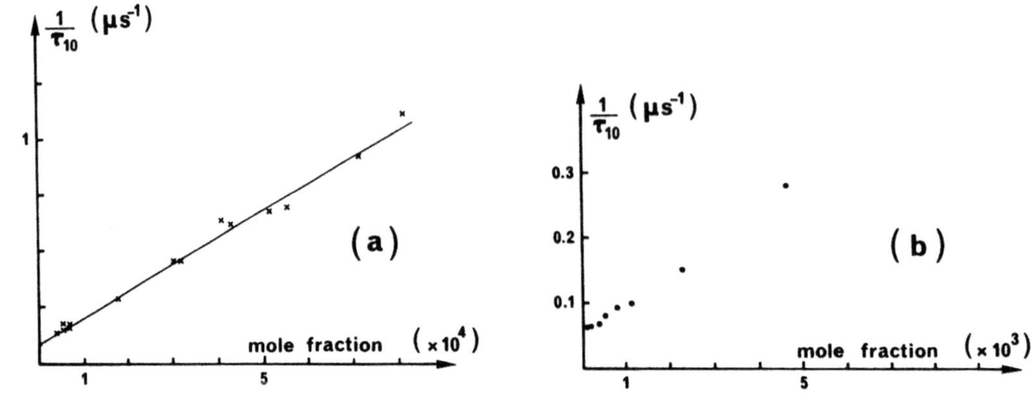

Fig.2 : Inverse relaxation time $1/\tau_{10}$ for v=1 of HCl in liquid xenon at 163 K (a), versus mole fraction, displaying a linear concentration-dependence and a non-zero intersection for zero concentration. The lower curve (b) displays the same variation in the solid matrix at 158 K. Notice the different units used.

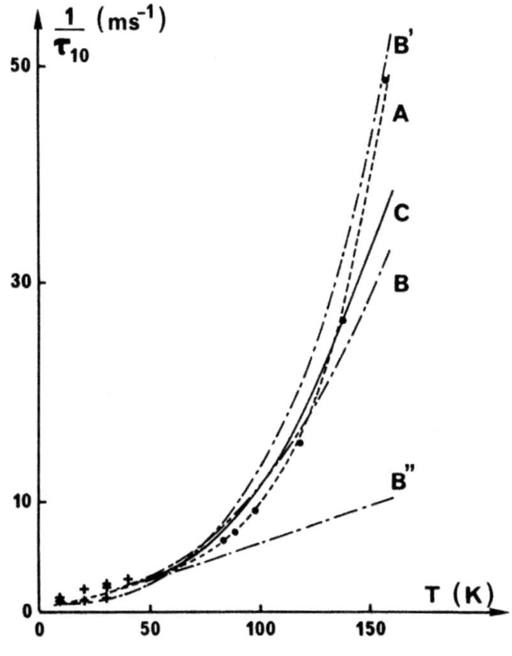

Fig.3 : Theoretical interpretation of the variation of the inverse relaxation time $1/\tau_{10}$ for HCl diluted in xenon versus temperature. The dotted line (A) represents the matching of the experimental points by a vibration to rotation mechanism [9]. The dashed dotted lines follow a vibration to translation and rotation mechanism [10] : B with one delocalised phonon (10 cm^{-1}) and one local phonon (45 cm^{-1}), B' with one delocalised phonon (20 cm^{-1}) and two local phonons (2 x 45 cm^{-1}), B" with only one delocalised phonon (10 cm^{-1}). The full line (C), shows the result of a binary collision model.

considered as isolated in the matrix. Fig.3 gives the variation of the inverse relaxation time with the temperature (points). Our data extends smoothly the low-temperature measurements of Young and Moore [5] (crosses in fig.3), but the temperature-variation of the relaxation time becomes more pronounced when the fusion point is approached.

Interpretation

We briefly summarize the main theoretical approaches to vibrational relaxation in dense phase [2]. Usually, the starting formula is the golden rule : in the frequently justified weak coupling limit, a molecule in the vibrational state v, submitted to the perturbing Hamiltonian H_B from the bath, relaxes to the state v'

$$\frac{1}{\tau_{vv'}} = \frac{2\pi}{\hbar} \sum_{e,f} \rho_e |\langle v,e|H_B|v',f\rangle|^2 \quad \delta(E_{vv'}-E_{f,e}) \qquad (2)$$

where the summation over initial (e) and final (f) bath states is executed, ρ_e

being the density of initial bath states. This formulation can directly be applied to solid matrices, since the quantum states can be given a reasonable form for all the degrees of freedom (vibration, rotation, translation).

But in liquids, as translational states cannot be expressed, binary approaches are followed. The Isolated Binary vibrational Interaction [6,2] (IBI) approach starts form a correlation function expression equivalent to the golden rule :

$$\frac{1}{\tau} = \frac{1}{\hbar^2} \int_{-\infty}^{\infty} dt \, \exp(i\omega t) \, \langle H_B(0) \, H_B(t) \rangle \tag{3}$$

this formulation is much easier to use than the Fermi golden rule, since a semi-classical interpretation of the correlation function $\langle H_B(0)\, H_B(t)\rangle$ is possible. If, moreover, the correlation function can be replaced by its binary approximation and if the dynamics are dominated by strong interactions occurring at short time on a limited potential range, the following basic formula is reached :

$$\frac{1}{\tau} = \rho \; k^{gas} \; \frac{g(R^*)}{g^{gas}(R^*)} \tag{4}$$

so that a link is made between the relaxation time τ in the dense phase (density ρ) and the pair distribution function g in this phase taken at the distance R* characteristic of the deexcitation process.

The latter approach, that is also the simpler to apply, is the Isolated Binary Collision approximation [6]. Symbolizing the molecules as hard spheres, the relaxation rate is then proportional to the collision rate ν :

$$\frac{1}{\tau} = P\nu \tag{5}$$

where P is the probability of relaxation per hard collision. In some way, it may be identified with eq.4 when pair distribution functions are taken for hard spheres [7].

In the case of a binary mixture (i of concentration $x \ll 1$, diluted in j) and with $k_{ij}^{gas} \gg k_{ij}^{gas}$, the IBI approach (eq.4) can be generalised when the self-relaxation rate is not diffusion limited.

$$\frac{1}{\tau} = x \, \rho \, k_{ii}^{gas} \, \frac{g_{ii}(R^*)}{g_{ii}^{gas}(R^*)} + \rho \, k_{ij}^{gas} \, \frac{g_{ij}(R'^*)}{g_{ij}^{gas}(R'^*)} \tag{6}$$

where the crossed pair distribution function g_{ij} is taken at a distance R'* characteristic of the relaxation of species i by j. Our measurement of the relaxation of HCl (i) diluted in liquid Xe (j) just above the fusion point is contained within this framework, and effectively, a linear variation of $1/\tau_{10}$ with x is observed giving a ratio of the pair distribution function $\simeq 1.4$. This magnitude is very reasonable on the repulsive part of the potential where relaxation takes place [2]. We can accept that the relaxation of HCl in liquid Xe, in agreement with most previous studies in liquid phase, is well described by the IBI approach.

In the solid phase, the Golden rule expression (eq.2) can be directly evaluated in principle. Only prototypic systems, such as the transfer to phonons [8] or to the rotation alone [9] were first treated. For hydrides such as HCl, the transfer is strongly in favour of rotation due to the large rotational quanta, so that pure transfer to rotation will be a better model than to translation. From the comparison of the results given by two theories (ref 9 and 10) of vibrational transfer mainly to the rotations, with the experimental results of relaxation in HCl/Xe, we see on fig.3 that the variation with the temperature can be reproduced quite well over the whole temperature-range in matrices up to the fusion point. With the theory by Metiu [9] considering a transfer to rotation, as well as with the results of Ref. 10 considering that several phonons participate to the transfer (clearly the straight line obtained with a single accepting phonon is no longer acceptable at high temperature). This agreement does not shed insight on the problem,since in the theory developed by Freed and Metiu, this variation is given by the rotational populations, while in the theory by Berkowitz and Gerber it is given by the populations of the phonons completing the energy mismatch.

Our observation of very similar values for τ_{10} in liquid and solid phases shows that the relaxation is quite insensitive to the site symmetry, and this cannot be easily accounted for by golden rule approaches that give a great importance to the symmetry of the crystal.

In contrast with golden rule theories,the binary approaches explain easily our complete set of data. The continuity of the relaxation time at the fusion point is easily understood,since the pair distribution functions are not noticeably altered on the repulsive part of the potential. Moreover, as shown by the full line on fig.3, a binary collision model explains relaxation rate in HCl dissolved in Xe. Other results on the relaxation of higher levels,or previously obtained by other groups at low temperature, [5] confirms this binary interpretation.

Conclusion

Vibrational relaxation of hydrogen chloride, studied in the liquid and solid phase, tends to show that the deexcitation process may be described in a similar way in all the phases, gas, liquid and solid. Both the temperature-dependence behaviour and the continuity of the relaxation rate accross the solid/liquid phase-transition do suggest the usefulness of binary interaction models.

We suggest here that the elementary deexcitation process can be known from an isolated collision (for example by a study in molecular beams or in a gas) and that the application to high densities only needs scaling with the density using some model. The methodology is that of the theorical interpretation of relaxation in liquids. The reason why scaling is sufficient comes from the necessity of strong interactions to relax a vibrational quantum.

References

1. H. Dubost in Inert gases, the Chemical Physics Aspects, M.L. Klein ed. Springer Series in Chemical Physics, 1984.
2. J. Chesnoy and G.M. Gale, Ann.Phys. Fr, 6, 893 (1984)
3. J. Chesnoy, Chem.Phys.Letters, 114, 220 (1985)
4. F. Legay in Chemical and biochemical Applications of Lasers II, C.B. Moore ed, Academic Press, 1977
5. L. Young and C.B. Moore, J.Chem.Phys., 81, 3137 (1984)
6. P.K. Davis and I. Oppenheim, J.Chem.Phys., 57, 505 (1972)
7. C. Delalande and G.M. Gale, J. Chem.Phys., 71, 4531 (1979)
8. J. Jortner, Mol.Phys., 32, 379 (1976)
9. K.F. Freed and H. Metiu, Chem.Phys.Lett., 48, 262 (1977)
10. M. Berkovitz and R.B. Gerger, Chem.Phys., 37, 369 (1979)

Ultrafast IR Spectroscopy Studying Nearly-Free Induction Decay

H.-J. Hartmann, K. Bratengeier, and A. Laubereau

Physikalisches Institut, Universität Bayreuth, D-8580 Bayreuth, F.R.G.

In recent years time-resolved coherent Raman scattering techniques have been used for numerous studies of molecular dynamics in the electronic ground state. In this brief article a picosecond IR method is reported, taking advantage of the different selection rules and higher efficiency of electric dipole coupling as compared to the Raman interaction. Our technique represents the analogue to time-resolved CARS. The time-resolved IR spectroscopy described here is based on the coherent propagation of low-intensity picosecond pulses.[1] The case of moderately short samples is considered, where nearly-free induction decay (NFID)[2,3] of the resonantly interacting molecules is observed, providing direct information of the dephasing time T_2 of the investigated molecular transition.

Our time-domain IR spectroscopy has been applied to various vibration-rotation transitions of HCl gas in mixtures with Ar buffer gas. Working at a total pressure of 3 bar, homogeneous collision-broadening is adjusted, while Doppler broadening is completely negligible. The chosen pressure-range, on the other hand, necessitates picosecond time-resolution. The measuring system is depicted in Fig. 1. Single picosecond pulses, generated by a Nd:glass laser system enter a double-pass parametric generator-amplifier system. High quality tunable infrared pulses (duration $t_p \simeq 2$ ps) and narrow frequency width ($t_p \times \Delta\nu \simeq 0.7$) are efficiently produced at the "signal" (ω_S) and "idler" (ω_{IR}) frequency positions by stimulated parametric amplification in a 3 cm $LiNbO_3$ crystal. The "idler" pulse serves for the propagation phenomenon; it is resonantly tuned to the molecular transition of interest and provides information on the molecular dynamics via the coherent reshaping effect. The parametric pulse at the "signal" frequency serves as probing pulse and operates the parametric light gate for the pulse shape analysis. Measuring the up-conversion emission of the light gate with a photomultiplier versus delay time t_D between infrared and probing pulse, the IR pulse intensity is measured as a function of time. The detection scheme achieves sub-picosecond time resolution of 0.2 ps and a high sensitivity: measurements of the pulse wings can be carried out over a dynamic range of 10^7. Using a second IR light gate, the shape and amplitude of the input IR pulse are simultaneously measured in a reference experiment.

An example for the observed signal transients is presented in Fig. 2. The system HCl:Ar with natural isotope abundance of ^{35}Cl and ^{37}Cl is studied. The infrared pulse is resonantly tuned to the R(2) transition frequency of $H^{35}Cl$ at 2944.9 cm^{-1}.[4] The corresponding transition of $H^{37}Cl$, which is shifted by approximately 2 cm^{-1} to smaller frequency, also interacts near-resonantly with the incident pulse. The time-integrated conversion signal $C(t_D)$ representing the intensity of the infrared pulse as measured with the light gate technique is plotted on a logarithmic scale versus delay time t_D of

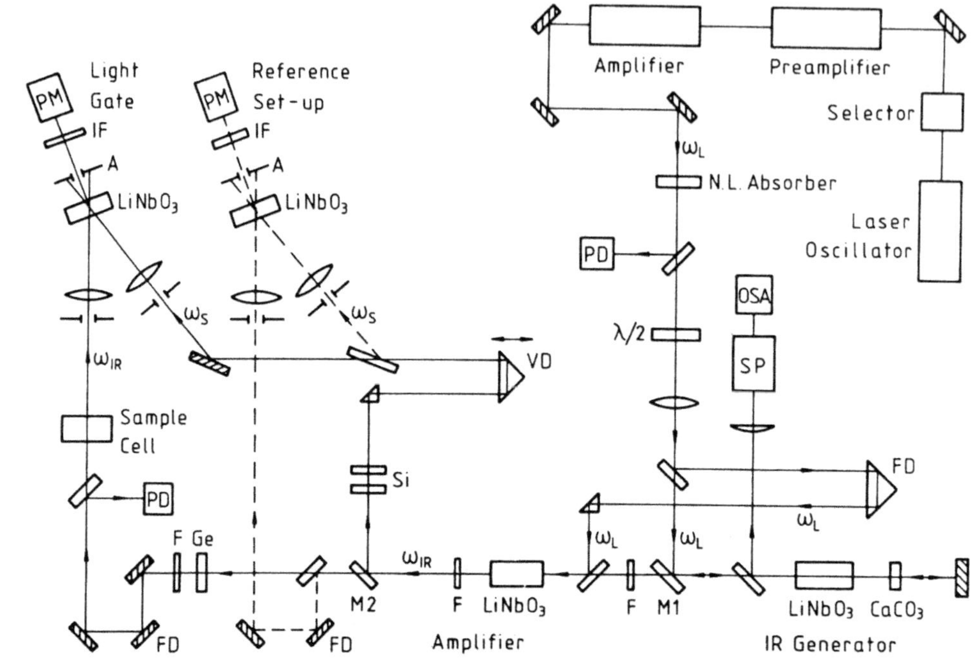

Fig. 1: Schematic of experimental system for ultrafast IR spectroscopy. Photodetector PD, mirror M, spectrometer SP, infrared multichannel analyser OSA, blocking filters Ge and Si, filters F and IF, variable delay VD, fixed delay FD, aperture A, photomultiplier PM

the probing pulse. C(O) = 1 marks the maximum signal measured with evacuated sample cell (transmission = 1). For the data of Fig. 2 with total pressure 3 bar and HCl partial pressure of 26 mbar we estimate a value of $\alpha\ell \simeq 1.5$. Here $\alpha = \alpha_1+\alpha_2$ denotes the sum of the conventional peak absorption coefficients α_i of the two isotopic species, and $\ell = 1.3$ cm is the sample length. Around $t_D = 0$ the conversion signal of the transmitted pulse (open circles, solid line) closely follows the rapid build-up and immediate decay of the input (full points, dotted line). For larger time values, $t_D \gtrsim 10$ ps, a drastic change occurs. More than three orders below the maximum, the trailing pulse wing displays a beating phenomenon, which is superimposed on an approximately exponential slope. From the oscillations of the signal curve the beating time $T_{is} = 15.3 \pm 0.3$ ps is directly measured corresponding to an isotopic line-splitting of $\Delta\omega/2\pi c = 2.18 \pm 0.04$ cm^{-1}. This value favourably compares with spectroscopic data and theoretical estimates. The envelope of the beating maxima yields the time-constant $\tau = 30$ ps. This value represents the decay time of NFID and differs from $T_2/2$. Applying an analytic correction formula[3] or fitting the semi-classical theory of coherent pulse propagation to the experimental data (solid line), we obtain the desired molecular time-constant $T_2 = 50 \pm 5$ ps. We also note that the observed intensity level of the pulse wing compares favourably with the calculated curve for the estimated value of $\alpha\ell$; i.e. quantitative agreement between theory and experiment is found.

The observed time-dependence with exponential decay is equivalent to a spectral line of Lorentzian shape with a half-width (HWHH in units of cm^{-1}) of $\Gamma = 1/(2\pi c T_2)$. The dephasing process originates

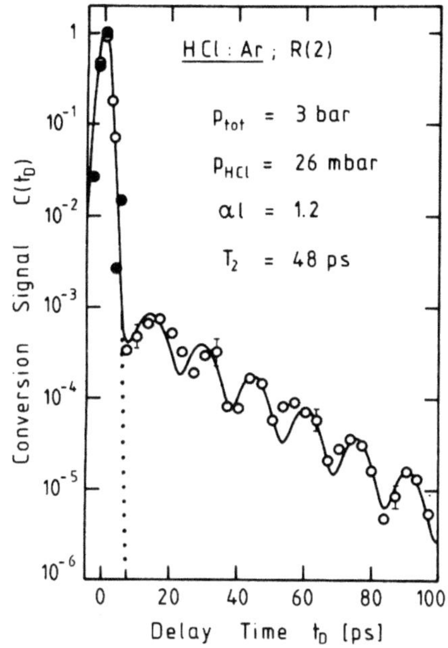

Fig. 2:

Picosecond dephasing of the R(2) transition of HCl:Ar at p_{HCl} = 26 mbar and 3.0 bar total pressure; conversion signal $C(t_D)$ representing the intensity of the input pulse (full points, broken line) and of the transmitted pulse (open circles, solid curve) versus delay time; theoretical curves

from binary HCl:HCl and HCl:Ar interactions. Results for the foreign gas broadening of various R-transitions by Ar collisions are depicted in Fig. 3. On account of the linear pressure-dependence, the experimental T_2-numbers were normalized to p_{tot} = 1 bar and $p_{HCl} \rightarrow 0$, yielding the dephasing time $T_{2,Ar} = (2\pi c\ \Gamma_{Ar})^{-1}$ due to foreign gas interaction. In the Figure the reciprocal time-constant $1/T_{2,Ar}$ is plotted (see left ordinate scale). The corresponding Lorentzian linewidth Γ_{Ar} is indicated by the right-hand scale. The hatched and open rectangles in Fig. 3 respectively represent the scatter of published IR and FIR linewidth data (see literature quoted in Ref. 4); the experimental error of the conventional spectroscopic measurements has been omitted. The results of rotational Raman scattering for J→J+2 transitions are also shown;[5] because of the different selecting rule of the Raman effect,the experimental points (◇) are displayed with an arbitrary shift of 0.5 to larger J values. The agreement with our time-resolved data is noteworthy. It is recalled that our time-resolved data and also the IR linewidth numbers contain a possible contribution of vibrational relaxation,in contrast to FIR spectroscopy and rotational Raman scattering.

The solid and broken curves in Fig. 3 represent the theoretical results of Ref. 6, considering rotational relaxation mechanisms, only. The solid curve is calculated for a purely theoretical interaction potential, while the broken curve refers to a semi-empirical HCl:Ar potential. The apparent agreement between theory and experiments in Fig. 3 and also between the experimental data involving purely rotational and rotation-vibration transitions shows that vibrational dephasing contributes little to the measured relaxation rates. There is evidence that apart from rotational population decay also pure rotational dephasing gives a significant contribution to the observed time-constants.[2]

The results for the isotopic beating effect are compiled in Fig. 4. The reciprocal beating time $1/T_{is}$ observed at 3 bar total pressure (open circles) is plotted versus rotational quantum number J. The

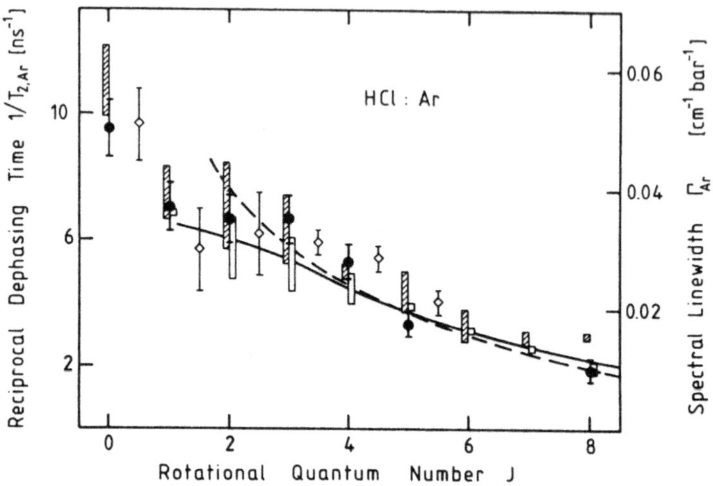

Fig. 3: Measured values of the reciprocal dephasing time $1/T_{2,Ar}$ versus rotational quantum number of the vibration-rotation transition (●); corresponding linewidth scale for foreign gas broadening Γ_{Ar}; IR-linewidth data (hatched rectangles), FIR-linewidth data (open rectangles) and rotational Raman linewidth (◇), are shown for comparison; theoretical curves

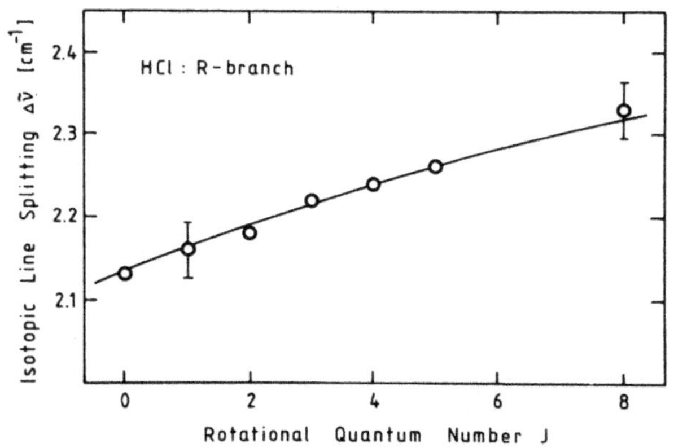

Fig. 4: Isotopic line splitting of $H^{35}Cl$ and $H^{37}Cl$ versus rotational quantum number of the vibration-rotation transitions (open circles); theoretical curve

corresponding isotopic line splitting $\Delta\tilde{\nu}$ is also indicated on the right-hand ordinate scale in Fig. 3. Experimental accuracy is typically .04 cm^{-1}. The solid curve represents the calculated frequency difference for the two isotopic species $H^{35}Cl$ and $H^{37}Cl$, taking into account the different isotopic masses and the rotational and centrifugal constants, B and D, respectively.[4] The good agreement with the experimental points demonstrates that our ultrafast infrared spectroscopy can directly measure frequency differences of ∿ 70 GHz in time domain.

In summary, we emphasize that we have applied coherent propagation of low-intensity picosecond pulses for a time-resolved IR spectrocopy. Working in the regime of nearly-free induction decay with moderately short samples, the spectroscopic information can be readily derived from the decay of the trailing pulse wing. The application of the phenomenon is facilitated by its linear intensity-dependence; i.e. the signal transients are intensity-independent apart from a scaling factor. Our transient IR spectroscopy complements coherent Raman techniques because of different selection rules; additional information is obtained.

References:

1 H.-J. Hartmann and A. Laubereau, Opt. Commun. 47, 117 (1983)
2 H.-J. Hartmann and A. Laubereau, J. Chem. Phys. 80, 4663 (1984)
3 H.-J. Hartmann, K. Bratengeier and A. Laubereau, Chem. Phys. Lett. 108, 555 (1984)
4 H.-J. Hartmann, H. Schleicher and A. Laubereau, Chem. Phys. Lett. 116, 392 (1985)
5 G.J.Q. van der Peijl, D. Frenkel and J. van der Elsken, Chem. Phys. Lett. 56, 602 (1978)
6 E.W. Smith and M. Girard, J. Chem. Phys. 66, 1762 (1977)

Dephasing in Liquid N_2 and in Binary Mixtures of N_2 with Liquid Argon

D. Brandt, H.J. van Elburg, B.L. van Hensbergen, and J.D.W. van Voorst

Laboratory for Physical Chemistry, University of Amsterdam, Nieuwe Achtergracht 127, NL-1018 WS Amsterdam, The Netherlands

In order to understand the contribution of several mechanisms to the dephasing time in liquid nitrogen as well as in binary mixtures with liquid argon for a broad range of physical parameters, in terms of molecular dynamics calculations, we have gathered a large number of data along the liquid-gas coexistance curve by spontaneous polarized Raman scattering from the vibrational transition in N_2 molecules /1/.

In terms of the mole fraction of nitrogen, x_{N2}, the composition of the mixtures was 0.75(2), 0.50(1), 0.30(2) and 0.055(10) respectively. The measured linewidths were converted to Raman linewidths assuming a Lorentzian lineshape. The Raman linewidth and the dephasing time were assumed to be related according to $\delta\omega_{Ram}\tau_D = 1$.

The data of pure liquid N_2 already reported in the literature /2, 3, 4/ agree well within experimental error with our measurements on pure liquid N_2. Also the data reported for a few mixtures /4, 5/ were in full agreement with our results. Our experiments also revealed that, throughout the temperature-range from the normal boiling point up to and beyond the critical point, the lineshape has a (mainly) Lorentzian character. Near the critical point a deviation can be expected /6/ and indeed is observed, but only a minor change is involved.

The results are presented as a linear plot of the dephasing time as a function of temperature. An example is shown in figure 1 for x = 0,5. The dephasing time was also plotted as a function of mole-fraction for a constant temperature-difference ΔT with respect to the appropriate critical temperature of the mixture (fig. 2). It

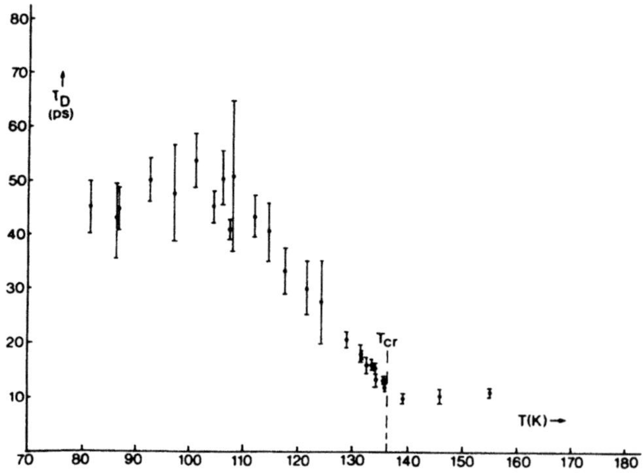

Fig. 1:

Dephasing times as a function of temperature for x_{N_2} = 0,50

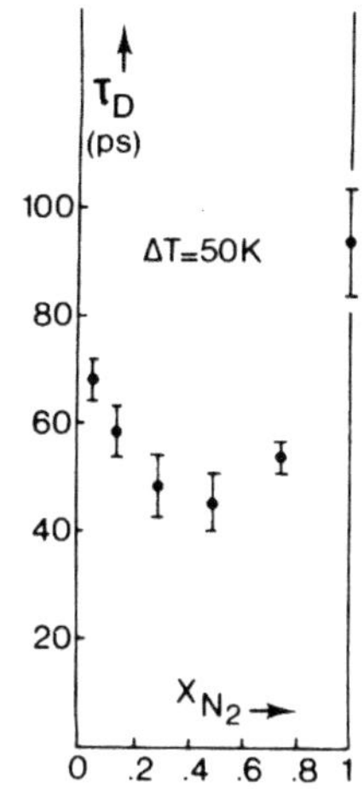

Fig. 2:

Dephasing times as a function of molefraction at a constant temperature-difference with the respective critical points

appeared that for all ΔT the curve could be fitted to a polynomial of degree two. The Raman linewidth in mixtures of intermediate molefraction increases with respect to the width in pure argon and nitrogen environments. This might suggest, although much less restricted, an explanation in the sense of a model given by Knapp and Fischer /7/ and extended to a spectral diffusion. First, in a mixture, even for a fixed number of Ar and N_2 as nearest neighbours, many different configurations can exist, each with its own average frequency shift due to the difference in intermolecular potential between the reference molecule with respect to Ar and N_2 respectively. Secondly, Quentrel and Brot /8/ have demonstrated for liquid nitrogen that there exists a correlation lasting over several picoseconds between the nearest neighbours that were present at a certain time around a reference molecule. To the fluctuating vibrational frequency thus contribute (i) rapid fluctuations due to rotational reorientational collisions with the neighbouring solvent molecules and (ii) a slower modulation caused by the change in number of nearest neighbours, composition and configuration of the solvent layer. The difference in time-scale between these two motions probably is sufficient to regard them as uncorrelated. The situation can be characterized by a weakly or coupled not too slow relaxation mechanism as mentioned by Oxtoby /9/. Especially at the intermediate molefractions the features of this model appear. Then there is a large variety in composition and configurations, and thus in local average vibrational frequencies as compared to the more homogeneous environments present in a pure liquid or dilute solution.

If a slower modulation of the N_2 vibrational frequency exists, it can only be traced by a coherent transient, i.e. the two-pulse Hahn type-echo. That in principle a Raman photo-echo in a vibrational degree of freedom is possible, was shown for gaseous nitrogen /10/. The essential of this experiment is a phase-matching for a frequency box CARS configuration for (i) the two pumping as rephasing laser frequencies, and (ii) the probe laser pulse and the coherent anti-Stokes scattering, in such a way that there is also a definite phase-matching for pumping and rephasing pulse as well as for the probe pulse and the anti-Stokes pulse. This prohibits a three frequency CARS generation influencing the population during the rephasing pulse, and gives a sensitive alignment (by four frequency box CARS) for the echo experiment.

Soon after the conference, we succeed in finding a Raman echo in a x = 0,5 mixture of N_2 and Ar /11/. These echo s inhibit two curious features (i) that the echomaximum was not found at two times the delay time between pumping and rephasing pulse, but precedes that point and (ii) that the echo amplitudes as a funtion of the delay time decays non-exponential. This is in full agreement with a model for two independent modulating processes, differing in magnitude and time-scale as compared to the time-scale of the experiment as given by Loring and Mukamel /12, 13/. In fact the coherent transient response seems to support the interpretation given under I.

Up to now we have completed molecular dynamics calculations for pure N_2 and for various points on the gas-liquid coexistance curve /14/. The calculations were more according to the formalism given by Oxtoby et al. /15/. As can be seen from fig. 3, the agreement of the MD calculation in the fast modulation limit with the experimental data is very satisfactory. Fig. 4 shows that and the force on the bond, and the vibration rotation coupling as well as their cross-correlations contribute significantly.

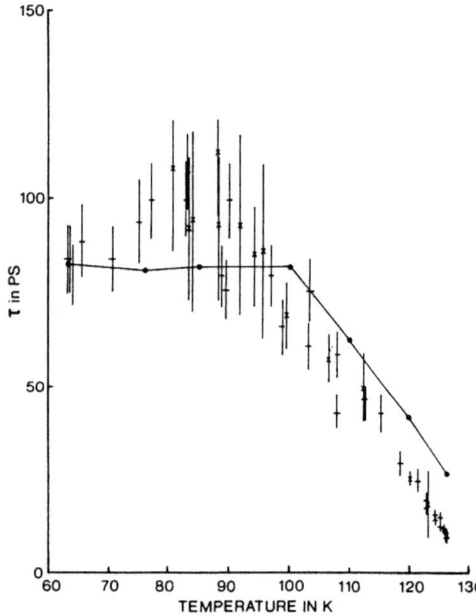

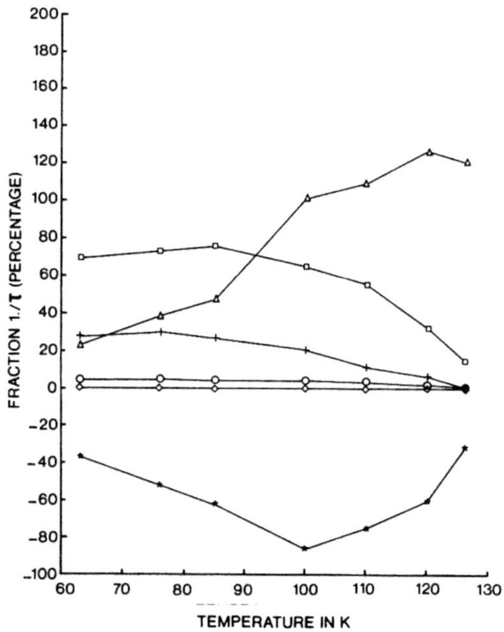

Fig. 3:

Dephasing time of N_2 in liquid nitrogen from Molecular Dynamics calculations compared with the experimental data

Fig. 4: Percentual contributions to the dephasing time

□ Force on the bond F_1 (from first term in Taylor expansion of the intermolecular potential)

○ Force on the bond F_2

+ Cross correlation F_1 - F_2

Δ Vibration rotation coupling VR

x Cross correlation F_1 - VR

◇ Cross correlation F_2 - VR

References

1 H.J. van Elburg and J.D.W. van Voorst, to be published

2 M.J. Clouter and H. Kiefte, J. Chem. Phys. 66, 173 (1977)

3 A. Laubereau, Chem. Phys. Lett. 27, 600 (1974)

4 H.M.M. Hesp, J. Langelaar, D. Bebelaar and J.D.W. van Voorst, Phys. Rev. Lett. 39, 1376 (1977)

5 S.A. Akhmanov, F.N. Gadjiev, N.I. Korolev, R. Yu. Orlov, Appl. Opt. 19, 859 (1980)
6 M.J. Clouter and H. Kiefte, Phys. Rev. Lett. 52, 763 (1984)
7 E.W. Knapp and S.F. Fischer, J. Chem. Phys. 76, 4730 (1982)
8 B. Quentrel and C. Brot, Phys. Rev. A12, 272 (1975)
9 D.W. Oxtoby, J. Chem. Phys. 74, 5371 (1981)
10 V. Brueckner, E.A.J.M. Bente, J. Langelaar, D. Bebelaar and J.D.W. van Voorst, Opt. Commun. 51, 49 (1984)
11 D. Brandt, A. Draijer, J. Pieterse and J.D.W. van Voorst, to be published
12 R.F. Loring and S. Mukamel, Chem. Phys. Lett. 114, 426 (1985)
13 R.F. Loring and S. Mukamel, to be published in J. Chem. Phys.
14 B. van Hensbergen and J.D.W. van Voorst, to be published
15 D.W. Oxtoby, D. Levesque, J.J. Weis, J. Chem. Phys. 68, 5528 (1978)
16 D. Levesque, J.J. Weis, D.W. Oxtoby, J. Chem. Phys. 72, 2744 (1980)

High Pressure Brillouin Scattering on Liquid Carbon Tetrachloride and Benzene

A. Asenbaum

Institut für Experimentalphysik der Universität Wien, Strudlhofgasse 4, A-1090 Wien, Austria

Brillouin spectra of liquid carbon tetrachloride and benzene have been measured at high pressures up to 1500 bars and at temperatures between 298 and 348 K. From the experimental spectra the energy vibrational relaxation time τ was determined as a function of pressure. At constant temperature τ decreases with increasing density and at constant density τ decreases with increasing temperature.

INTRODUCTION

From Brillouin scattering experiments on nonassociated molecular liquids, information about vibrational relaxation can be obtained (1-4). By measuring the Brillouin peak shift, both sound velocity and sound frequency can be determined, if the wave number is known, which depends on the index of refraction of the liquid under study and the scattering angle.

Further information can be obtained from the half-widths of the Brillouin peak, which corresponds to sound damping, and the halfwidth of the Mountain peak, which is indirectly proportional to the relaxation time. This relaxation time (5) can be interpreted as the time which is necessary to change the distribution of the vibrational degrees of freedom according to sound wave - induced changes of external (translational, rotational) degrees of freedom.

Brillouin scattering experiments on liquids under high pressure (3,6,7) allow measurements of the vibrational relaxation time as a function of density at constant temperature, where only the mean free path in the liquid is changed. Furthermore, measurements of the relaxation time as a function of temperature are possible at constant density, where mainly the transition probability between translational and vibrational degrees of freedom is changed, whereas the mean free path remains nearly constant.

EXPERIMENTAL

The light-scattering apparatus consisted of an argon ion laser in single-mode operation (λ=514.5 nm) with a power of 150 mW, an electronically stabilized five-pass Fabry-Perot interferometer as spectrometer (8), a cooled photomultiplier (FW 130) as detector, and a multichannel analyzer as data aquisition system.

The multiscaler was triggered by the Rayleigh peak to avoid broadening of the spectral components of the Brillouin spectra caused by drifts of the laser frequency and the misalignment of the Fabry-Perot mirrors. During the experiment a finesse between 25 and 30 was maintained.

The liquids, carbon tetrachloride and benzene, were contained in a high-pressure cell with three windows in a 90 degree scattering geometry. The high-pressure cell was surrounded by a thermostat to hold the temperature of the liquid sample constant within ±0.2 K. The scattering angle was measured to be 90 ± 20'. The high pressure system consisted of a hand pump up to 700 bars and a hand screw press up to 2000 bars. Pressure was measured with an error of ±12 bar. The relaxation times were determined within ±6%.

RESULTS AND DISCUSSION

Figure 1 shows two experimental Brillouin spectra at 298K of carbon tetrachloride at 150 bars and 1200 bar respectively.

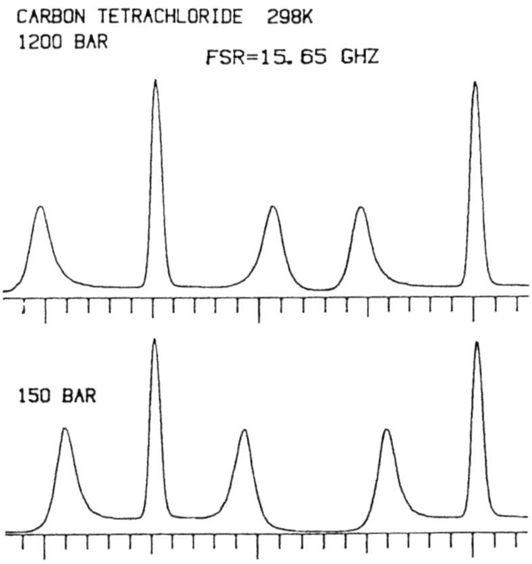

Fig.1 Brillouin spectra of carbon tetrachloride at 298K and at 150 bars and 1200 bars respectively. The finesse is 27, the FSR is 15,65 GHz.

At carbon tetrachloride the vibrational relaxation time was measured at pressures up to 1250 bars at 298K (3), at pressures up to 1500 bars and the temperatures 298, 323 and 348 K (6) and as a function of temperature at constant density (9). Figure 2 shows the relaxation time of carbon tetrachloride as a function of pressure at three different temperatures.

At benzene the relaxation time was measured at pressures up to 1300 bars and at the temperatures 298, 313, 323, 343K, whereas earlier papers (10,11) exist on the hypersound velocity at high pressures. Figure 3 shows the relaxation time of all but the lowest vibrational modes of benzene as a function of pressure at constant temperature.

From fig.2 and 3 follows, that τ decreases with increasing pressure , whereas at constant pressure τ decreases with increasing temperature.

Assuming that isolated binary collisions (IBC) models are

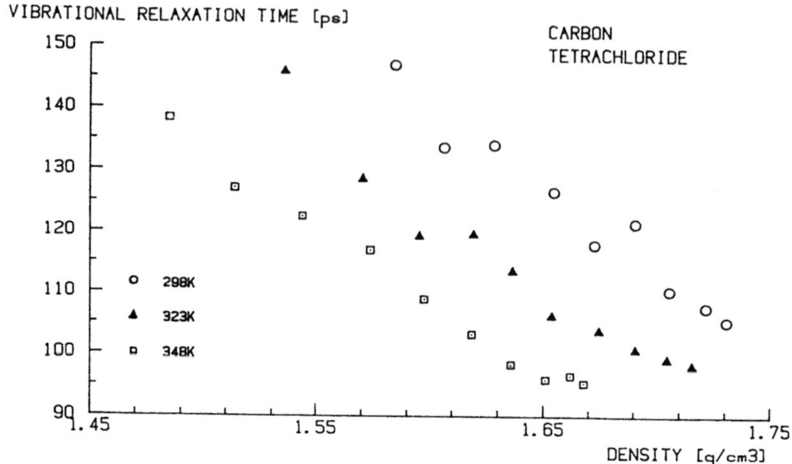

Fig.2 Vibrational relaxation time of carbon tetrachloride at a function of density at 298, 323, 348K.

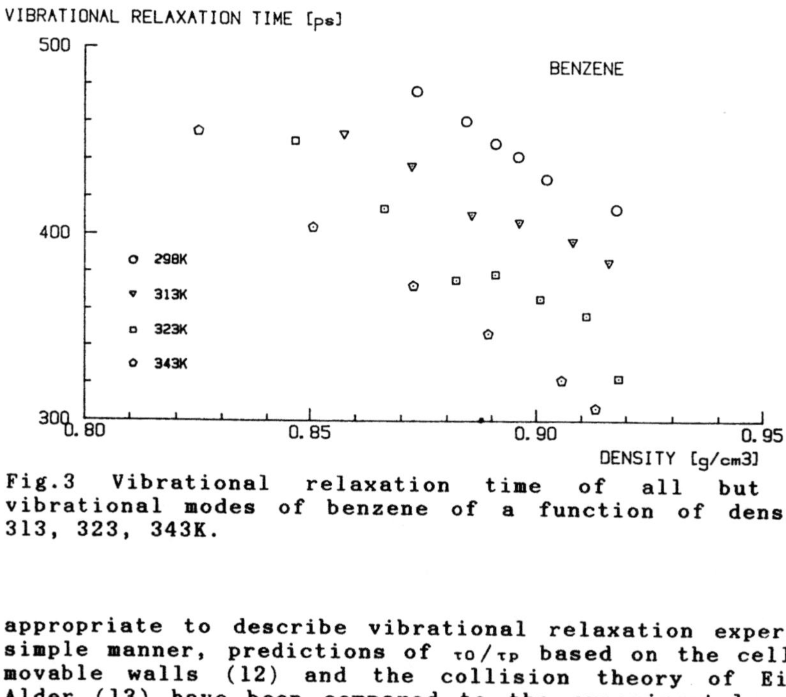

Fig.3 Vibrational relaxation time of all but the lowest vibrational modes of benzene of a function of density at 298, 313, 323, 343K.

appropriate to describe vibrational relaxation experiments in a simple manner, predictions of τ_0/τ_P based on the cell model with movable walls (12) and the collision theory of Einwohner and Alder (13) have been compared to the experimental values (7,9). For the hard sphere diameter of the molecules, temperature-dependent values derived from transport coefficients were chosen (benzene σ=5.12A (14,15), carbon tetrachloride σ=5.25A (16)). It could be shown (7), that the model of Einwohner and Alder (13) gives a reasonably good fit to the experimental data, compared to the cell model with movable walls (12). Nearly the same result was obtained by a recent Brillouin scattering experiment on carbon tetrachloride (9), at which the vibrational relaxation time was measured as a function of temperature at a constant density of 1.622 g/cm^3.

ACKNOWLEDGEMENTS

The author wishes to thank Professor Dr. P. Weinzierl for the interest in this work and Professor Dr. M. Sedlacek for helpful discussions.

1 R.D.Mountain, J.Res.NBS 70A, 207 (1966)
2 W.H.Nichols and E.F. Carome, J.Chem. Phys. 49, 1000 (1968)
3 M.Sedlacek, Z.Naturforsch. 29a, 1622 (1974)
4 A.Asenbaum and M.Sedlacek, Adv.Mol.Relax.Proc. 13, 225 (1978)
5 K.F.Herzfeld and T.A.Litovitz, Absorption and Dispersion of Ultrasonic waves (Academic, New York, 1959).
6 A.Asenbaum and H.D.Hochheimer, J.Chem.Phys. 74, 1 (1981)
7 A.Asenbaum and H.D.Hochheimer, Z.Naturforsch. 38a, 98 (1983)
8 J.R.Sandercock, J.Sci.Instrum. 9, 566 (1976)
9 M.Sedlacek and M.Wöhrl, to be published
10 J.H.Stith, L.M.Peterson, D.H.Rank, and T.A. Wiggins, J.Acoust. Soc.Amer. 55, 785 (1974)
11 F.D.Medina and D.C. O'Shea,J.chem.Phys. 66,1940 (1977)
12 W.M.Madigosky and T.A.Litovitz, J.Chem.Phys.34,489 (1961))
13 T.Einwohner and B.J.Alder, J.Chem.Phys. 49, 1458,(1968)
14 H.J.Parkhurst and J.Jonas, J.Chem.Phys.63, 2698 (1975)
15 H.J.Parkhurst and J.Jonas, J.Chem.Phys.63, 2705 (1975)

Part III

Vibrational Dynamics in Solids

Vibrational Energy Relaxation in Solid Benzene[1]

R.M. Hochstrasser

Department of Chemistry, University of Pennsylvania,
Philadelphia, PA 19104, USA

1. CARS Measurements of Vibrational Relaxation in Solids

Molecular crystal vibrational transitions have certain unique characteristics which allow their dynamics to be explored using coherent methods. This is made possible principally as a result of the averaging of the inhomogeneous distribution by vibrational energy-transfer [1,2]. Consider, as in Figure 1, the two vibrational states of a molecule, v=0 and v=1, between which Raman transitions can occur. If the vibrational excitation is localized, then in the aggregate, Figure 1 center, a distribution of transitions would be observed. This distribution arises because the crystal is not perfect as a result of its finite extent, strains and so on. Thus the translational symmetry is not exact and a distribution of transition frequencies exists. This distribution can be observed in the lineshape of transitions in dilute mixed crystals. The existence of vibrational energy-transfer or delocalization of the excitation changes the picture depending on the relative values of the root mean square deviation in frequency, D, and the vibrational exciton bandwidth β_V. It turns out that D/β_V is very small compared with unity for many vibrational transitions, such that classically the pure coherence decay is altered from its static form $\exp\{-D^2t^2\}$, to a dynamical, or motionally averaged form $\exp\{-(\gamma + D^2/\beta_V)t\}$ where γ is the vibrational relaxation time. Thus, whenever $\gamma \gg D^2/\beta_V$ the effects of the coherence loss due to the inhomogeneities can be neglected. In effect, the timescale of this coherence loss is increased from $1/D$ to β_V/D^2.

In the case of benzene the Raman transition exciton bands have widths of ca. 1 - 3 cm^{-1} whereas the D-values are a few tenths of a cm^{-1}, thus the vibrational transitions are significantly narrowed by the energy-transfer. A more detailed quantum mechanical calculation of the coherence loss time yields a slower

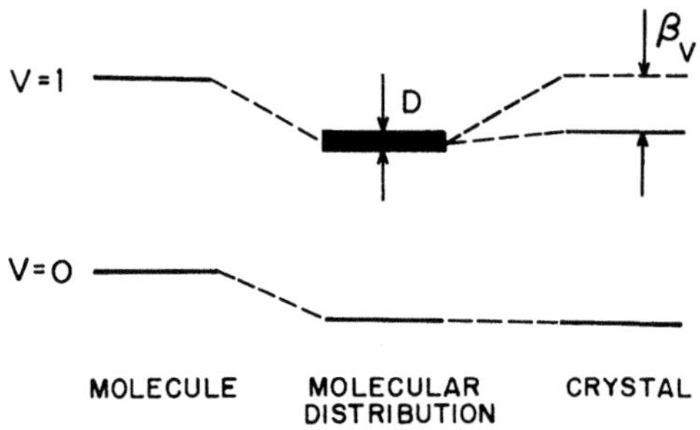

Fig.1 Effect of energy-transfer on vibrational states. Left: Molecular states. Center: Distribution of transitions of spread D resulting from strains, defects, impurities, etc. Right: Spread of levels into a band of width β_V, thereby suppressing the effect of the inhomogeneous distribution.

[1]This research was supported by grants from NIH and NSF.

than exponential decay [3] which for benzene predicts decay on the timescale of greater than 10 ns. Thus we should not expect these crystals to exhibit such inhomogeneous influences on subnanosecond coherence decays. The measured times should therefore reflect population decay only.

There are various types of population decay: (i) pure vibrational relaxation (ii) impurity trapping of the excitations (iii) trapping at surfaces and defects of an otherwise pure system. In (ii) both chemical and isotopic impurities can be important, but experiments can be derived to eliminate such effects [4]. The contribution of surface trapping can be diminished by using large crystals excited in the bulk. This is one significant advantage of Raman pumping, which requires a substantial pathlength. Trapping at defects cannot be satisfactorily eliminated and its contribution, though expected to be small, requires further study. These factors allow us to conclude that the decay of the vibrational coherence in our experiments is measuring the population, or T, decay of the vibrations.

2. Vibrational Relaxation in Benzene

As we have noted elsewhere [4] the vibrational relaxation rates of the fundamentals of benzene do not correlate well with the number of available relaxation channels. However, the rates generally increase with the total vibrational energy content as indicated in Figure 2. The benzene crystal is sufficiently harmonic that we can expect $\Delta v = \pm 1$ selection rules to determine the relaxation pathways where Δv indicates changes in both the internal and the external, relative motion, degrees of freedom. If this is correct, then the variations of rate with energy in Figure 1 have an obvious interpretation in terms of the internal anharmonicities of the benzene modes.

The anharmonicity of the benzene vibrations must increase with increasing energy content. Thus it could be considered that at low energy, say less than 1000 cm^{-1}, the modes are essentially harmonic on the relevant timescales of relaxation. In this "restricted access" region the existence of $\Delta v = \pm 1$ possibilities is not sufficient for a rapid relaxation, and a significant coupling must also be sought. In the higher energy, or "free access" region, coupling may occur to many of the levels that are separated by less than one lattice vibrational quantum, as a result of the molecular anharmonicities becoming significant. At a certain state density still in the region of the fundamentals, the internal harmonic modes

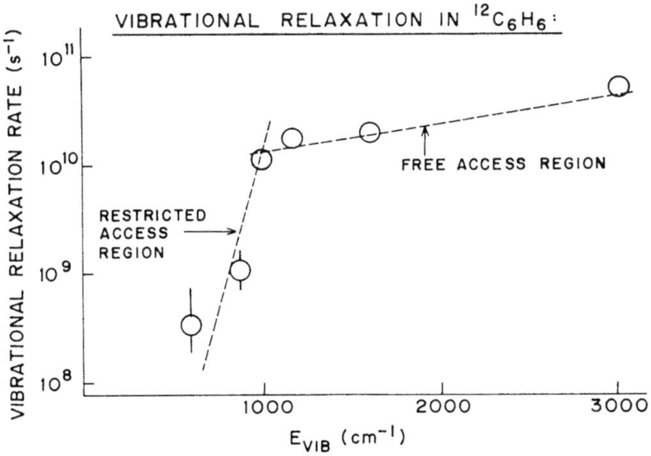

Fig.2 Vibrational relaxation rates of some Raman active modes in benzene. The dashed line is intended to draw the eye to the two regions of behavior of rate versus total vibrational energy. (Data from Ref. [4])

might be sufficiently mixed to allow relaxation into many channels while still maintaining the $\Delta v = 1$ selection rule for external modes. Internal mode mixing was previously invoked to describe the relaxation of CH modes in liquids [5] specifically in that case in terms of their coupling to CH bending motions.

It will be interesting to discover by studies of the infrared active modes of benzene whether the trend of Figure 2 is upheld. As for other systems, the basic ideas should carry over. Of course, significant differences may be expected in the energy at which the anharmonic effects become dominant. For example, in larger molecules where there are many lower frequency modes, the onset of the free access region may be at quite low total energy. The lowest frequency modes may then be too close to the lattice modes for there to be a significant restricted region. It is interesting that the trends indicated here for benzene crystals are not remarkably different from those seen for benzene molecules undergoing collisions with isopentane [6]. In that case the excitation of some of the combinations involving ν_6, ν_{16} and ν_1 in the electronically excited state yields relaxation times that are not very sensitive to mode type or to total energy content. In fact, one could imagine a similar curve drawn through these gas phase data as is drawn as a dashed line in Figure 2. It would be of great interest to have results for ground state relaxation resulting from benzene-benzene collisions, and for liquid benzene.

3. Mechanisms of Energy Relaxation [7-9]

The vibrational energy redistribution in a molecular crystal can arise in a variety of ways. Vibrational excitation may be degraded into lattice motions and new internal mode excitations on the same molecule or on different molecules. In a crystal with only one molecule in the asymmetric unit these two processes are indistinguishable. However, by effective use of isotopic impurities which involve slight changes in frequency and normal mode composition, it should be possible to sort out the various pathways.

Crucial to a proper understanding of energy relaxation in molecular crystals is the correct description of the initially excited state. An important ground rule is that in laser experiments there are generally many molecules within the excitation volume. Even at the lowest measurable trap concentrations there are also many traps per excitation volume. This led us to suppose that a proper theory of relaxation, when this is fast compared with dephasing, should include a distribution of traps in the volume that is commensurate with the concentration. Finally the observed relaxation should be an average over all such distributions. The basic physics of this approach concerns the difference in the rate of energy transfer to a localized impurity from localized compared with extended states. For example, a trap equidistant from two donors over which an excitation is delocalized becomes excited twice as fast than if the excitation was localized on one donor only -- assuming that the excitation amplitudes on the two donor molecules are equal in the former case. It is not difficult to show [7] that the optical, infrared or Raman excitation produces just such initial states having the amplitudes equal on all sites of the host lattice such as pictured in Figure 3. The energy relaxation process can then be regarded as the collapsing of the initially prepared distribution into the traps by nearest neighbor coupling.

It should be remembered that even in a pure crystal there are traps for the vibrational energy corresponding to the other vibrational levels of the host molecules. Thus the process of energy redistribution is quite complex in general, especially when the excitation amplitudes and phases are considered at each stage of the relaxation. For example, the initially excited states having equal amplitude on each site may decay to vibrational excitations that involve all possible phasing of the excitation. It remains to discover experimentally whether any coherence is transferred in such processes.

If it is now assumed that there is a time-independent rate-constant describing the relaxation in this picture, two qualitatively different contributions arise. The

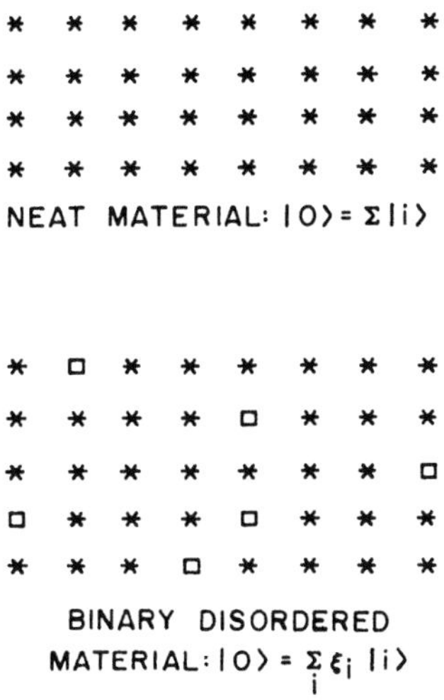

Fig.3 Representation of the states |0⟩, initially excited by pulses shorter than the dephasing time in solids. The asterisks correspond to equal excitation amplitudes. Upper: The initial state is a superposition of equal site amplitudes |i⟩. Lower: The squares represent impurity centers, such that in a binary disordered system the site amplitudes are equal to ζ_i, where ζ_i = 1 or 0 depending on whether or not |i⟩ is a host molecule.

rate-constant contains terms proportional to the concentrations squared which we have termed coherent energy flow, and those proportional to the product of the concentrations of host and guest which represent incoherent energy flow [7]. Nonmonotonic behavior of the rate versus concentration is therefore predicted by this approach and has been observed for the 991 cm^{-1} level of benzene [4,8]. This general approach to relaxation in solids predicts that different rates will be observed depending upon whether or not the dephasing of the excitation amplitudes occurs before relaxation. Experiments have not yet been devised to show such effects and only the case where the dephasing is slower than relaxation has so far been studied.

References

1. R. M. Hochstrasser and I. I. Abram in Light Scattering in Solids, J. Birman, H. Cummins and K. Rebane eds., Plenum NY (1979).
2. P. L. DeCola, R. M. Hochstrasser and H. P. Trommsdorff, Chem. Phys. Letts. 72, 1 (1980).
3. I. I. Abram and R. M. Hochstrasser, J. Chem. Phys., 72, 3617 (1980).
4. J. Trout, S. Velsko, R. Bozio, P. L. DeCola and R. M. Hochstrasser, J. Chem. Phys. 81, 4746 (1985); F. Ho, W. S. Tsay, J. Trout, S. Velsko and R. M. Hochstrasser, Chem. Phys. Letts., 83, 5 (1981).
5. A. Fendt, S. F. Fischer and W. Kaiser, Chem. Phys. 57, 55 (1981).
6. K. Y. Tang and C. S. Parmenter, J. Chem. Phys. 78, 3922 (1983).
7. S. Velsko and R. M. Hochstrasser, J. Chem. Phys. 82, 2180 (1985).
8. S. Velsko and R. M. Hochstrasser, J. Phys. Chem. 89, 2240 (1985).
9. S. Velsko and R. M. Hochstrasser, J. de Physique, in press.

Time-Resolved Raman Spectroscopy of Laser-Heated Semiconductors

D. von der Linde

Universität-GHS-Essen, Fachbereich Physik, D-4300 Essen 1, F.R.G.

1. Introduction

Laser processing of semiconductors or metals usually starts with the deposition of a certain amount of laser energy in a thin surface layer of the material. With laser fluences of typically 0.1 to 1 J/cm^2, and light penetration depths of about 10 nm --characteristic of strong optical absorption-- the density of absorbed energy can reach 10^5 to 10^6 J/cm^3. If the laser pulses are sufficiently short such that thermal conductivity does not play a role during heating, the surface temperature can reach very high values, e.g., with the Dulong-Petit heat capacity one obtains a rough estimate of 10^4 to 10^5 K for rise of the surface temperature. Thus the temperatures attainable with rather modest laser energies may readily exceed the melting and boiling points of any known material.

An additional interesting aspect of pulsed laser surface heating is that extremely steep temperature gradients (10^9 K/cm) and extremely high rates of heating and cooling (up to 10^{14} K/s) can be achieved /1/. The exciting possibilities of laser processing have stimulated new activities in many fields of materials research /2/, a recent example being laser annealing /3/ of radiation damage which occurs in semiconductors as an undesired by-product of ion implantation during the fabrication of microelectronic devices.

In many applications of laser materials processing familiar phase-transitions such as melting and evaporation play an important role. They may occur, however, under conditions and on time-scales that have never been investigated before. For the understanding of these phase-transitions time-resolved measurements of the surface temperature are quite important. Vibrational Raman spectroscopy can -- under certain circumstances-- provide valuable information on the temperature of laser-irradiated surfaces.

This article describes time-resolved Raman temperature measurements on crystalline silicon with a view of determining the nature of the phase-transition which takes place in pulsed laser heating of silicon, when the laser energy exceeds a certain threshold. This transition has been identified as an essential step in pulsed laser annealing of crystal damage. Speculations /4/ that the phase transition is not ordinary thermal melting of the surface have stimulated an intense debate and detailed experimental investigations involving quite a variety of measuring techniques. Measurements of the lattice temperature at the transition point are crucial for answering questions concerning of the physics of the phase-transitions.

2. Nanosecond Laser Heating of Silicon

As a rather direct manifestation of the phase-transition during laser heating of silicon one observes an abrupt increase of the optical reflectivity /5/. An

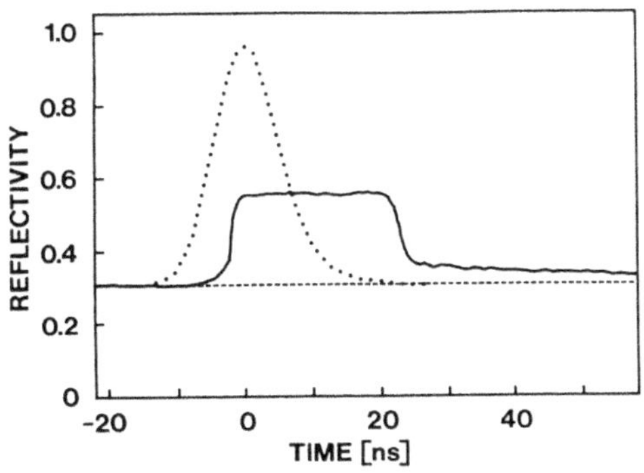

FIG. 1: Solid curve: change of the optical reflectivity at 488 nm during irradiation with a 10 ns laser pulse of 0.65 J/cm^2 at 532 nm. Dotted curve: laser pulse shape. Dashed line: initial reflectivity of the silicon crystal.

example is given in Fig. 1 which shows the time-resolved reflectivity of a silicon crystal irradiated by a 10 ns laser pulse (wavelength 532 nm). The change of the reflectivity of the laser-heated surface area is probed with a cw argon ion laser beam (wavelength 488 nm). It is seen that at the beginning of the laser pulse (dotted curve) the reflectivity increases gradually from the initial value of 31 % to about 38 %. At this point, however, a sudden jump to a value of 58 % is observed, which persists for 25 ns. A careful analysis indicates that the optical properties during the high reflectivity phase (HRP) are identical with those of liquid silicon, suggesting that the sudden rise and the subsequent drop of the reflectivity signify, respectively, melting and resolidification of the surface. The challenge is, of course, to measure the surface temperature right before the transition occurs, and to compare the result with the equilibrium melting temperature of silicon, T_m = 1685 K. Optical techniques including Raman scattering can readily provide the high time resolution that is required. In addition, Raman scattering gives structural information useful in deciding the disputed nature of the transition. Raman scattering is expected to vanish, if the transition is an ordinary melting process, i.e. involving a transition from a covalently bonded crystalline state to a metallic liquid in which optical phonon Raman scattering is absent.

3. Raman Temperature Measurements

The Raman spectrum of crystalline silicon is characteristic of the diamond lattice structure: there is only one triply degenerate optical phonon at the center of the Brillouin zone, the vibrational frequency ω_p corresponding to 520 cm^{-1} (at room temperature). The shift and the changes of the width of the optical phonon line with temperature are well known /6,7/, and it is possible to infer the lattice temperature from measurements of the line position and width.

The intensity ratio R of the Stokes-to-anti-Stokes scattering can also be used for a determination of the lattice temperature. R is given by $R = C(T)\exp(\hbar\omega_p/kT)$ where C(T) is an important temperature-dependent correction which involves the optical constants and the Raman scattering cross-sections /8/.

There are, however, a number of complications of Raman temperature measurements in silicon under pulsed laser conditions. For example, the Raman line will be influenced and possibly obscured by competing processes such as electronic scattering from photoexcited carriers /9/, thermal radiation /10/, and laser-induced stress /11/. If the temperature is to be inferred from measurements of the Stokes-anti-Stokes ratio, the correction factor C(T) must be precisely known over an extended range of temperatures. A general difficulty of Raman scattering from hot silicon is the fact that as the temperature increases the scattering intensity decreases quite substantially, mainly because of the strong increase of the optical absorption at 532 nm /12/. The problem is enhanced by the presence of very large temporal and spatial temperature gradients, which require high temporal and spatial resolution to avoid undesired temperature-averaging which would generally lead to an underestimate of the temperature. These complications are believed to explain the inconclusive results of previous Raman temperature measurements. The difficulties have been extensively discussed in the literature /8,12,13/. Fortunately, these problems are now overcome or can be adequately accounted for.

4. Experimental

For laser-heating and Raman scattering we use the same 10 ns laser pulse from a frequency-doubled, passively Q-switched Nd-YAG laser operating in a single longitudinal and transverse mode. The samples are standard (100) wafers of crystalline silicon, which must be rapidly raster-scanned to avoid multiple exposure (multiple melting) of the same surface area. Scattered light from the laser pulse is collected in a backward scattering geometry. To ensure homogeneous heating conditions, we use only the central portion of the illuminated area defined by the 90 % of maximum intensity contour (diameter 150 µm) of the Gaussian laser beam. As mentioned previously, a cw laser probe beam at 488 nm is focussed to a spot diameter of 20µ in the center of the active area for precise monitoring of the onset and duration of the HRP, which indicates the phase-transition. With incident and scattered light polarized along (100) and (010), respectively, Raman scattering from the longitudinal (LO) component of the Γ_5 zone-center optical phonon is detected. Thermally-induced uniaxial stress would reduce the O_h cubic to a D_{2d} tetragonal local symmetry, causing a splitting into a Γ_4 singlet and a Γ_5 doublet. In our scattering geometry, only the singlet can be observed, and complications due to the lifting of the degeneracy by stress are avoided.

Scattered light is passed through a double monochromator and detected by a fast photomultiplier (1P28). The photomultiplier signals are transferred to a waveform digitizing oscilloscope and then processed by a computer. We use two different modes of data processing: (i) a simple arithmetical summation of the individual photomultiplier signals of each laser pulse. Events are recorded until the waveforms representing the Stokes or the anti-Stokes scattering intensity are faithfully restored; (ii) individual processing of each photomultiplier signal to determine the time of the signal maximum, which can be localized with a precision of better than 100 ps. One then counts the number of events in certain time slots which are formed by subdividing the total observation time into a suitable sequence of time intervals. It has been established that the ultimate time resolution in this mode is 100 ps.

The choice between the two different methods offers a trade-off between the speed of data accumulation and the actual time resolution, which is several nanoseconds in the direct summation mode. The Raman signals range typically from about five single photon events per pulse to about one event in a hundred pulses. Recording of a complete Raman waveform requires at least a thousand photons. With the 20 Hz pulse repetition rate of the Nd-YAG laser, the data accumulation times for 1000 photons range from 50 s to about 1.5 hours.

An example of a typical scattering signal is depicted in Fig. 2 which shows a histogram representing the number of anti-Stokes scattering events in consecutive

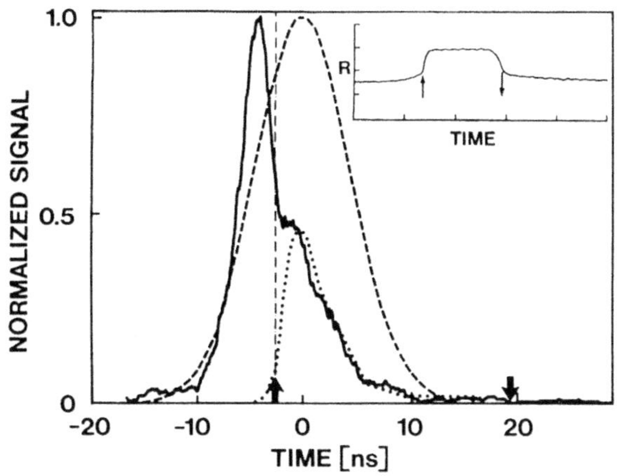

FIG 2: Total signal near the anti-Stokes wavelength (547 nm) with a frequency resolution of 200 cm^{-1}. Dashed curve: laser pulse (0.7 J/cm^2) at 532 nm. Dotted curve: thermal radiation. Insert: reflectivity curve indicating the beginning and the end of the HRP.

1 ns time intervals. The total number of detected single photon events is one thousand. The dashed line indicates the laser pulse, and the insert shows the accompanying reflectivity measurement. In this case the energy of the laser corresponds to 2.3 times the threshold energy for the onset of the phase-transition, and the arrows pointing up and down indicate, respectively, the beginning and the end of the high reflectivity phase. Note that the anti-Stokes Raman signal decreases strongly when the jump of the reflectivity occurs. However, it is seen that the signal does not vanish completely. A detailed spectral analysis shows that residual scattering signal during the HRP is in fact due to thermal radiation /10/. The dotted curve in Fig. 2 shows the contribution of thermal radiation to the total signal, indicating that during the HRP the entire observed signal is indeed caused by thermal emission. The Raman signal goes to zero within a time given by the time resolution of the system.

5. Time-Resolved Raman Spectra

Time-resolved Raman spectra /14/ are measured in a point-by-point fashion by stepping the monochromator through wavelength increments of 0.2 nm. Figure 3 shows an example of a series of Raman spectra (frequency resolution of 10 cm^{-1}) corresponding to a heating pulse energy of 0.31 J/cm^2, slightly less than the threshold of the phase-transition. The labels 1 to 4 indicate the time with respect to the laser pulse which is depicted in the insert. Spectrum 1 corresponding to the beginning of the laser pulse peaks at 520 cm^{-1}, the room-temperature line position of the optical phonon. Spectra 2,3 and 4 representing later times with more energy deposited in the material clearly show a distinct line broadening and a shift of 20 cm^{-1} to lower frequencies. The spectra are normalized with respect to the instantaneous laser intensity; unity corresponds to the maximum of spectrum 1.

From a detailed computer simulation of the laser heating process it has been established that the observed changes of the Raman spectra with time can be fully explained by a thermal model. These calculations take into account the complete space and time variation of the temperature, the temperature-dependence of the optical and thermal constants and of the position and the width of the optical phonon Raman line, thermally induced stress effects, and finally, the spectral response function of the monochromator and the impulse response of the photomultiplier.

For a 10 ns laser pulse of 0.31 J/cm^2, the surface temperature is calculated to be 350, 720, and 1400 K for -6 ns, 0, and +5 ns, respectively. The corresponding calculated Raman spectra given by the solid curves in Fig. 3 show that there is good agreement between the experimental data and the line position and width of the calculated spectra. In particular, spectrum 4 is seen to be characteristic of a surface temperature of about 1400 K, indicating that with 0.31 J/cm^2 the temperature rises, indeed, very close to the melting point. It should be noted that curve 1 also takes into account a Fano-type interference between Raman scattering from the optical-phonon mode and the quasi-continuum of intervalence band excitations due to free carriers. Electronic Raman scattering from photoexcited free holes (3×10^{19} cm^{-3}) explains the weak asymmetry of the spectrum at t = -10 ns. At higher temperatures (later times), interference with electronic Raman scattering has been neglected in the calculations.

The Raman scattering cross-section has been taken to be constant; if the actual minor increase of the cross-section with temperature is taken into account the agreement between the measured spectra and the calculated spectra could be further improved.

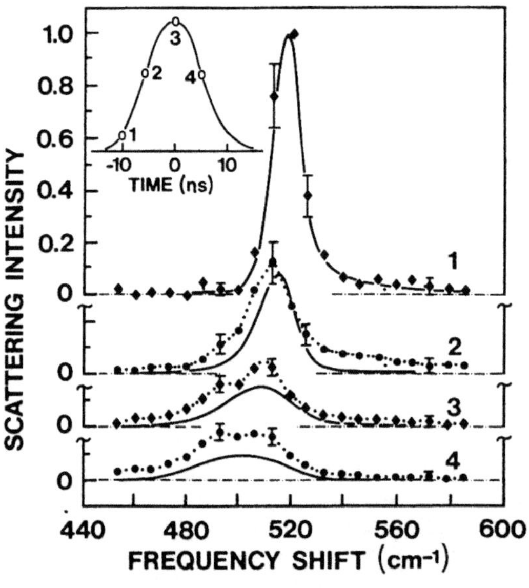

FIG. 3: Time-resolved Raman spectra just below the melting threshold (laser energy 0.3 J/cm^2). Insert: laser pulse with numbers indicating the times of the four spectra. The solid lines are calculated. Dotted lines: guide to the eye connecting the experimental points.

6. Measurements of the Stokes/anti-Stokes Ratio

For the determination of the temperature from the Stokes/anti-Stokes ratio /14/ the bandwidth of the monochromator is increased to 200 cm^{-1} such that the total spectrally integrated Stokes or anti-Stokes Raman band is measured. From the time dependent Raman signal of the Stokes and the anti-Stokes measured in this way one obtains the ratio R(t), and, since the correction function C(T) is now well known. the surface temperature as a function of time. High time resolution is important in this measurement, because the temperature increases by several hundred degrees in a few nanoseconds.

In Fig. 4 the temperature obtained from the measured ratio R(t) is plotted as a function of time. The dotted curve, and the upper and lower dashed line represent, respectively, the laser heating pulse, room temperature (300 K), and the melting temperature of silicon (1685 K). Data for three different laser energies are shown. For all energies, the experimental points indicate room temperature at the beginning of the pulse. For 0.2 J/cm^2 (squares), a temperature maximum around 900 K is reached at the end of the laser pulse, whereas just below threshold at 0.31 J/cm^2 (circles), the Raman temperature data come very close to the melting point. Well above threshold at 0.7 J/cm^2 (diamonds), the temperature is seen to rise very rapidly up to the melting point. Note that T_m is reached at the same time as the transition to the HRP is observed (arrow in Fig. 4) in the accompanying reflectivity measurement. The duration of the reflectivity enhancement was measured to be 30 ns in this case.

The solid curves in Fig. 4 represent the results of the temperature calculations with the laser energy being treated as an adjustable parameter. The best fit of the experimental data is obtained with slightly greater energy. Nevertheless, the agreement with the measured temperature is good, and there is no indication of any serious discrepancy with the thermal melting model.

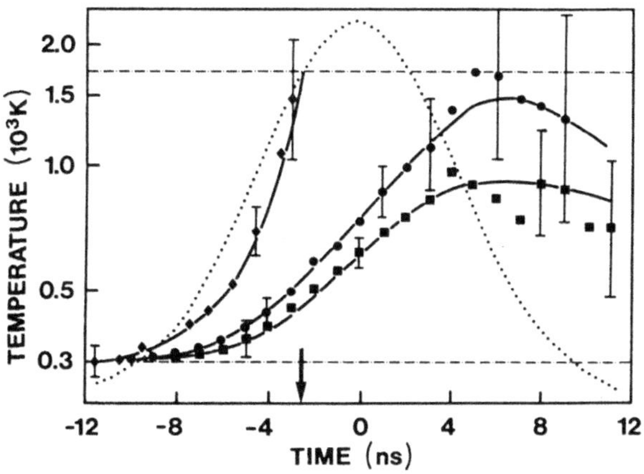

FIG. 4: Temperature as a function of time from the measured Stokes /anti-Stokes ratios. Three different laser energies: 0.7 J/cm^2 (diamonds), 0.31 J/cm^2 (full circles), and 0.2 J/cm^2 (squares). Upper and lower dashed line: melting temperature and room temperature respectively. Arrow: onset of the HRP.

7. Conclusions

The following conclusions can be drawn from the experiments. The Raman spectra have exposed time-dependent line shifts and broadenings which are fully consistent with the temperature rise predicted by the thermal model. The observation that optical-phonon Raman scattering disappears abruptly when the transition to the HRP takes place is consistent with a structural change from the fourfold covalent crystalline atomic coordination to a metallic liquid. From the measured Stokes/anti-Stokes ratio R the detailed temperature evolution during the heating pulse has been obtained. Because T is proportional to log R, the experimental error is large for $T > \omega_p/k$. The excellent quantitative agreement between the measured and the calculated temperature may be somewhat fortuitous in view of the interference with electronic scattering, which could lead to additional error in the determination of T. However, the essential point is that from the present Raman data there is no evidence whatsoever of any substantial disagreement with a simple thermal model of the laser heating process.

1. N. Bloembergen, in Laser-Solid Interactions and Laser Processing, e.d. S. D. Ferris, H. J. Leamy, and J. M. Poate (AIP, New York, 1979)
2. D. Turnbull, Metallurgical Transactions 12A, 695 (1981)

3. See, e.g., Laser Annealing of Solids, ed. by J. M. Poate and J. W. Mayer, (Academic Press, N.Y. 1982)

4. J. A. Van Vechten, in Laser and Electron Beam Processing of Materials, ed. by C. W. White and P. S. Peercy (Academic Press, N.Y. 19801)

5. D. H. Auston, C. M. Surko, T. N. C. Venkatesan, R. E. Slusher, and J. A. Golovchenko, Appl. Phys. Lett. 33, 437 (1978)

6. T. R. Hart, R. L. Aggarwal, and B. Lax, Phys. Rev. B1, 638 (1970)

7. M. Balkanski, R. F. Wallis, and E. Haro, Phys. Rev. B28, 1928 (1983)

8. A. Compaan and H. J. Trodahl, Phys. Rev. 29, 793 (1984)

9. F. Cerdeira, T. A. Fjeldy, and M. Cardona, Phys. Rev. B8, 4734 (1973)

10. M. Kemmler, G. Wartmann, and D. von der Linde, Appl. Phys. Lett. 45, 159 (1984)

11. G. E. Jellison and R. F. Wood, Mat. Res. Soc. Symp. Proc. 23, 153 (1984)

12. G. E. Jellison and F. A. Modine, J. Appl. Phys. 53, 3745 (1982)

13. G. E. Jellison, D. H. Lowndes, and R. F. Wood, Phys. Rev. B28, 3272 (1983)

14. G. Wartmann, M. Kemmler and D. von der Linde, Phys. Rev. B30, 4850 (1984)

Picosecond Vibrational Dynamics of Hydrogen Bonded Solids: Phonons and Optical Damage

J.R. Hill, T.J. Kosic[1], E.L. Chronister[2], and D.D. Dlott

School of Chemical Sciences, University of Illinois at Urbana-Champaign, 505 South Mathews Avenue, Urbana, IL 61801, USA

A variety of simple and complex biological molecules form H-bonded solids. We have studied vibrational dynamics in amino acid, peptide, and protein crystals [1]. Ps CARS was used to measure the lifetime of various optical phonons (ω = 30 - 1500 cm^{-1}) in H-bonded crystals, and in a number of cases surprisingly long phonon lifetimes (many ns) were observed. Because there are many long-lived phonons in these crystals, we can study how the phonon dynamics are altered by the phonon frequency, the crystal structure, the temperature, or the presence of impurities [1].

The experiments reported in this paper were performed in two distinct regimes. With "low" laser intensities, excited vibrations have small amplitudes and intermolecular restoring forces are nearly linear (the small non-linearities cause phonon decay). With "high" laser intensities the amplitudes become large, restoring forces are highly non-linear, and laser destruction of the crystal is observed.

PsCARS experiments on crystal phonons are typically performed in the nearly linear regime, where phonons are excited in an impulsive manner and allowed to freely relax. A complete knowledge of the phonon dynamics, coupled with the fluctuation-dissipation theorem,is sufficient to determine completely the equilibrium fluctuations of the molecules in the crystal. Other dynamical processes which involves phonons, for example electron or exciton dressing and hopping, or thermal conductivity, can be understood in a similar manner. On the other hand, the dynamics of solid-state chemical reactions, or crystal destruction which involves laser excited shock waves,cannot be described merely in terms of the crystal phonon states.

Acetanilide was chosen for the experiments because it forms H-bonded crystals with long-lived optical phonons, and it can be zone-refined to high purity [1]. The molecular structure is shown in Fig. 2.

The experimental apparatus is shown in block form in Fig. 1. A continuously pumped, mode-locked and Q-switched Nd:YAG laser is used to synchronously pump two tuneable dye lasers. The psCARS experiment uses 20ps dye laser pulses at ω_1 and ω_2 to coherently excite phonons at $\omega = \omega_1 - \omega_2$ [1]. Anti-Stokes emission stimulated by a delayed ω_1 pulse is detected by a photomultiplier versus pulse delay. The laser repetition frequency is variable up to 1kHz maximum. Damage experiments are performed by irradiating the crystal with intense 85ps green and tuneable pulses at intensities ranging from .1 to 2 GW/cm^2. A photodiode (PD) before and after the sample measures the transmission of the green pulse, which decreases when damage occurs. The computer controls the damage pulses with an electronic shutter.

---

1 Current address: Hughes Aircraft Corporation, El Segundo California 90245

2 Current address: Chemistry Department, University of California at Los Angeles, California 90024.

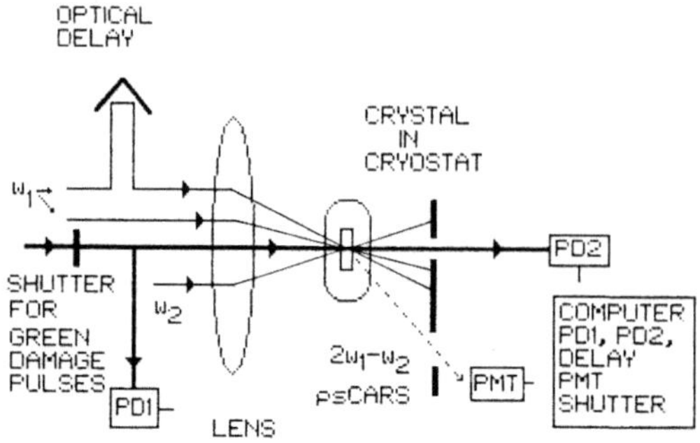

Figure 1: Two Nd:YAG pumped ps dye lasers at ω_1 and ω_2 perform the ps CARS experiment. The green (532nm) damage pulse is controlled by the computer. Optical damage is monitored by two photodiodes (PD).

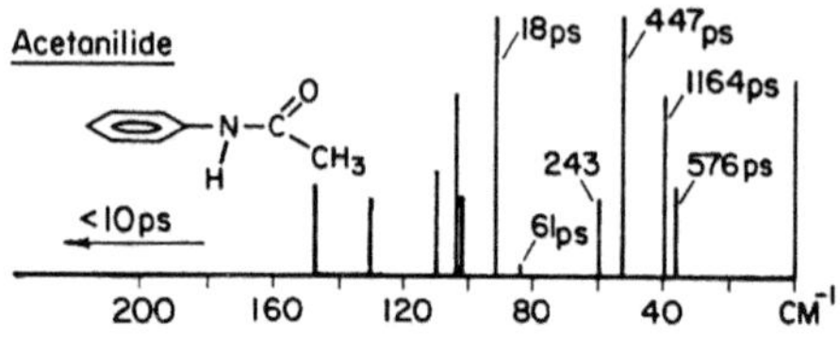

Figure 2. Picosecond CARS simulated spectra of low-temperature acetanilide crystal [1]. The 1000 cm^{-1} vibron (not shown) has T_1 = 25ps.

Figure 2 shows a psCARS "spectrum" of optical phonons in low-temperature acetanilide [1]. The vertical lines represent intense CARS resonances of the crystal, with the height proportional to the signal intensity. Half-wave plates were used to optimize this emission for each phonon. The psCARS decay was obtained by sweeping the optical delay line for each phonon. In low-temperature pure crystals, the psCARS decay is due entirely to the phonon lifetime [1,2], which is indicated for each phonon in the Figure. Phonons above 100 cm^{-1} decay faster than 10ps with the exception of the 1000 cm^{-1} in-plane phenyl ring stretch. In acetanilide, as well as several other simple H-bonded crystals, the phonon [Ω=30-100 cm^{-1}] decay rate is found to increase as Ω^4, where Ω is the phonon frequency. This dependence is characteristic of the process where each optical phonon spontaneously fissions, via cubic anharmonic interactions, into two counterpropating acoustic phonons of frequency $\Omega/2$ [1].

The optical damage experiments are illustrated in Fig. 3. During the incubation period, typically 1K-50K pulses (irradiation for ca. 100ns to 5µs), the crystal remains transparent. When a critical local concentration of defects is generated, destruction begins [3]. During the destruction period, typically 10 pulses (ca. 1ns irradiation), optical damage in the form of a small inclusion in the bulk of the crystal appears and spreads until a cavity approximately the diameter of the laser beam (25 microns) is observed [4]. With further irradiation, carbonized material appears. In most cases the initial damaged region expands with each successive laser pulse, but in about 10% of the cases oscillating behavior with alternating phases of growth and contraction are observed [4]. During all these phases, appreciable absorption of the laser pulses (>10%) was not observed.

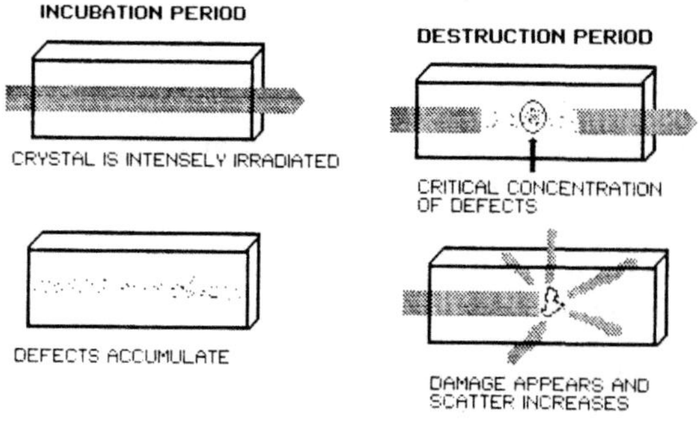

Figure 3. Schematic diagram of the accumulated damage process.

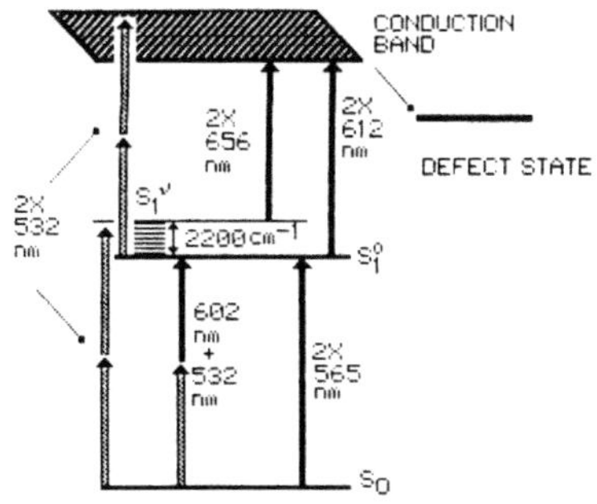

Figure 4. Energy level diagram for acetanilide.

We have used a new technique--ps probe pulse destruction spectroscopy--to verify that the defects result from 2-photon absorption to S_1 followed by 2-photon ionization [4]. Figure 4 is the relevant energy-level diagram for acetanilide [4]. Two-photon absorption of the green damage pulse creates vibrationally excited S_1^v. The lifetime of S_1^v is less than 10ps [1], and repopulation of vibrationless S_1 occurs on the 25ps timescale, so the subsequent 2-photon ionization during the 85ps pulse may occur from a distribution of vibrationally hot S_1 molecules.

In our experiments, we damaged crystals with simultaneous green and tuneable pulses, and with the tuneable pulse delayed 770ps from the green pulse [4]. This delay was chosen because it is shorter than the S_1 lifetime, and longer than the pulse duration or the S_1^v lifetime. The mean number of laser pulses for damage to occur for the various experiments is given in Table 1 [4]. The data is the average of about 50 damage events. The most efficient damage occurs with 560nm and green. Both wavelengths can 2-photon excite and ionize acetanilide. Simultaneous pulses are more efficient than delayed because 560nm + 532nm absorption is possible when the pulses are simultaneous. The 566nm pulse can 2-photon ionize acetanilide, but can only 2-photon excite S_1 in combination with 532nm, so destruction is less efficient unless the pulses are simultaneous. The 612 and 604 nm pulses cannot excite S_1 except by a 3-photon process, but both can in principle 2-photon ionize S_1. Since S_1 has a lifetime much longer than the delay, these results are nearly independent of delay. Overall, 612nm is more efficient at 2-photon ionization than 604nm.

TABLE 1: MEAN NUMBER OF PULSES (THOUSANDS) TO DAMAGE CRYSTAL [4]

WAVELENGTH	SIMULTANEOUS	TUNEABLE PULSE DELAY = 770PS
532+612 nm	20K	21K
532+604 nm	32K	40K
532+566 nm	19K	39K
532+560 nm	8K	13K

Although these experiments show that the defects are a result of ionization of acetanilide, they do not allow us to characterize the defect state itself. The defect state may involve H-bonds ruptured by the ionization, fragmented molecules, or molecular ions. At low temperature, the defects persist at least overnight. We have not yet been able to detect absorption or fluorescence from the defect state itself.

It is known to be difficult to observe accumulated defects [3]. We discovered that the accumulation of defects can be monitored by psCARS. The optical phonon lifetime decreases as defects accumulate. One data set is shown in Fig. 5. The lifetimes of two phonons were studied while the sample was irradiated with intense green damage pulses. The crystal was destroyed at about 30 μsec of irradiation. The two phonons have different initial lifetimes, and the rate of defect scattering is found to be the same for both. At present we cannot use this method to determine the defect concentration because the

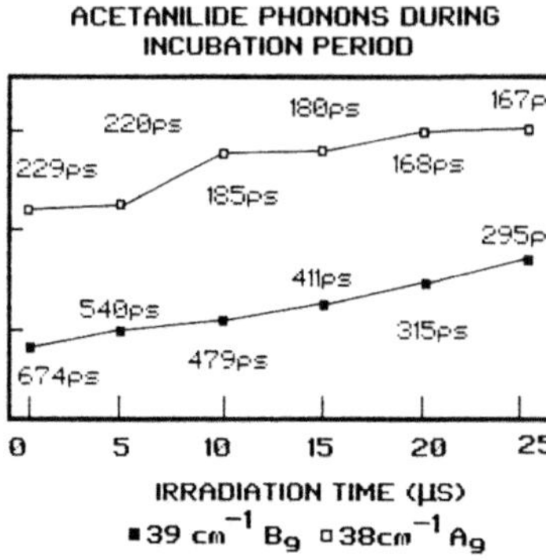

Figure 5. The low-temperature phonon lifetimes in acetanilide decrease as defects accumulate.

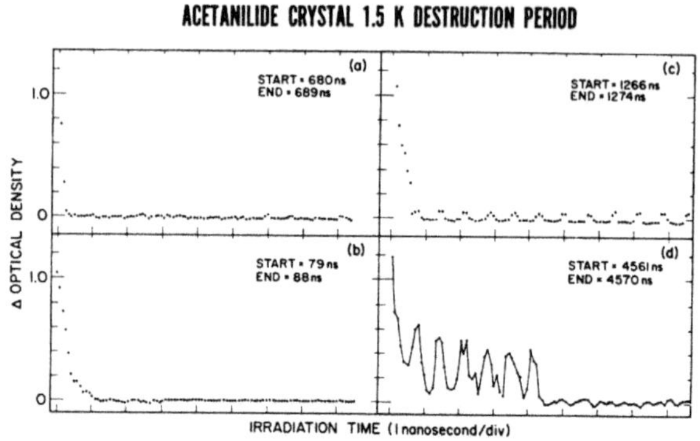

Figure 6. The destruction period is monitored by the transmission of ps pulses. The 10ns period before destruction is shown. In (d), points are connected by lines for clarity.

phonon-defect cross-section is not known. The effect of impurities on phonon lifetimes was previously studied by comparing ℓ-alanine to disordered crystals where 60% of the exchangeable protons were substituted with deuterium [1]. In the impure ℓ-alanine, the phonon lifetimes decreased to roughly the same degree as in the irradiated acetanilide. This result indicates that either the defect concentration prior to destruction is in the tens of percent, or the defects scatter phonons more strongly than partially deuterated molecules.

Damage occurs when a region with a critical defect concentration is prepared. This is defined as the concentration, averaged over many events, required to initiate damage on 50% of the pulses [3]. The damage may be the results of heating and melting, or optoelastic shock waves [3]. An analysis of the rapid heating and cooling is complicated by the crystal being below the Debye temperature, since the heat capacity and thermal conductivity are strongly temperature-dependent. We predict [4] that even with complete absorption of the pulse,the T-jump is from 10 to ~40K with complete cooling between pulses. Since the measured absorption was far less than complete, the inclusion appears to be formed by shock waves rather than melting.

In previous studies of accumulated optical damage of solids [3], it has been proposed that the mechanism of defect production and accumulation was entirely different than the mechanism of crystal destruction. We studied the duration of the destruction period with the two-color damage pulse combinations given in Table 1. Within experimental error we find that the same conditions which decrease the duration of the incubation period also decrease the duration of the destruction period [4], leading us to the conclusion that the mechanism of defect formation is the <u>same</u> as the mechanism of crystal destruction. If this is indeed the case, percolation theory [5] is a natural model for this process, since the continuous accumulation of defects will, at some critical concentration, create large, connected damaged regions of the crystal, resulting in a sudden onset of crystal destruction.

When accumulated damage is studied using ns lasers [3], it is found that the damaged region is created in just one laser pulse. By contrast, with ps pulses we are able to create a small region which expands with subsequent pulses. Figure 6 shows the transmitted intensity of the last 100 pulses before burnup in four cases. At the left of each data set, the OD rises to 1-1.5. This is typical of an inclusion with no black material. It is clear that the formation of the inclusion occurs in two distinct modes: the continuous growth mode in 6a and 6b, and the oscillating growth-collapse mode in 6c and 6d. The oscillating mode appears about 10% of the time. This could only occur if the laser pulses can anneal as well as damage the crystal [4].

We also studied the effect of large, non-equilibrium vibrational populations on the rate of damage. With intense pulses the 1000 cm^{-1} phenyl ring stretching mode was excited to an occupation number of $n \sim 10^{-4}$, corresponding to an effective temperature $T^* \sim 160K$ during the duration of the pulses. In this case it took about twice as many pulses to damage the crystal as under identical conditions without vibrational excitation. This interesting result shows that vibrational excitation can be used to control the rate of a solid state reaction, possibly by annealing the defect regions.

This research is supported by the National Science Foundation (USA). DDD is an Alfred P. Sloan fellow.

1. Thomas J. Kosic, Raymond E. Cline, Jr., and Dana D. Dlott: Chem. Phys. Lett. 103, 109 (1983); J. Chem. Phys. 81, 4932 (1984).
2. K. Duppen, B. H. Hesp, and D. A. Wiersma: Chem. Phys. Lett. 79, 399 (1981); Claire L. Schosser and Dana D. Dlott: J. Chem. Phys. 80, 1369 (1984).
3. A. A. Manenkov, G. A. Matyushin, V. S. Nechitailo, A. M. Prokhorov, and A. S. Tsaprilov: Opt. Eng. 22, 400 (1983); Larry D. Merkle, Michael Bass, and Randall T. Swimm: Opt. Eng. 22, 405 (1983).
4. Thomas J. Kosic, Jeffrey R. Hill, and Dana D. Dlott: in preparation.
5. See e.g. J. M. Ziman: Models of Disorder (Cambridge University Press, 1979).

Time-Resolved Two-Color Optical Coherence Experiments

K. Duppen, D.P. Weitekamp, and D.A. Wiersma*

Picosecond Laser and Spectroscopy Laboratory,
Department of Chemistry, University of Groningen, Nijenborgh 16,
NL-9747 AG Groningen, The Netherlands

Coherent Raman experiments have been shown to be very effective in studying the vibrational dynamics of molecules in the condensed phase (1). When in the excitation process two different frequencies are being used, vibrational mode selectivity can be achieved, and quite often electronic resonances can be used to enhance the Raman scattering cross-section (2). It is also well known that such coherence Raman experiments can be performed in either the frequency - or time-domain and that the choice of technique depends on the particular problem. These techniques are therefore complementary in the same way as holeburning and photon echoes for the study of electronic transitions. We note that time-domain experiments are mandatory for those cases where we wish to study a transient species.

In the past decade we have been deeply involved with the question of the nature of optical and more recently vibrational dephasing in molecular mixed crystals. For reasons of experimental convenience, the large organic molecule pentacene has been the test object in many of these studies. In this paper we wish to focus on our recent four-wave mixing and two-color photon echo experiments on the mixed crystal system of pentacene in naphthalene. We start with a discussion of recently performed time-resolved coherent Stokes Raman scattering (CSRS) experiments. In Fig. 1 the pulse cycle (a), the level scheme (b,c) and the generation and detection scheme (d) for these experiments are shown. In insert b and c the lowest level is the electronic ground state, the next level a selected vibrational level of the ground state, the third level the electronically excited state and the upper level the corresponding vibrational level in the excited state. The optical absorption spectrum of pentacene in naphthalene at two different temperatures is shown in Fig. 2. The upper spectrum was taken at 1.5 K and the lower one at 77 K

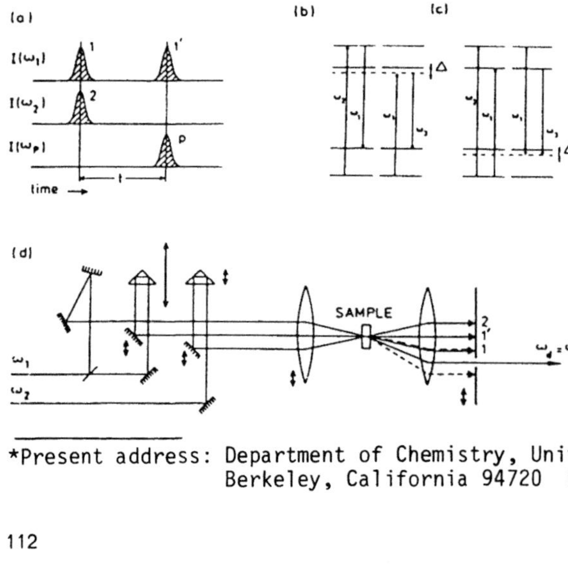

Fig. 1. Pulse cycle (a), level schemes (b,c) and schematic set-up (d) for the time-delayed CSRS experiments. The pulses at ω_1 and ω_2 are amplified pulses from a synchronously pumped tandem dye-laser system.

*Present address: Department of Chemistry, University of California at Berkeley, Berkeley, California 94720 USA

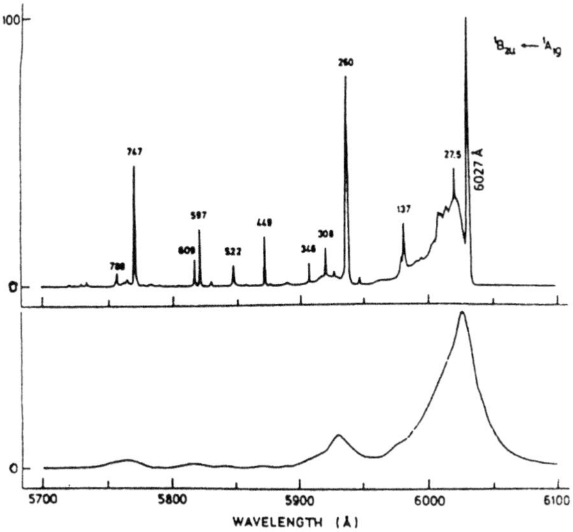

Fig. 2. Absorption spectrum of pentacene in naphthalene at 1.5 K (upper) and at 77 K (lower trace).

showing that at this latter temperature the absorption linewidth is close to 100 cm^{-1} and that most of the vibrational structure is washed out. In our experiments we have selected the 747 cm^{-1} excited state vibrational mode with a 756 cm^{-1} mode as ground state partner, as inferred from frequency-domain four-wave mixing experiments (3).

Two type of CSRS experiments were performed: in the first one, scheme 1b, the intensity of the delayed Stokes signal monitors the decay of the ground state vibrational coherence. At low temperature an exponential decay was found with a time-constant of 51 ± 1 psec, in good agreement with the results of previous time-resolved CARS experiments on the same mode (4). The temperature-dependence of the vibrational homogeneous lineshape, as derived from the vibrational coherence lifetime (T_2) via the relation $\Delta\nu_h = (\pi T_2)^{-1}$, is plotted in the lower curve in Fig. 3. In the second type experiment, scheme 1c, the excitation pulses at t = 0 prepare vibrational coherence in the electronically excited state. The delayed Stokes signal measures in this case the width of the vibrational transition in the upper state. Figure 4 shows two typical traces obtained with a different intensity of the probe beam. At low probe intensity, the CSRS signal peaks at t = 0; when the probe intensity is increased by a factor of 10 an apparent shift

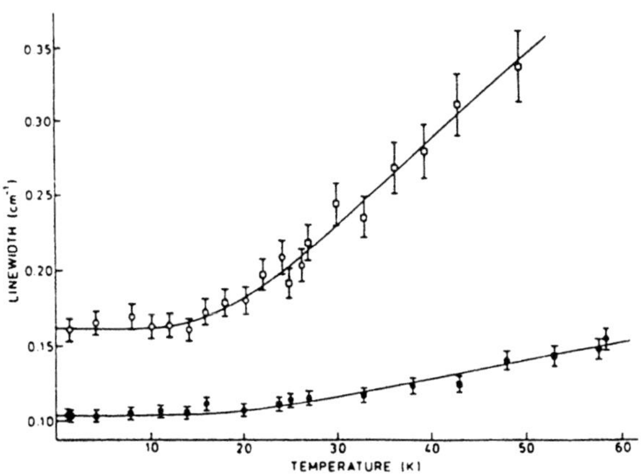

Fig. 3. Plots of the homogeneous vibrational width for the 747 cm^{-1} mode in the excited state (upper plot) and 756 cm^{-1} in the ground state (lower plot). The solid lines are fits to an assumed exponential activation of the vibrational T_2 with parameters discussed in the text.

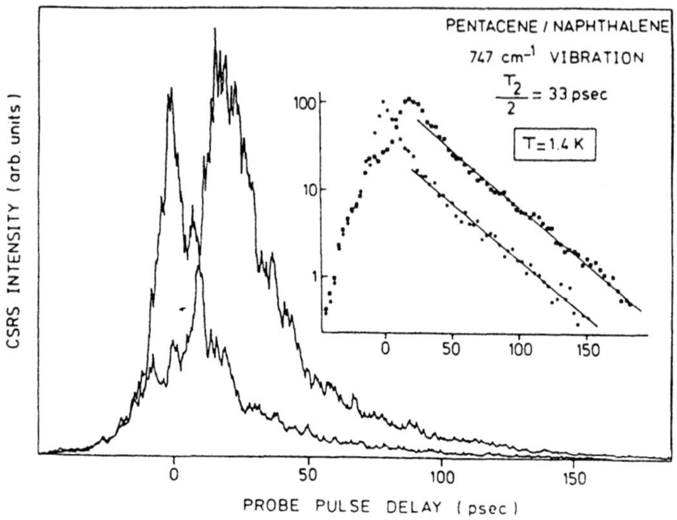

Fig. 4. Time-delayed CSRS signal, according to scheme c of Fig. 1, for different intensities of the probe beam.

of the signal maximum, at a 20 psec delay of the probe pulse, is observed. We explain this phenomenon as being due to interference of the probe pulse, at $t = 0$, with the creation of an optimal vibrational coherence (5). It is gratifying to note that in either case the same decay time of 33 psec is measured, again in excellent agreement with earlier photon echo experiments involving this vibrational level (6). The measured temperature-dependence of the deduced homogeneous width is plotted in the upper curve of Fig. 3.

Before proceeding with a discussion of the results obtained, it is worth noting that in the frequency-domain version of CSRS, as described in the $\chi^{(3)}$ formalism, the excited state vibrational resonance is only allowed in the case that "pure dephasing" occurs in the system. These excited state resonances have been coined PIER-4 (7) and DICE (8) in the literature. Dick and Hochstrasser (9) showed in a recent non-perturbational treatment of frequency-domain CSRS that these extra resonances also become allowed at higher field strength, thus can be field-induced. In a non-perturbational treatment of time-resolved four-wave mixing by Weitekamp et al. (4) it was shown that at any field strength a *transient* vibrational coherence occurs in the excited state, even in the absence of pure dephasing. This is an interesting case where in the excitation process a transient coherence appears which is damped under steady state conditions at low-field strength.

We now return to a discussion of the results displayed in Fig. 3. In an attempt to offer an explanation of the observed temperature effect on the vibrational resonance in the ground and excited state, we have fit the displayed curves to the following expression for the vibrational T_2: $T_2^{-1} = \frac{1}{2}\,\tau_v^{-1} + T_2^{*-1}\exp(-\hbar\omega_0/kT)$. This equation for the vibrational T_2 shows that we have identified the low-temperature decay constant with the population relaxation time; which is an accepted fact (6). From the fits indicated by the solid lines in Fig. 3, we obtain the following: in the ground state $T_2^* = 58$ ps and $\omega_0 = 56$ cm^{-1} while in the excited state these parameters become $T_2^* = 13$ ps and $\omega_0 = 52$ cm^{-1}. We immediately stress the fact that these parameters are fitting parameters and that without additional evidence the "activation energies" can not be assigned to a degree of freedom of either the pentacene molecule or the naphthalene host. It is interesting to note however that the naphthalene host crystal has an optical phonon at 57.5 cm^{-1}, well within the error range of the measured activation energy. Whether or not this naphthalene optical phonon is involved in the vibrational dephasing process could possibly be sorted out by studying these vibrational line shapes in a deuterated naphthalene host crystal where the lowest optical phonon drops to 54 cm^{-1}.

Recent theoretical (10) and experimental (11) work on vibrational dynamics in pure molecular crystals seems to indicate that a large fraction of the homogeneous width of a vibrational transition at any temperature is due to population relaxation induced by cubic vibron-phonon anharmonicity. An alternative interpretation of the observed activation then invokes the participation of a number of vibrational doorway states with a frequency difference such that it is within the range of the host phonon density of states and with an appreciable cubic anharmonic interaction (6, 10). In order then to draw any conclusion regarding the mechanism of vibrational dephasing of guest molecules, it is necessary to study the population relaxation dynamics as a function of temperature. As it turns out, this is not an easy matter to accomplish. In fact we have, so far, not been able to obtain useful relaxation data at temperatures where the temperature-induced part of the vibrational lineshape is appreciable. For the sake of completeness we will discuss some of the unsuccessful experiments performed and suggest possible improvements. One attempt to study vibrational population relaxation in pentacene at elevated temperature involved a two-color grating scattering experiment. In this experiment a spatial population grating was created at the vibronic transition with frequency ω_1 by crossing two time-coincident beams in the sample, while the probe beam was tuned to the resonance at frequency ω_2 (as in Fig. 5b). At low temperature, where the vibronic transitions do not overlap, this experiment was quite successful in measuring the vibrational decay. At higher temperatures however, above 30 K or so, the scattering efficiency of the selected vibrationally hot transition interfered with light scattered from the grating in the ground state and with the one relaxing into the electronically excited state, leading to probe-delay independent signal in this temperature-range. A similar type grating experiment involving the corresponding ground state mode was for the same reason unsuccessful. It seems that a better chance of success in these experiments could be created if the grating probe pulse were applied at a frequency far removed from the strongly-allowed pure electronic transition, e.g. at a combination band. The draw back of this "solution" is that the transition moment of such a combination band will be very much reduced, which will lead to a reduced scattering efficiency. Finally we wish to emphasize the close connection that exists between these spatial two-color grating experiments and two-color photon echo experiments. In the photon echo case the excitation pulses at ω_1 are not time-coincident which

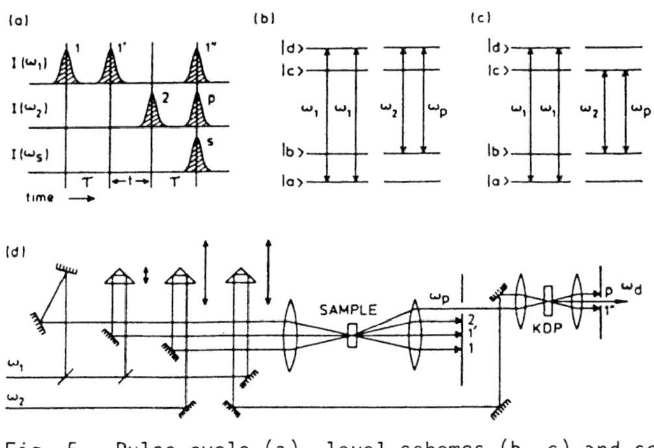

Fig. 5. Pulse cycle (a), level schemes (b, c) and schematic set-up (d) for the generation of two-color photon echoes. Note that all pulses in schemes b and c are applied at different times, and that the probe pulse at ω_p is mixed externally with the photon echo.

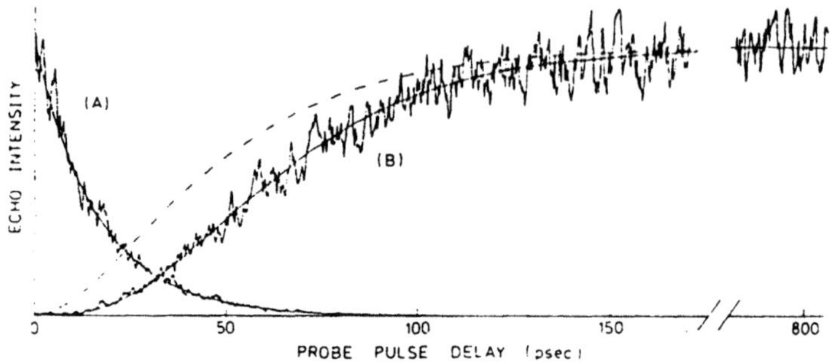

Fig. 6. Two-color stimulated picosecond photon echo signals. In (A) the connected echo is measured, see Fig. 5b, in (B) the relaxed echo is measured according to scheme c in Fig. 5. Note that in the latter case the echo transition does not share any level with the initially excited transition.

leads to a "frequency grating" rather than a spatial grating. The probe pulse then generates from the ground state and/or excited state an echo which can be detected by optical mixing (12). As was emphasized earlier, such an echo can be created from any state that is involved in the optical pumping cycle, provided that the inhomogeneous broadenings at the different optical transitions are correlated. Typical results of such stimulated echo experiments are shown in Fig. 6 whereby the intriguing fact is noted that the echo indeed can be generated from a state that acts as the acceptor for the initially prepared frequency grating (13). In the case of the connected two-color stimulated echo (C2CSE) shown in Fig. 5b, the echo, as a function of probe delay, measures the population decay from the upper level. With a relaxed two-color stimulated echo (R2CSE) the rate of transfer into the initially unexcited level |c > is measured (Fig. 5c). However, the intensity of these echoes is also dependent on the *optical* T_2 of the electronic transitions involved;this method, with the currently available psec pulses, is also not suited to probe the population dynamics at higher temperature. It's hard to win in this game.

In summary: we have been able to measure, using time-delayed CSRS, the temperature dependence of the vibrational lineshape of the same vibration in the ground and excited state of pentacene in naphthalene. Presently it is not clear whether the observed activation is dominated by thermally induced population relaxation or by dephasing effects involving host optical phonons or even higher frequency librations of the guest. What is urgently needed is a sensitive and selective method to measure the temperature-dependence of the population flow in molecular solids.

References

1 A. Laubereau and W. Kaiser: Rev. Mod. Phys. 50, 607 (1978)
2 N. Bloembergen, H. Lotem and R.T. Lynch: Indian J. Pure Appl. Phys. 16, 151 (1978)
3 P.L. Decola, J.R. Andrews, R.M. Hochstrasser and H.P. Trommsdorff: J. Chem. Phys. 73, 4695 (1980)
4 K. Duppen, D.P. Weitekamp and D.A. Wiersma: J. Chem. Phys. 79, 5835 (1983)
5 D.P. Weitekamp, K. Duppen and D.A. Wiersma: Phys. Rev. A 27, 3089 (1983)
6 W.H. Hesselink and D.A. Wiersma: J. Chem. Phys. 74, 886 (1981)
7 Y. Prior, A.R. Bogdan, M. Dagenais and N. Bloembergen: Phys. Rev. Lett. 46, 11 (1981)
8 J.R. Andrews and R.M. Hochstrasser: Chem. Phys. Lett. 82, 381 (1981)
9 B. Dick and R.M. Hochstrasser: Chem. Phys. 75, 133 (1983)
10 R. Righini: Chem. Phys. 84, 97 (1984)
11 O.C.L. Schosser and D.D. Dlott: J. Chem. Phys. 80, 1394 (1984)
12 W.H. Hesselink and D.A. Wiersma: Chem. Phys. Lett. 56, 2 (1978)
13 K. Duppen, D.P. Weitekamp and D.A. Wiersma: Chem. Phys. Lett. 108, 551 (1984)

High Sensitivity Time-Resolved Study of the Coherence and Parametric Instabilities of Two-Phonon States

G. Gale, F. Vallée, and C. Flytzanis

Laboratoire d'Optique Quantique, Ecole Polytechnique,
F-91128 Palaiseau Cédex, France

Coherent picosecond excitation and probe techniques have recently been extended [1-4] to the selective investigation of the dynamics of vibrational overtones and multiphonon states in condensed media. These states, which involve the creation and annihilation of several phonons, are essential for the understanding and the description of large amplitude vibrational motion in condensed matter and the anharmonic forces that come into play there. Through the study of their time evolution, much insight can also be gained about phonon breakdown and large wave vector phonon dynamics.

A central problem here is the coherence and population evolution of these multiphonon states. For single phonon states these processes are usually associated [5] with the transverse (T_2) and longitudinal (T_1) relaxation times respectively, but for the multiphonon states the situation is far more complex because of the compound character of these excitations and the additional relaxation channels that this introduces ; more importantly the anharmonic interactions, although much weaker than the harmonic ones, under certain conditions introduce profound changes in the multiphonon spectrum [6,7,8,9,10].

These questions were addressed for the first time in refs 1-3 in the case of the two-phonon states using time-resolved nonlinear optical techniques and in particular CAHORS (Coherent Anti-Stokes Higher Order Raman Scattering). Here we summarize the picture that emerges out from these studies concerning the coherent evolution of these states, and also present a simple description of a class of parametric instabilities that may occur there. The latter may allow to generate high density of large wave vector phonons and probe their dynamics.

To fix our ideas we recall that in the harmonic approximation the excited energy of a lattice with two phonons in branches σ' and σ" with wave vectors $\underline{k}'$ and $\underline{k}''$ respectively is given by [9,11]

$$\hbar\Omega(\underline{k}) = \hbar\{\omega(\underline{k}') + \omega(\underline{k}'')\} \qquad (1)$$

with $\underline{k} = \underline{k}' + \underline{k}''$. The optically (infrared or Raman) accessible two-phonon states form a quasi continuum with $\underline{k} = 0$ or $\underline{k}' = -\underline{k}''$ and $\underline{k}'$ lies anywhere within the first Brillouin zone. The bandwith of this quasi continuum is equal to the sum of the widths W of the individual phonon branches ; W is a measure of the intermolecular coupling and phonon localization. When the anharmonicity is switched on the phonons interact with each other through the third and fourth order terms, $h^{(3)}$ and $h^{(4)}$ respectively, in the usual expansion of the lattice hamiltonian [9,11]

$$h = h^{(0)} + h^{(3)} + h^{(4)} \qquad (2)$$

If the strength Γ_B of the fourth order term $h^{(4)}$, which is essentially intramolecular, is much larger than the intermolecular coupling W a localized bound two-phonon state [12] or bi-phonon splits off the free two-phonon continuum (see Fig. 1) with narrow linewidth and nonnegligible oscillator strength borrowed from the two-phonon quasi-continuum which is considerably reduced ; the residual quasi-continuum contains the free two-phonon states. In addition, if such a bound (or quasi-

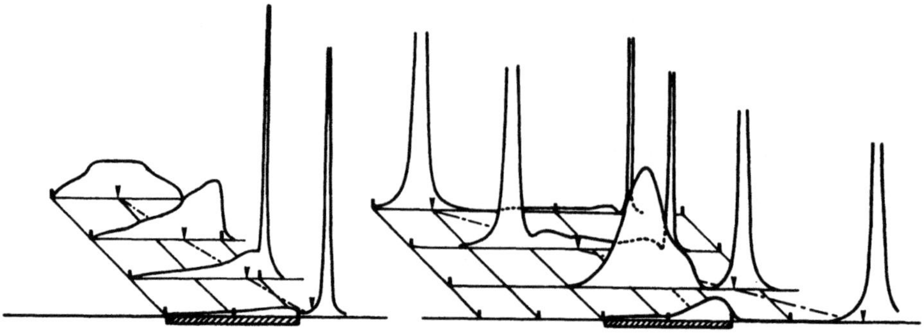

Fig. 1. Left ; dependence of two-phonon spectra on anharmonic strength Γ_B (increasing from left rear to right front). Right ; production of two bound states as a one-phonon line approaches the free two-phonon band. In both cases the free band position is shaded.

bound) two-phonon state of energy Ω_B is nearly degenerate [6] with a single phonon state of energy $\hbar\omega_1$ a hybridization mediated through the third order term $h^{(3)}$ may occur resulting in a double peak (Fermi doublet [12]) whose two components split off on either side of the free two-phonon residual continuum (see Fig. 1) ; the relative strength of the two components depends on the ratio $\beta/(\Omega_2 - \omega_1)$ where $\hbar\beta$ is the strength of the third order term $h^{(3)}$[6,14,15].

Following refs 1-3 let us introduce the expectation value of the optically accessible two-phonon coordinate

$$<q_+q_->=\mathrm{Tr}\rho q_+q_- \quad (3)$$

where q_+ and q_- are the single phonon coordinates with wavevector $\underline{k}$ and $-\underline{k}$ and ρ is the density matrix operator in the optically accessible two-phonon state space which consists of the ground, the bound two-phonon and the quasi continuum of the free two-phonon states, $|\psi_o>$, $|\psi_B>$ and $|\psi_F>$ respectively ; for simplicity we restricted ourselves to the phonons of a single optic branch. The dynamics of the bound two-phonon state which is singled out in the coherent excitation process in CAHORS, enter the calculation through the expectation value

$$<Q>_B \equiv \mathrm{Tr}\rho_B q_+q_- \quad (4)$$

where ρ_B is the density matrix operator in the subspace spanned by $|\psi_o>$ and $|\psi_B>$. Note that because of different interactions with the bath $<Q>_B \neq <q_+>^o<q_->$ while for the expectation values in the free two-phonon subspace $<q_+q_->_F \equiv \mathrm{Tr}\rho_F q_+q_- \simeq <q_+><q_->$. In the following we shall omit the brackets <> but keep in mind the real physical content of these coordinates. By explicitly introducing the bound two-phonon coordinate Q_B and its frequency Ω_B we actually take into account the main effect of the fourth order term $h^{(4)}$ and are left with a residual term of the form

$$h^{(4)}_{BF} = \Gamma_{BF}Q_Bq_+q_- \quad (5)$$

the coherent interaction term between the bi-phonon and the free two-phonon states ; any other term in $h^{(4)}$ can be included in the damping of the two-phonon state. The term (5) plays a crucial role both for the transfer of coherence from the bound two-phonon state to the free two-phonon states and the parametric instabilities that may occur then.

The work of refs 1-3 revealed that depending on the strength of the two anharmonic terms in (2) the loss of coherence of the bound two-phonon states mainly occurs through two channels

- an intrinsic [1,2] one where the bound or quasi bound two-phonon state internally loses its coherence to the free two-phonon states from which it is formed through the term $h^{(4)}$ and the later are subsequently dissolved into single phonons : this mechanism which essentially leads to a temperature independent behavior occurs whenever the bound or quasi bound two-phonon state is near or on top of the quasi-continuum of the free two-phonon spectrum a common situation when only phonons of a single phonon branch are involved. The model has been described in ref. 1 where using Green's function techniques it was found that this mechanism in general leads to an exponential decay of the coherence with a lifetime $T_B' = \Gamma_B'^{-1}$

$$\Gamma_B' \simeq \nu_o \tag{6}$$

where ν_o is the density of free two-phonon states calculated where the overlap with the bound two-phonon state is maximal.

- an extrinsic [3] one where the frequency of the bound (or quasibound) two-phonon state, which is essentially intramolecular, fluctuates in time as a result of the intermolecular degrees of freedom or other low-lying states ; this mechanism which leads to a temperature-dependent behavior (characteristically a T^2 dependence is dominant whenever the bound (or quasibound) two-phonon state is far removed from the quasi continuum as a consequence of a mutual repulsion with a closely lying single phonon state through the $h^{(3)}$ term. It was shown in ref. 3 that this mechanism too in general leads to an exponential decay of the coherence with a damping constant

$$\Gamma_B'' \sim kT^2 \tag{7}$$

One generally expects that the intrinsic mechanism is dominant at low temperatures or in crystals with one or two simple molecules per unit cell, while the extrinsic one prevails in crystals with many molecules per unit cell or large and easily deformable molecules. The two extreme situations are exemplified [2,3] with the CS_2 and CO_2 crystals respectively, while N_2O crystal constitutes [4] an intermediate case, since the Fermi resonance is less pronounced than in CO_2 but more so than in CS_2.

In all these studies the time-resolved technique CAHORS (Coherent AntiStokes Higher Order Raman Scattering) (see Fig. 2) was used [1-4] which exploits the narrow linewidth of the compound states and the appreciable oscillator strength that is concentrated there from the large joint density of pairs of phonons all over the Brillouin zone. More precisely, it was shown there that these compound states can be coherently driven with high efficiency by two time-coincident picosecond pulses of frequencies ω_L and ω_S respectively with $\omega_L - \omega_S = \Omega_B$ and the coherent decay of this state is subsequently followed by delayed ($t = t_D$) coherent antiStokes scattering at $\omega_p + \Omega_B$ of a probe pulse at frequency ω_p. The experimental setup is described elsewhere ; the high sensitivity of our experimental system,

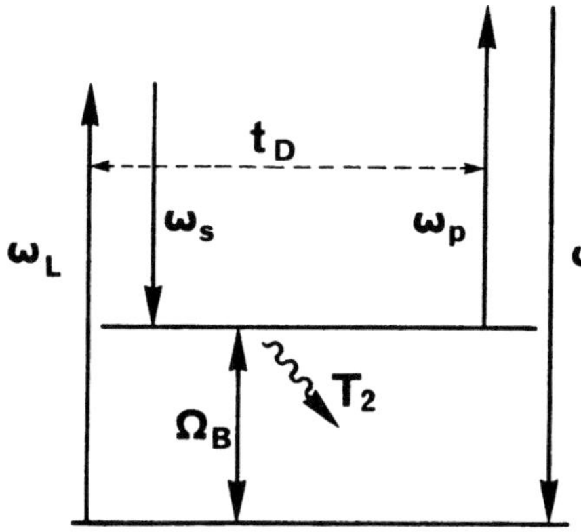

Fig. 2. Coherent excitation and probing schema for two-phonon bound states in solids and vibrational overtones in liquids. Ω_B is the bound state frequency and T_2 its coherent decay time

which allows a dynamic measurement range of 10^9 for a strong Raman scatterer, means that previously inaccessible weak Raman features may now be studied and that the coherent evolution of narrow Raman lines in a cluttered region of the spectrum may be isolated.

The description of the excitation and probing stages in the CAHORS technique were given in refs 1-3 both for a bound two-phonon state and a Fermi doublet.

The first demonstration of this technique was done [2] on the $2\nu_2$ bound two-phonon state of crystalline CS_2 located at Ω_B = 801 cm^{-1}. This state is only coupled with the continuum of free two-phonon states. An exponential decay roughly independent on temperature was observed with T_B = 14 ± 3 ps in agreement with the assumption of an intrinsic mechanism ; this crystal contains only two molecules per unit cell and thus a small number of intermolecular modes. The coherence life time T_B is longer than the one expected if independent free phonon states only were involved in the two-phonon state.

The CO_2 crystal illustrates [3] the other extreme case. This crystal has four molecules per unit cell, and there is a large number of low-frequency modes representing the rotational, librational and translational intermolecular motions, and this introduces fluctuations in the intermolecular motion. Furthermore, a Fermi resonance occurs between the symmetric Raman active stretching mode ν_1 and the first overtone $2\nu_2$ of the doubly degenerate infrared-active bending mode as the overtone frequency Ω_2 falls very close to the unperturbed ω_1 fundamental frequency. As a consequence a double peak in the vibrational spectrum appears, whose two components at frequencies Ω_+ = 1383 cm^{-1} and Ω_- = 1276 cm^{-1} split well off on either side of the two-phonon continuum. Exponential decay is maintained for both components over the whole temperature considered from 9 to 245 K where CO_2 is liquid (see Fig. 3a). It was also found that the two components exhibit a strong temperature-dependence, and the corresponding curves are parallel (see Fig. 3b) ; the temperature-dependence is quadratic in T in accordance with (7) obtained for the extrinsic case.

Finally, in the case of N_2O an exponential decay was obtained for the $2\nu_2$ state which is only weakly coupled with the ν_1 mode and is located at 1165 cm^{-1}. The coherence lifetime T_B was found nearly insensitive to temperature, T_B = 11 ps, confirming that the intrinsic mechanism is operative here. However, some complications arise because of an inherent orientational disorder related to the lack of inversion symmetry of the molecule N_2O (both forms NNO and ONN can occur [4]).

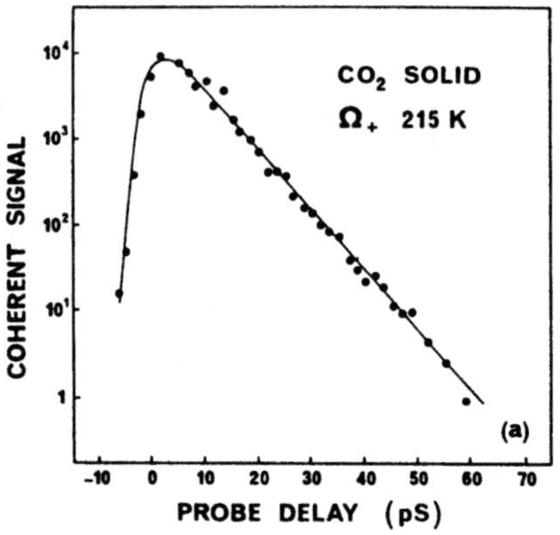

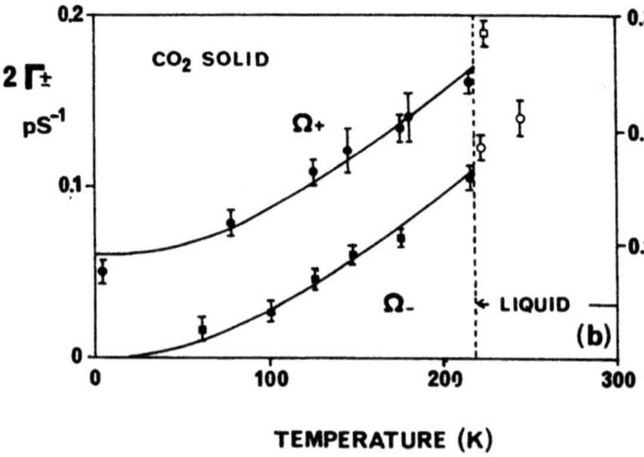

Fig. 3(a). Coherent anti-Stokes signal vs probe delay for the Ω_+ line of CO_2 solid at 215 k. (b). Variation of the relaxation rates $2\Gamma_+$ and $2\Gamma_-$ with temperature for the Ω_+ line (circles) and the Ω_- line (squares) fo the Fermi doublet in CO_2

The important point to notice from these investigations is that the bound two-phonon states may maintain its coherence for relatively long times when the intrinsic mechanism is operative ; if the free two-phonon states into which this coherence is transferred can retain it long enough, under certain conditions, this long overlap in time may engender [1] parametric instabilities. Below we wish to address this important point in a more quantitative way.

The coherent excitation with two intense fields allows one to generate coherent bound two-phonon states with wave vector $\underline{k} \simeq 0$ and large amplitude. Their amplitude can be viewed as a field amplitude satisfying an inhomogeneous wave equation. Since the overlap $<\psi_F/\psi_B>$ of the bound and free two-phonon states can be substantial, the bound two-phonon can break down with finite probability into two free phonons coherently propagating with wave vectors $\underline{k}'$ and $\underline{k}''$ such $\underline{k}' + \underline{k}'' = 0$ and amplitudes q' and q'' that satisfy the one-phonon propagation equation with a source term. The source term is related to $<\psi_F/\psi_B>$ whose square also measures the probability of transferring the coherence from the bound to the free two-phonon amplitudes with total wave vector conserved or $\underline{k}' + \underline{k}'' = \underline{k}$.

A similarity with the parametric amplification and oscillation in three wave interactions in nonlinear optics can be drawn as follows : the overlap $<\psi_F/\psi_B>$ corresponds to the second order susceptibility $\chi^{(2)}$ and Q_B, q_+ and q_- corresponds to the field amplitudes of the pump (p), signal (s) and idler (i) respectively [16]. The phonon propagation equation can be linearized within the envelope approximation similar to the one used in nonlinear optics. This striking analogy pointed in ref.1 can be the starting point of nonlinear phonon optics.

We present now a simple description of a class of parametric instabilities which are of relevance in many interesting situations when the intrinsic mechanism for coherence transfer is operative. We will assume that no Fermi resonance occurs or $|\beta/(\Omega_2 - \omega_1)| << 1$, so that the third order term $h^{(3)}$ only introduces a mere renormalization of the one-phonon spectrum [10]. Furthermore, since the main effect of $h^{(4)}$ is accounted for by the formation of the bound two-phonon state, as stated previously, the coherent interaction between the bound two-phonon amplitude Q and the single phonon amplitudes q_+ and q_- is mediated through the residual term (5) where q_+ and q_- are the amplitudes of the two phonons of opposite wave vectors that interact most strongly with the bound two-phonon state. Any other term in $h^{(4)}$ will be absorbed in the damping constants.

The vibration couples [1,2] with the light fields in the coherent excitation stage through

$$h_e = -\frac{1}{2}\alpha E^2 \tag{8}$$

where is the molecular polarizability, $E = E_L \cos\omega_L t + E_S \cos\omega_S t$ and $\omega_L - \omega_S \simeq \Omega_B$. We assume as in refs 1-2 that the relevant optic mode is Raman inactive and expand α

$$\alpha \simeq \alpha_o + \frac{1}{2}\alpha'' Q_B \tag{9}$$

where $\alpha'' = \partial^2\alpha/\partial q^2$ is the second order Raman tensor. From (8) the effective interaction hamiltonian is

$$h_e'' = -\frac{1}{4}\alpha'' Q_B E^2 \tag{10}$$

Using (5) and (10) the equations of motion of the coordinates are then given by

$$\ddot{Q}_B + \frac{1}{T_B}\dot{Q}_B + \Omega_B^2 Q + \Gamma_{BF} q_+ q_- = \frac{1}{4}\alpha'' E^2 \tag{11}$$

$$\ddot{q}_+ + \frac{1}{T_+}\dot{q}_+ + \omega_+^2 q_+ + \Gamma_{BF} Q_B q_- = 0 \tag{12}$$

$$\ddot{q}_- + \frac{1}{T_-}\dot{q}_- + \omega_-^2 q_- + \Gamma_{BF} Q_B q_+ = 0 \tag{13}$$

For a first approximation the term $\Gamma_{BF} q_+ q_-$ in (11) can be disregarded (it can be absorbed in the damping) and in E^2 only the cross term that oscillates at frequency $\omega_L - \omega_S \simeq \Omega_B$ will be retained [1,2]. For simplicity we may also safely assume that $\omega_+ = \omega_- = \omega_o$ and $T_+ = T_- = T$ for phonons of the same branch and opposite wave vectors. Introducing then $q_\Sigma = q_+ + q_-$ and $q_\Delta = q_+ - q_-$ in (11-13) one has

$$\ddot{Q}_B + \frac{1}{T_B}\dot{Q}_B + \Omega_B^2 Q_B = F\cos\Delta\omega t \tag{14}$$

$$\ddot{q}_\Sigma + \frac{1}{T}\dot{q}_\Sigma + \omega_o^2 q_\Sigma + \Gamma_{BF} Q_B q_\Sigma = 0 \tag{15}$$

and similarly for q_Σ, $F = \frac{1}{2}\alpha'' E_L E_S$ and $\Delta\omega = \omega_L - \omega_S \simeq \Omega_B$. Solving for Q_B in (14) one obtains $Q_B = \mathrm{Re}\,\{Q_o e^{\cdot i\Delta\omega t}\}$ where

$$Q_o = \frac{1}{4}\,\frac{(\alpha'')^2 E_L E_S}{\Omega_B^2 - \Delta\omega^2 + i\Delta\omega\, T_B^{-1}} \tag{16}$$

and replacing it in (15) one gets

$$\ddot{q}_\Sigma + \frac{1}{T}\dot{q}_\Sigma + \omega_o^2 (1 + \mu\cos\Omega_o t)\, q_\Sigma = 0 \quad \text{with} \tag{17}$$

$$\mu = \frac{\Gamma_{BF} Q_o}{\omega_o^2} << 1 \tag{18}$$

where it will be assumed that $\Delta\omega T_B > 1$; in general we may set

$$\Omega_o = 2\omega_o + \varepsilon \tag{19}$$

where $\varepsilon << \omega_o$ is the frequency mismatch between the bound and free two-phonon states. Equation (17) is that of the parametric oscillator which has been extensively studied in the literature [17] ; its solution may be sought in the form [17]

$$q_\Sigma = a(t)\cos(\omega_o + \tfrac{1}{2}\varepsilon)t + b(t)\sin(\omega_o + \tfrac{1}{2}\varepsilon)t \tag{20}$$

where we neglect all harmonics of $(2\omega_o + \varepsilon)$. We also assume that a(t) and b(t) are slowly varying and we seek solutions of the form exp(st). To first order in ε the conditions for parametric instability [17] (or amplification, $s>0$) is

$$Q_o > Q_{th} = \frac{\omega_o}{\Gamma_{BF}}\left(\varepsilon^2 + \frac{1}{T^2}\right)^{1/2} \tag{21}$$

which together with (16) introduce a threshold value for $E_L E_S$. The fields required to reach this condition are usually large but manageable in certain cases ; clearly for $\varepsilon >> \frac{1}{T}$ the threshold condition simply becomes

$$Q_o > \frac{\omega_o}{\Gamma_{BF}}\,\varepsilon \tag{22}$$

which is easier to satisfy.

The occurrence of parametric instabilities has some major implications :

- large densities of one-phonon states can be coherently driven even when they do not couple directy with the fields (Raman inactive modes)

- the parametric process favors creation of large wave vector phonons and their dynamics may be probed optically ; indeed since in general Ω_B overlaps or lies below the lower part of the free two-phonon quasi continuum, the frequency mismatch ε is smaller [11] for pairs of phonons at the edge of the Brillouin zone than in the center

- new effects appear when the polariton character of the phonons is also taken into account [7]

- when a structural phase-transition occurs, the point density of two-phonon states changes and according to (6) also T_B will change, which may affect the parametric instability ; this could be the case in the NH_4Cl crystal [18]

- one can have transition to chaotic behavior since the set of equations exhibit such a behavior.

- all the above considerations can be extended to other types of elementary excitations in particular acoustic phonons, magnons or plasmons ; the case of magnons is of particular interest [18] since bound two-magnons states care large oscillator strength.

The authors wish to thank Amhed Mokhtari for discussions on the problem of parametric instabilities.

1 - C. Flytzanis, G.M. Gale and M.L. Geirnaert, in Applications of Picosecond Spectroscopy to Chemistry, edited by K.B. Eisenthal (Reidel, Higham, Mass. 1984)p.205

2 - M.L. Geirnaert, G.M. Gale and C. Flytzanis, Phys. Rev. Lett. 52, 815 (1984)

3 - G.M. Gale, P. Guyot-Sionnest, W.Q. Zheng and C. Flytzanis, Phys. Rev. Lett. 54, 823 (1984)

4 - F. Vallée, G.M. Gale and C. Flytzanis to be published in Chemical Physics Letters

5 - see for instance A. Laubereau and W. Kaiser, Rev. Mod. Physics 50, 608 (1978)

6 - J. Ruvalds and A. Zawadowski, Phys. Rev. B2, 1172 (1970)

7 - V.M. Agranovich and I.I. Lalov, Zh. Eksp. Teor. Fiz. 61, 656 (1971) [Sov. Phys. JETP 34, (1972)]

8 - M.V. Belousov, in Excitons, edited by E.I. Rashba and M.D. Sturge (North-Holland, Amsterdam, 1982) p. 772

9 - see for instance, Lattice Dynamics and Intermolecular Forces edited by S. Califano (Academic Press, N.Y., 1975)

10 - J.C. Kimball, C.Y. Tong and Y.R. Shen, Phys. Rev. B23, 4946 (1981)

11 - J.H. Reisland, The Physics of Phonons (John Wiley, London, 1973)

12 - M.H. Cohen and J. Ruvalds, Phys. Rev. Lett. 23, 1378 (1969)

13 - F. Fermi, Z. Phys. 71, 250 (1931)

14 - F. Bogani, J. Phys. C11, 1283 and 1269 (1978)

15 - F. Bogani and P.R. Salvi, J. Chem. Phys. 81, 4991 (1984)

16 - Y.R. Shen, Principles of Nonlinear Optics, (John Wiley, N.Y. 1984)

17 - see for instance, L. Landau and E. Lifshitz, Mechanics (Pergamon Press, London, 1978) p. 80

18 - V.S. Gorelik, G.G. Mitin and M.M. Sushchinski in Theory of Light Scattering in Condensed Matter edited by B. Bendow, J. Birman and V. M. Agranovich (Plenum, N.Y. 1975) p. 109

19 - W. Hayes and R. London, Scattering of light by crystals (John Wiley - Interscience N.Y. 1978)

Two-Phonon Bound States in Molecular Crystals

F. Bogani

Dipartimento di Fisica and Centro Interuniversitario Struttura della Materia, Unità di Firenze, Largo E. Fermi, 2, I-50125 Firenze, Italy

Two-phonon transitions in molecular crystals form an important class of anharmonic processes, and the possible existence of a two-phonon bound state was pointed out a long time ago by GUSH et al. [1]. Since then a lot of work has been done on the argument: two-phonon bound states and resonances have been detected in several molecular (e.g. CO_2, N_2O, HCl, HBr, CS_2, etc.) and ionic crystals (e.g. NH_4Cl, HIO_3, $LiNbO_3$). A theory explaining the formation of a phonon bound state was first proposed by COHEN and RUVALDS [2] and AGRANOVICH [3]. A more complete theory, allowing a quantitative comparison with the experimental data, has been recently developed by the author and coworkers [4-6]. For a recent review of the argument see the paper of AGRANOVICH in [7].

Molecular crystals of small molecules (e.g. CO_2, HCl, etc.) represent a good subject of study for high anharmonic processes in solids for several reasons. They can be fairly well described as a collection of molecular units frozen at the lattice sites and held together by long-range dipolar and quadrupolar forces. Their spectrum can be separated in two regions: the high energy region ($\geqslant 500$ cm^{-1}) corresponding to the intramolecular modes, each with small dispersion ($\leqslant 100^{-1}$) due to the long-range forces, and the low-energy region ($\leqslant 200^{-1}$) corresponding to the external modes [8]. Our knowledge, both experimental and theoretical, of the harmonic lattice dynamics is good, and most of the molecular constants, such as the intramolecular anharmonic constants, are well known from gas phase measurements.

On this basis we can construct a reliable and relatively simple theory for the anharmonic processes. The basic model Hamiltonian for the combination band $w_i + w_j$ is [4]

$$H = \tfrac{1}{2} \sum_{jls} (\dot{Q}^2(jls) + w_j^2 Q^2(jls) + \sum_{m's'} M_j(ls,l's')Q(jl's')Q(jls)) + K \tag{1}$$

where Q(jls) is the molecular mass-weighted normal coordinate for the j internal model; l and s the cell and site indexes respectively, w_j the j-mode frequency of the isolated molecule, $M_j(ls,l's')$ the intermolecular interaction between the internal j-modes of the molecules at sites (l,s) and (l's'), and

$$K = \sum_{ij} X_{ij} Q^2(ils)Q^2(jls) \tag{2}$$

accounts for the intramolecular anharmonic coupling between the two modes. The two-phonon spectra can be obtained from the Fourier transform of the retarded two-phonon Green function (GF), i.e. from the elements of the matrix G defined as

$$\langle\langle Q^+(jl_1s_1)Q^+(jl_2s_2) \; ; \; Q^-(il_1's_1')Q^-(jl_2's_2') \rangle\rangle_w \tag{3}$$

By standard techniques one obtains for the renormalized two-phonon Green funtion G a matrix Dyson equation

$$G = g + gXG \tag{4}$$

where the matrix g is the harmonic two-phonon GF and X is the anharmonic coupling matrix. The formal solution of the Dyson equation is

$$G = (I-gX)^{-1}g \qquad I_{ij} = \delta_{ij} \tag{5}$$

and corresponds to summing up an infinite series of diagrams like

We see from (5) that the two-phonon GF can have poles corresponding to

$$\det|I-gX| = 0 \tag{6}$$

besides the poles of the harmonic GF. An analytical expression for the matrix elements of G can be easily obtained in some cases, depending on the structure of the matrix X.

The simplest situation is that of a combination band between a undispersed (Raman active) mode and a dispersed (IR active) internal mode (e.g. the combination band of the symmetric and antisymmetric stretching modes in CO_2). The relevant GF can be exactly evaluated, and we get for the renormalized two-phonon density of states A, proportional to ImG,

$$A(w) = n(w)\left[\left(1 - X\int_p \frac{n(z)dz}{w-z}\right)^2 + \left(\pi X\, n(w)\right)^2\right]^{-1} \tag{7}$$

where n(w) is the harmonic two-phonon density of states . The internal anharmonicity can strongly deform the two-phonon spectrum, and gives rise to resonances if the first term in the denominator vanishes. If this zero falls outside the two-phonon continuum (n(w) = 0) we have the appearance of a two-phonon bound state and for the latter

$$A(w) = X^{-2}(df/dw)_0^{-1}\,\delta(w-w_0) \tag{8}$$

where f and w_0 are defined by the equations

$$f(w) = 1 - X\int \frac{n(z)dz}{w-z} \qquad f(w_0) = 0 \tag{9}$$

The parameter determining the shape of the spectrum and the appearance of the bound states is fundamentally the ratio between the anharmonic coupling constant and the width of the two-phonon continuum: if this parameter is big enough we have the formation of a bound-state and strong deformation in the observed spectra. The renormalized density of states A as a function of the anharmonic coupling constant is shown in fig. 1. The experimental and computed spectra for the w_1+w_3 combination band of CO_2 are shown in fig. 2.

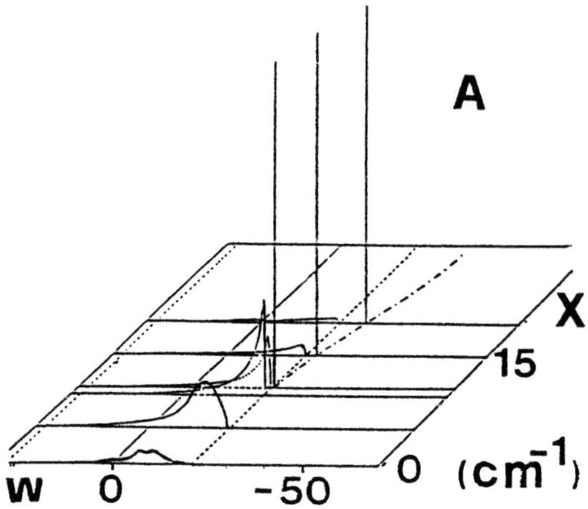

Fig. 1. The renormalized density of states as a function of the anharmonic coupling constant in CO_2.

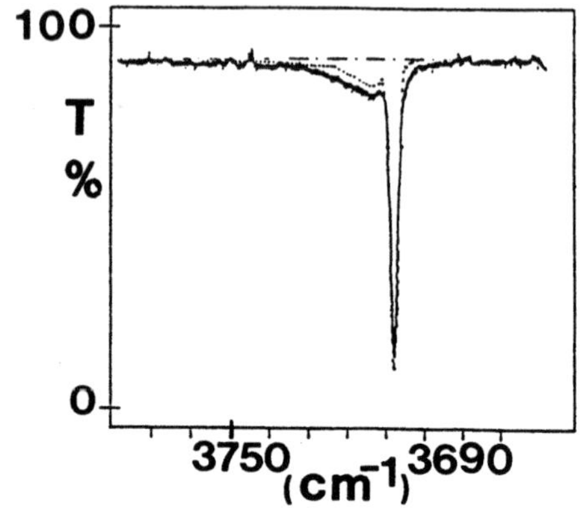

Fig. 2. The experimental and computed (dotted line) spectra for w_1+w_3 in solid CO_2.

If the two-phonon band falls near a fundamental mode, the anharmonic coupling between the modes is strongly enhanced by the Fermi resonance: a typical case is offered by CO_2 where the frequency of the symmetric stretching and that of the overtone of the bending mode are nearly coincident. In this case the situation is more involved, and we have to sum up the two infinite series

An analytical solution is still possible and the basic GF looks like

$$G = \left(w-w_1-X\int_p \frac{n(z)dz}{w-z} + i\pi Xn(w)\right)^{-1} \tag{10}$$

where w_1 is the bare frequency of the fundamental mode, X is the anharmonic coupling constant and n(w) the harmonic two-phonon density of states. We have the appearance of one or two bound states corresponding to the zeroes outside the two phonon continuum of the function:

$$f(w) = w - w_1 - X\int_p \frac{n(z)dz}{w-z} \tag{11}$$

The situation is depicted in fig. 3 for CO_2. In fig.4 the experimental and computed Raman spectra are shown.
If the combination mode is IR active, polariton dispersion has to be taken into account, giving rise to additional features [7].

Until now the theory accounts fairly well for the experimental spectra, but some points need better understanding. These bound states represent collective non-linear excitations of the crystal, and should be associated with classical non-linear waves (similar to soliton waves): a clear description in these terms is

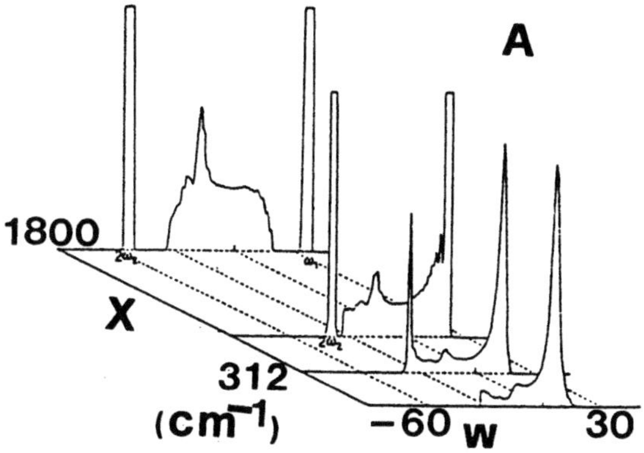

Fig. 3. The renormalized density of states for the Fermi resonant w_1 and $2w_2$ band in solid CO_2.

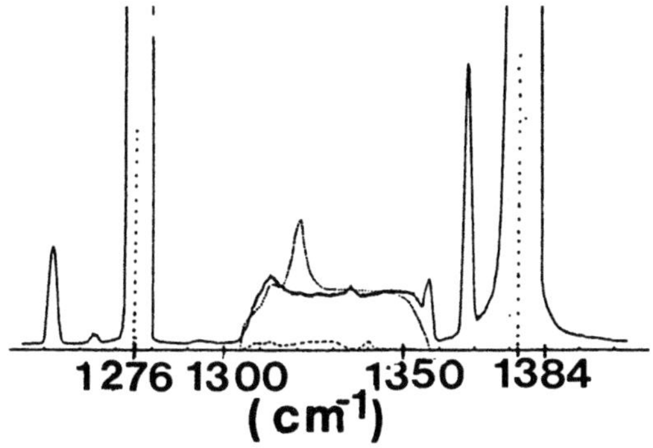

Fig. 4. The experimental and computed spectra for the Fermi resonant w_1 and $2w_2$ band in CO_2.

still lacking [7]. Moreover, we have to better understand how these bound states give rise to localized or propagating modes inside the crystal [9]. In this respect the time-resolved study of these bound states is relevant. Recent measurements have been done by the FLYTZANIS group with time-resolved CARS on a two-phonon resonance ($2w_2$ in CS_2 [10]) and on a two-phonon bound state (w_1 and $2w_2$ in CO_2 [11]) and by CALIFANO et al. with high resolution Raman spectroscopy [12]. For the resonance in CS_2 the decay time is found to be $\sim$14 ps at 160°K; in CO_2 both measurements give for the bound states a decay time of $\sim$ 40 ps for one component and $\geq$ 2650 ps for the other at 4°K.

A theory for the decay mechanism of these bound states is still lacking, but a line of work can be sketched. For the two-phonon resonance, corresponding in (10) to $f(w_0) = 0$ and w_0 inside the two-phonon continuum, assuming for the resonance a laurentian lineshape, we get for the half-width

$$\Gamma = 2\pi \, X n(w_0)(df/dw)_0^{-1} \tag{12}$$

In this simple case the resonance is directly coupled to the two-phonon continuum and as a consequence the lifetime is short and nearly independent from the temperature. A more complicated situation arises for the bound states: the decay and dephasing mechanisms involve the anharmonic coupling of the internal and external modes. In this case cubic anharmonic terms like V(11e) and V(22e), coupling two stretching or bending modes to an external one, are mainly responsible for the damping of the bound states. Due to the thermal motion of the molecules around the lattice sites a branch of the ladder can decay and the infinite series breaks down.

The contribution of this process can be easily evaluated for CO_2 at least as far as the V(11e) term is concerned. It can be shown that in this case the GF turns out to be

$$G = (w - w_1 - X \int \frac{n(z)dz}{w-z} + \Gamma(w))^{-1} \tag{13}$$

$$\Gamma(w) = N^{-1} \sum_{kj} S(k)(w - w_1 \mp W(jk))^{-1} \begin{Bmatrix} 1 + Z(W) \\ Z(W) \end{Bmatrix} \tag{14}$$

where S(k) is a rather complicated k-dependent form factor proportional to the Fourier transform of $V(11e)^2$, Z(W) is the Bose occupation number, w_1 is the bare stretching frequency and W(k) the frequency of the j external mode. An analogous, but more complicated expression can be written for the processes due to V(22e) involving the bending modes. Looking at the external phonon dispersion curves of CO_2 [13] and taking into account the energy and momentum conservation rules, we see that the proposed process should be effective. The frequency of the bare w_1 mode falls near 1039 cm^{-1} and the decay of the bound states involves the destruction of a lattice phonon at ~ 62 cm^{-1} for W_- and the creation of a lattice phonon at ~ 46 cm^{-1} for W_+. They correspond to an acoustic X and R phonon at the boundary of the Brillouin zone respectively. The corresponding temperature-dependence for the width Γ and the experimental data from [11] are reported in fig. 5.

Dephasing processes, not due to isotopes or to defects, involve higher anharmonic processes and give rise to a non-linear dependence of Γ from the occupation number Z. They should be effective at low temperatures for the long-living W_- mode as shown by the asymmetry of the line in the high-resolution spectrum (see fig. 6 and |12|).

A detailed evaluation of these processes requires further work and can lead to a deeper understanding of the nature of these bound states.

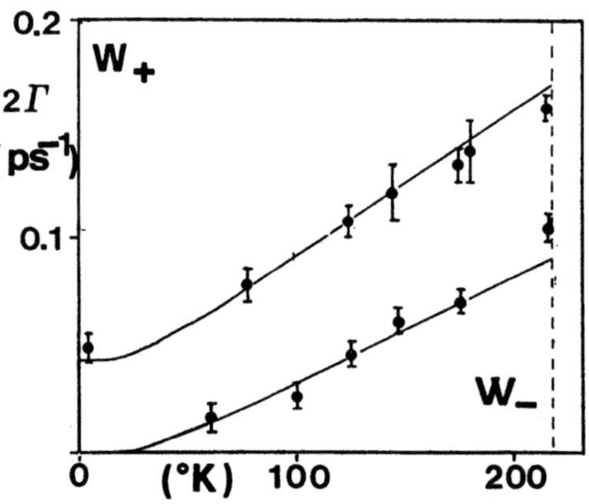

Fig. 5. Temperature-dependence of the width for the Fermi resonant bound states in CO_2 |11|. The continuous line is the fit to the data with (14).

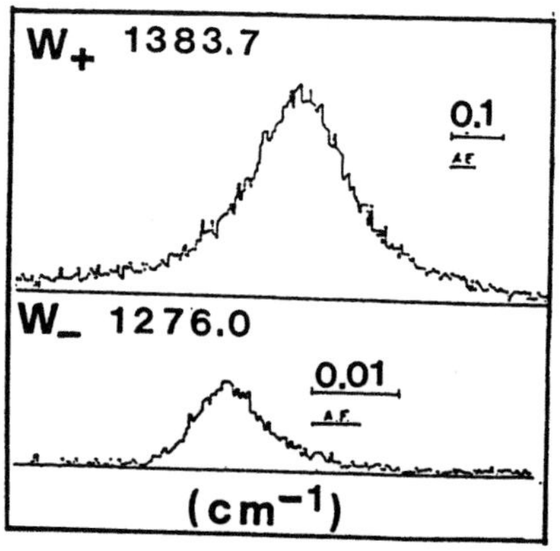

Fig. 6. High-resolution Raman spectra for the Fermi diad in solid CO_2 at 4°K |12|.

1. M.P. Gush, W.F. Hare, E.J. Allin and H.C. Welsh: Phys. Rev. 106, 1101 (1957)
2. M.H. Cohen and J. Ruvalds: Phys. Rev. Lett. 23, 1379 (1969)
3. V.M. Agranovich: Sov. Phys. Solid St. 12, 430 (1970)
4. F. Bogani: J. Phys. C11, 1283 and 1269 (1978)
5. F. Bogani, R. Giua and V. Schettino: Chem. Phys. 88, 375 (1984)

6. F. Bogani and P.R. Salvi: J. Chem. Phys. 81, 4991 (1984)
7. V.H. Agranovich in "Spectroscopy and Excitation Dynamics of Condensed Molecular Systems" ed. by V.M. Agranovich and R.M. Hochstrasser, North-Holland, Amsterdam 1983
8. S. Califano, V. Schettino and N. Neto: "Lattice Dynamics of Molecular Crystals", Springer, Berlin 1981
9. J.C. Kimball, C.Y. Fong and Y.R. Shen, Phys. Rev. B23, 4946 (1981)
10. M.L. Geirnaert, G.M. Gale and C. Flytzanis: Phys. Rev. Lett. 52, 815 (1984)
11. G.M. Gale, P. Guyot-Sionnest, W.Q. Zhehg and C. Flytzanis: Phys. Rev. Lett. 54, 823 (1985)
12. R. Ouillon, P. Ranson and S. Califano: to be published
13. B.M. Powell, G. Dolling, L. Piseri, P. Martel: Proceedings of the 5th IAEA Conference on Neutron Scattering. Grenoble 1972, 207 , IAEA - Vienna (1972).

Part IV

Photochemical Reactions

Resonance Raman Studies of Transient Intermediates in Photoreactions of Anthraquinone and Flavone Species

J.N. Moore[1], *D. Phillips*[1], *P.M. Killough*[2], *and R.E. Hester*[2]

[1]Davy Faraday Lab., The Royal Institution, London W1X 4BS, England
[2]Chemistry Department, University of York, York YO1 5DD, England

Introduction

Using two-colour pump-probe nanosecond pulsed laser techniques we have identified a number of transient species in the photoreactions of anthraquinone-2,6-disulphonate (AQS) and 7-hydroxyflavone (7-FOH) in solution through their resonance Raman (RR) spectra. The anthraquinone system is of interest due to its wide ranging applicability as a photosensitizer [1,2], while the flavone species provides an example of excited state proton-transfer reactions [3].

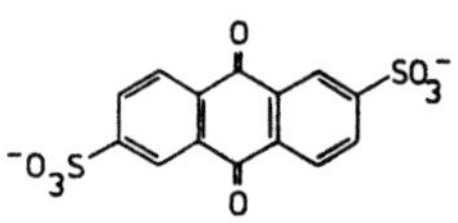

AQS

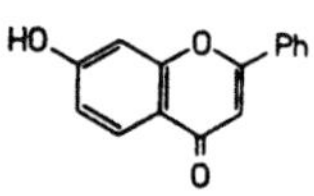

7-FOH

Transient absorption studies have revealed a complicated reaction scheme following absorption of u.v. light by AQS in aqueous solution, with up to five distinct species having been reported [4]. Efficient intersystem crossing from the short-lived excited singlet state to the triplet species, ^{3}AQS, is followed by reactions as shown in Figure 1.

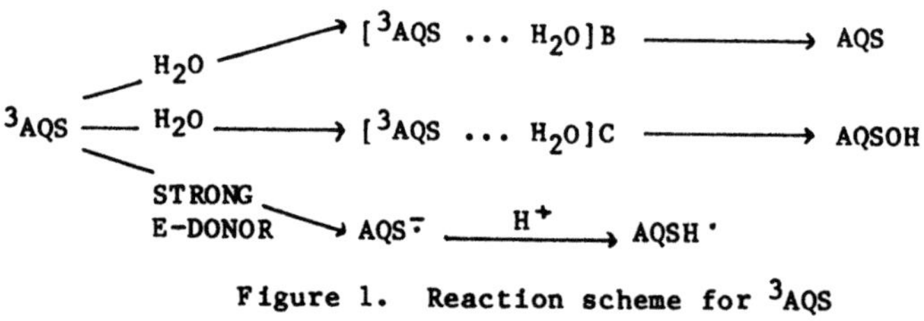

Figure 1. Reaction scheme for ^{3}AQS

In pure water, transient intermediates B and C are formed, these having been assigned as different complexes of ^{3}AQS with H_2O. Species C leads to a hydroxyanthraquinone product (AQSOH) but B does not. In the presence of high concentrations of a strong electron donor (eg. NO_2^- ion) the radical anion, AQS$^{\cdot-}$, is formed (as AQSH$^\cdot$ in acid solution). We report here results from time-resolved RR measurements aimed at further identification and characterization of several transient species in this photoreaction scheme.

Transient absorption data from the flavone system in methanol solution [3] have been interpreted in terms of the reaction scheme shown in Fig. 2.

Our aim here is to test this scheme through the application of the RR method and to characterize the ground state species I and II by their vibrational spectra.

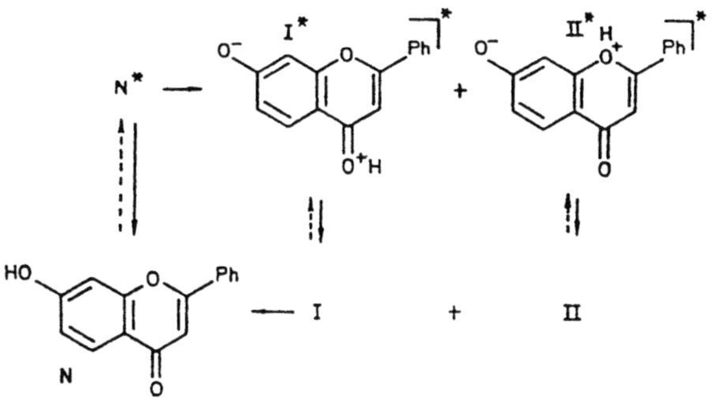

Figure 2. Photolysis reaction scheme for 7-FOH

Experimental

Two excimer lasers were used, one to pump the sample into a photoexcited state (337 or 351 nm for both AQS and 7-FOH, 10 ns pulsewidth), the other to drive a dye laser (370-540 nm, 10 ns) which, in turn, was used to probe the RR spectra of the reacting systems at various times after photolysis. Spectra were collected with an apparatus comprising a triple spectrometer and intensified diode-array detector (OMA). Typically, with laser repetition rates of 10 Hz, ca. 2500 pulse pairs were used to obtain each spectrum. Sample solutions were flowed through the loosely focused coaxial laser beams either in an open jet or through a quartz capillary tube. Solute concentrations were 5 x 10^{-3} M (in water) for AQS and 1 x 10^{-3} M (in methanol) for 7-FOH.

Results and Discussion

1. AQS. The best quality anthraquinone data were given by solutions containing 0.1 M $NaNO_2$ as the strong electron donor. Figure 3 shows typical RR spectra obtained under conditions (probe wavelength and delay time) appropriate to maximise the band intensities for the $AQS^{\bar{\bullet}}$ species (absorption max. at 400 and 510 nm) [5]. These spectra were given by solutions at pH 6. Lowering the pH to less than 3 and tuning the probe laser wavelength to 430 nm gave RR spectra from AQSH$^{\bullet}$ (abs. max. 400 nm).

In the absence of a strong electron donor, the ^{3}AQS triplet species is longer lived (ca. 500 ns, depending on solution conditions, and decays to form transients

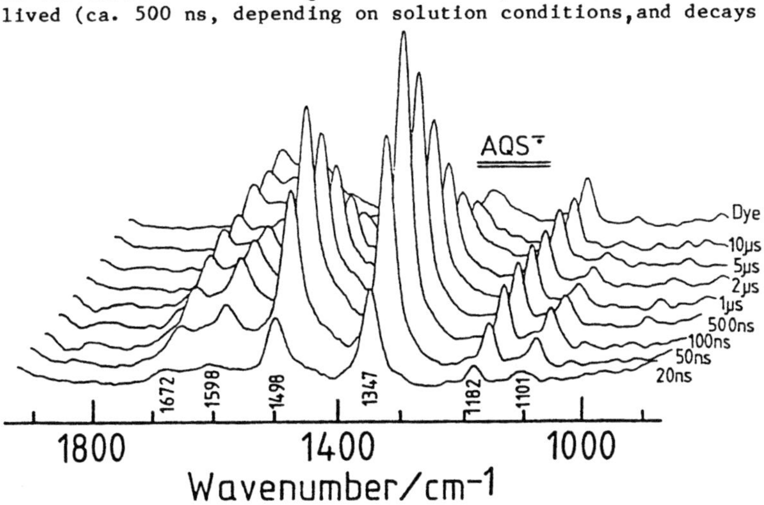

Figure 3. RR spectra of 5 x 10^{-3} M $AQS^{\bar{\bullet}}$ in 0.1 M $NaNO_2$ (pump 337 nm, probe 480 nm)

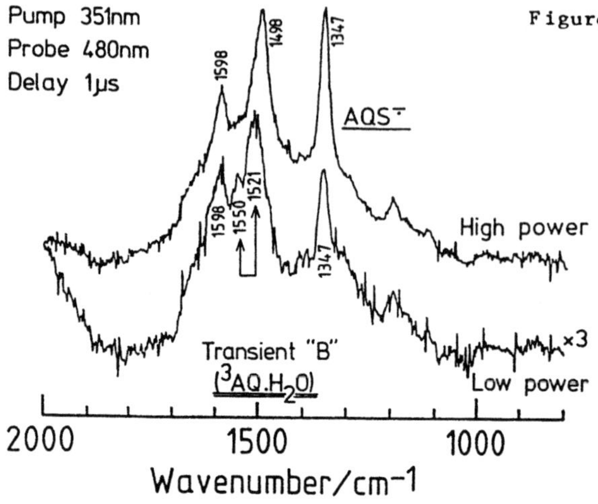

Figure 4. Conversion of transient "B" to $AQS^{\overline{\cdot}}$ at high laser power

B and C (see Fig. 1). RR spectra from each of these species have been obtained, although the fact that all three species occur together in solution and their absorption spectra overlap (abs. max. 390 and 480 nm for ^{3}AQS, 480 nm for B, 380 and 600 nm for C) places particular importance on the time-dependence of the RR band intensities in assigning the component spectra. A further problem encountered with transients B and C was their tendency to convert into the $AQS^{\overline{\cdot}}$ radical anion under the influence of the probe laser. This sensitivity necessitated the use of low laser irradiance, particularly at the wavelengths of the absorption maxima, as illustrated in Fig. 4. Raman bands determined for all the transient species shown in Fig. 1 are summarised in Table 1.

Table 1. RR bands from AQS and derived reaction intermediates

	AQS	$AQS^{\overline{\cdot}}$	AQSH$^{\bullet}$	B	C	^{3}AQS
ν_a(CO)	1681(i.r.)	1598(dp)	-	1596	(1590)	(1590)
ν_s(CO)	1672	1347	1606	1343	1347	1210
ν(CC)	1596	1498	1576	[1550	[1518	-
	-	-	-	[1514	[1483	(1490)
δ(CH)	1182	1182	1177	1179	-	1182
ν(CC)i.p.δ(CO)	-	1101	-	-	-	(1126)
ring	683	683	-	-	-	683

Key: (i.r.) = infrared; (dp) = RR band depolarized.

The assignment of vibrational bands to specific normal modes begins with the parent compound, AQS. There is considerable evidence [6] available for the band assignments given in Table 1. In relating the spectra from the photoreaction intermediates to that of the AQS parent, we also have made use of the assignments reported for the analogous but simpler benzoquinone (BQ), benzosemiquinone ($BQ^{\overline{\cdot}}$, BQH$^{\bullet}$) and hydroquinone (HQ) systems [7]. Coupling of the two carbonyl bond stretching modes in these centrosymmetric molecules (all except BQH$^{\bullet}$) leads to symmetric and asymmetric normal modes, ν_s(CO) and ν_a(CO), respectively, the former being Raman active but infrared forbidden and vice versa for the latter. However, our spectra of $AQS^{\overline{\cdot}}$ show an RR depolarized band at 1598 cm^{-1} (all other bands being polarized) which must arise from a ν_a(CO) mode, thus indicating some asymmetry in the $AQS^{\bar{\cdot}}$ structure. This may be due to ion pair formation, although the e.s.r. evidence from aqueous solutions does not support this [8]. However, the depolarized nature of the 1598 cm^{-1} band also indicates that the two carbonyl oscillators remain coupled and the $BQ^{\bar{\cdot}}$ comparison leads to the tentative assignments shown in Table 1.

The $\nu(CO)$ and $\nu(CC)$ bands shift to lower wavenumbers when AQS is photolysed (Tables 1 and 2). As for BQ, the effect is larger for $\nu(CO)$ than for $\nu(CC)$ but the $\nu_s(CO)$ shifts are uniformly larger for AQS than those reported for the analogous BQ species. Full reduction of BQ to HQ results in a -400 cm^{-1} shift while the ^3AQS spectrum shows a -462 cm^{-1} shift in $\nu_s(CO)$ from the parent AQS spectrum. The $\nu_s(CO)$ wavenumber values show the triplet ^3AQS species to be shifted much further from the parent AQS than is the radical anion, although this difference is not apparent in the $\nu(CC)$ values (see Table 2). This is consistent with the ^3AQS species being an $(n\pi^*)$ triplet rather than a $(\pi\pi^*)$ triplet [9]. Moreover, since the B and C spectra are more like that of $AQS^{\bar{\cdot}}$ than ^3AQS, these may be formulated as complexes with water molecules in a partial charge-transfer relationship to the ^3AQS, viz. $[H_2O(\dot{+}) \longrightarrow {}^3AQS(\dot{-})]$. Further evidence for this was found in the ease with which species B and C were converted to the radical anion, $AQS^{\bar{\cdot}}$, by the Raman laser.

Table 2. RR band shifts from the AQS spectrum

	$AQS^{\bar{\cdot}}$	$AQSH^{\cdot}$	B	C	^3AQS
$\nu_a(CO)$	-83	-	-85	(-91)	(-91)
$\nu_s(CO)$	-325	-66	-329	-325	-462
$\nu(CC)$	-98	-20	[-46 -82	[-78 -113	- (-106)

Since transient "C" leads to formation of α-AQSOH (see reaction scheme) but transient "B" does not, we might expect to find some indication in the RR spectra of differential C-C bond activation. The $\nu(CC)$ bands listed in Table 1 do indeed show larger shifts for "C" than for "B" (see also Table 2).

2. 7-FOH. Time-resolved RR spectra of the species resulting from photolysis of 7-FOH in methanol, after scaled subtraction of the strong solvent bands, are shown in Fig. 5. The strong bands seen at the shortest times after photolysis appear to decay non-exponentially, indicating possible accidental degeneracy of modes from two or more species with different lifetimes. The transient absorption data [3] yield lifetimes for species I and II (see Fig. 2.) of 65 μs and 380 ns, respectively, and a triplet species with 5 μs lifetime. All three species absorb at the probe laser wavelength used and thus all three may be expected to contribute to the RR spectra observed.

For the parent 7-FOH, by comparison with the previous discussion, we may tentatively ascribe the band at 1626 cm^{-1} to the $\nu(C{=}O)$ mode, 1457 cm^{-1} to $\nu(CC)$ and 1250 cm^{-1} to $\nu(C{-}OH)$. The shift of $\nu(C{=}O)$ down to 1571 cm^{-1} is much less than would be expected for an excited triplet species, and we therefore make the preliminary assignment of 1571 cm^{-1} to species I and/or II. Similarly, the strong transient Raman band at 1401 cm^{-1} may be associated with the second C-O bond stretching in I and/or II, since this is expected to become conjugated with the ring system following proton-transfer and thus lead to a higher frequency than in the 7-FOH parent. Further, more definite analysis requires data on the probe wavelength- and time-dependence of the RR band intensities.

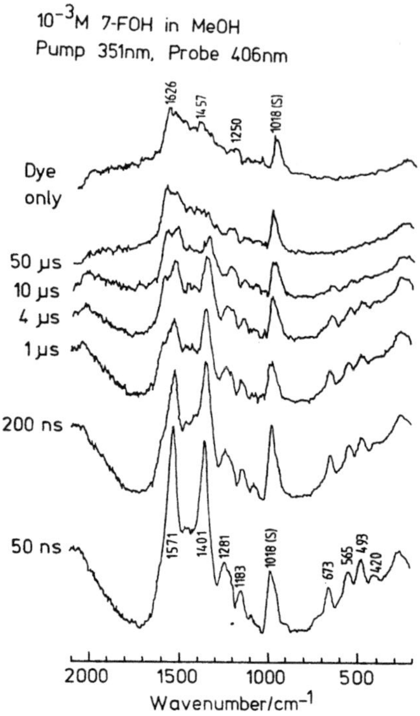

Figure 5. RR spectra of species from 7-FOH photolysis

Acknowledgements

We are grateful to Prof. M. Itoh for donating the 7-FOH sample and to the SERC for financial support.

References

1. J.M. Bruce in "The Chemistry of the Quinonoid Compounds", S. Patai (ed.), Wiley, N.Y., 1974.
2. I. Okura and N. Kim-Thuan, Chem. Lett., 1569 (1980).
3. M. Itoh and T. Adachi, J. Am. Chem. Soc. 106, 4320 (1984).
4. J.N. Moore, D. Phillips, N. Nakashima and K. Yoshihara, to be published; I. Loeff, A. Treinin and H. Linschitz, J. Phys. Chem. 87, 2536 (1983).
5. J.N. Moore, G.M. Atkinson, D. Phillips, P.M. Killough and R.E. Hester, Chem. Phys. Lett. 107, 381 (1984).
6. S.N. Singh and R.S. Singh, Spectrochim. Acta 24A, 1591 (1968); A. Girlando, D. Ragazzon and C. Pecile, Spectrochim. Acta 36A, 1053 (1980).
7. R.E. Hester and K.P.J. Williams, J. Chem. Soc. Faraday 2, 78, 573 (1982); R. Rosetti, S.M. Beck and L.E. Brus, J. Phys. Chem. 87, 3058 (1983); R.H. Schuler, G.N.R. Tripathi, M.F. Prebenda and D.M. Chipman, J. Phys. Chem. 87, 5357 (1983).
8. M.B. Hocking and S.M. Mattar, J. Mag. Res. 47, 187 (1982).
9. O.S. Khalil and L. Goodman, J. Phys. Chem. 80, 2170 (1976).

Resonance Raman Spectra and Assignments of Deuterated Stilbene Anion and Cation Radicals

C. Richter and W. Hub

Institut für Physikalische und Theoretische Chemie,
Technische Universität München, Lichtenbergstraße 4,
D-8046 Garching, F.R.G.

1. Introduction

The formation of ion radicals of trans-stilbene (TS) in photochemically-induced reactions with various substrates in polar solvents was established recently by TRRR spectroscopy /1,2/. These investigations revealed the reaction mechanisms of (i) the photoaddition of tert. alkylamines to TS /1/ and (ii) the photooxygenation of TS sensitized by cyanoanthracenes /2/.

The RR spectra of both trans-stilbene ion radicals excited in their strongest absorption band in the visible region at 480 nm are different with respect to the vibrational frequencies and the number of bands. In order to get structural informations from these spectra it is necessary to have reliable vibrational assignments. For this purpose and as a basis for a normal mode analysis of the ion radicals /3/ we have measured the RR spectra of the anion and the cation radicals of nine differently deuterated trans-stilbenes.

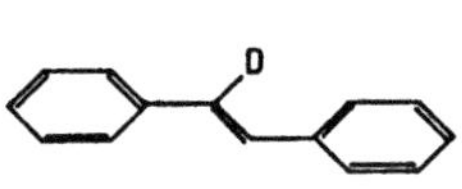

D01

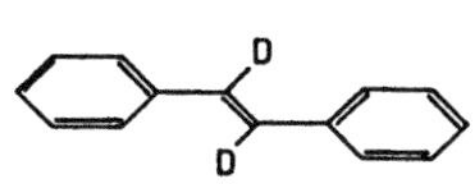

DAA

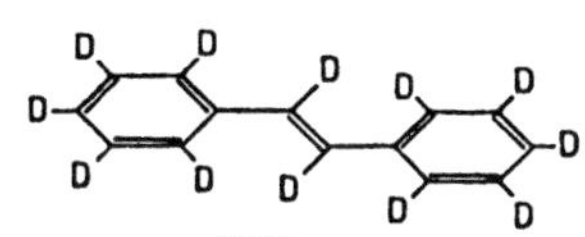

D12

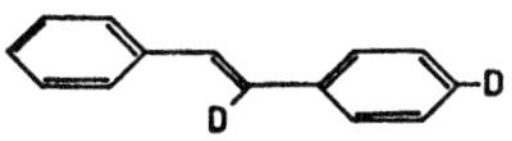

DAP

D44

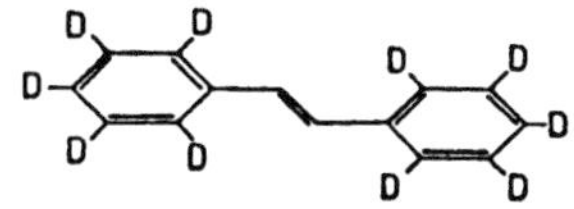

D10

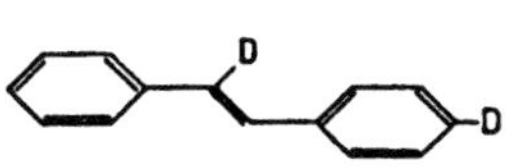

DAP'

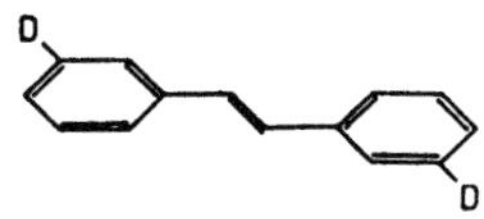

D33

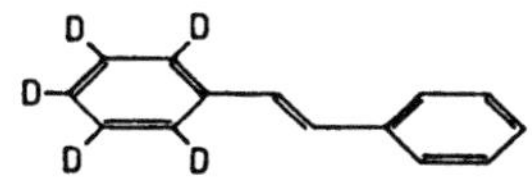

D05

2. Experimental

The ion radicals have been produced by the above-mentioned reactions in acetonitrile solutions at room temperature with a photolysis laser pulse at 337 nm and a probe laser pulse at 480 nm with gated multichannel detection (Δt=100 ns). Calibration of the raman bands was achieved by benzene Raman bands and are accurate to about 3 cm^{-1}.

3. Results and Discussion

Anion radicals. The RR spectra of $D00^-$, DAA^-, $D10^-$, and $D12^-$ are shown in Fig. 1a. It is obvious, that the band at 1577 cm^{-1} in $D00^-$ is shifted to lower frequencies upon ring deuteration, whereas the band at 1553 cm^{-1} in $D00^-$ is sensitive to deuteration in the vinyl group. Therefore we assign the 1577 cm^{-1} band to the ring stretching vibration and the 1553 cm^{-1} band to the vinyl stretching vibration. These assignments are further supported by the RR spectra of the partially ring deuterated species $D33^-$, and $D05^-$. In these spectra the position of the 1553 cm^{-1} band remains unshifted, whereas the 1577 cm^{-1} band is found at lower frequencies (see table 1). Deuteration of a single vinyl carbon atom leads to a shift of the vinyl

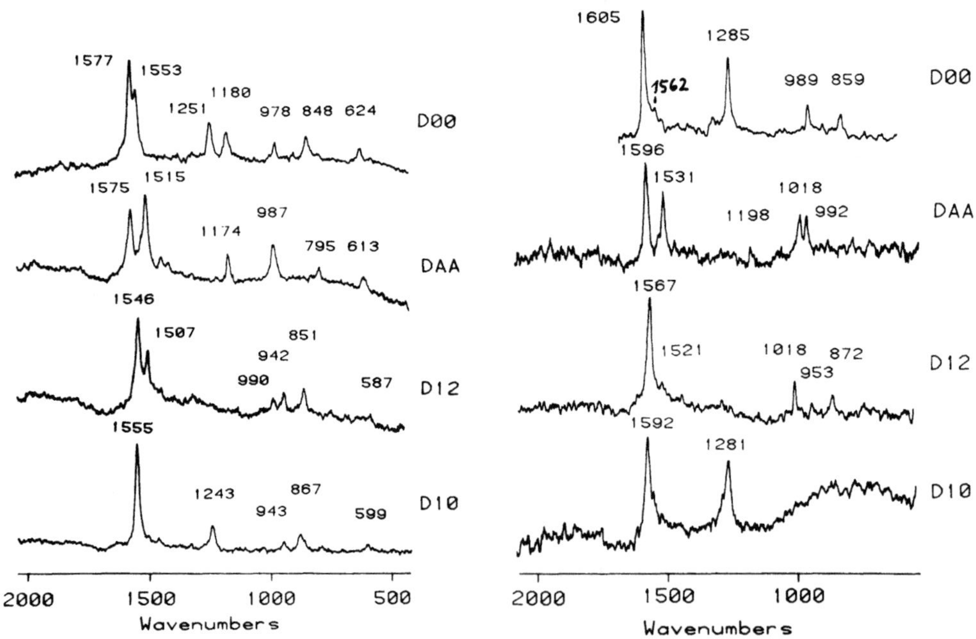

Figure 1a Figure 1b

RR spectra of the anion (Fig. 1a) and cation (Fig. 1b) radicals of deuterated trans-stilbenes (solvent bands subtracted)

Table 1. RR frequencies of deuterated <u>trans</u>-stilbene ion radicals

	TS	D00-	DAA-	D12-	D10-	D33-	D44-	D01-	DAP-	DAP'-	D05-
C=C str	1642	1553	1515	1507		1552	1553	1536	1531	1536	1554
C=C ring	1598	1577	1575	1546	1555	1572	1576	1578	1574	1577	1570
CH df vin	1325	1251	987	990	1243	1246	1249	1225	1223	1222	1243
								975	971	968	
CH df ring	1187	1180	1174	851	867	1177	1183	1178	1173	1178	1152
ring def	993	978	987	942	943	981	965				972
ring def	854	848	795				837	818	817	825	824
ring def	640	624	613	587	599	620	619	623	615	618	613

	TS	D00+	DAA+	D12+	D10+	D33+	D44+	D01+	DAP+	DAP'+	D05+
C=C str	1642	1562	1531	1521	1569	1572	1568	1577	1552	1552	
C=C ring	1598	1605	1596	1567	1592	1604	1602	1600	1596	1597	1600
CH df vin	1325	1285	1018	1018	1281	1284	1285	1257	1254	1259	1293
CH df ring	1187										
ring def	993	989	992	953		993	977	989	996	990	
ring def	854	859					859			845	842
ring def	640	628				629	629	624		629	622

stretching vibration from 1553 cm^{-1} to about 1535 cm^{-1} ($D01^-$, DAP^-, DAP'^-).

The frequency shifts in the C=C stretching region are accompanied by a strong variation of the relative intensities of the ring and the vinyl stretching bands. As a common feature of all anion radical spectra it is observed that the intensity of the vinyl C=C stretch is reduced as it approaches the frequency of the ring stretch. The single band at 1555 cm^{-1} in the $D10^-$ spectrum may therefore be explained by an effective suppression of the vinyl C=C stretch band intensity due to a low frequency difference between the two modes.

There are two bands in the $D00^-$ spectrum between 1100 and 1300 cm^{-1}, which may be assigned to CH bending modes because of the observed frequency shifts in the spectra of the isotopically labelled <u>trans</u>-stilbenes. The 1251 cm^{-1} band disappears upon deuterium substitution of both vinyl hydrogens (DAA^- and $D12^-$) and new bands are found around 990 cm^{-1}. In the spectra of $D01^-$, DAP^-, and DAP'^- the band is found at 1223 cm^{-1} with reduced intensity and a new band at about 970 cm^{-1} appears. All these bands are assigned to the vinyl CH deformation mode. Other substitution patterns do not influence the band position significantly.

The assignment of the 1180 cm^{-1} band in the DOO^- spectrum to the ring CH deformation mode is based on the shifts occurring upon ring deuteration ($D10^-$: ca. 867 cm^{-1}; $D12^-$: 851 cm^-) and the almost unaffected band positions in the other spectra.

Obviously a band assignable to the C-ring stretching vibration (which is at 1193 cm^{-1} in neutral TS) can not be found in the anion radical spectra.

The three medium intense bands below 1000 cm^{-1} in the DOO^- spectrum may be assigned to the ring deformation modes. The shifts of the anion radical frequencies upon deuterium substitution are comparable to those observed for the neutral trans-stilbene.

Cation radicals. In the cation radical spectra the assignment of the C=C stretching frequencies on the basis of the RR spectra reported here is not straightforward. At present most features of the cation radical spectra in the C=C stretching region can be explained by the assignment of the strong band at 1605 cm^{-1} in the DOO^+ spectrum to the ring C=C stretching mode. This explanation needs the assumption of a strong coupling between the two in-phase C=C stretching vibrations, leading to the suppression of the vinyl C=C stretching band intensity and in addition to a splitting of the C=C stretching frequencies. Both effects depend on the frequency difference between the vinyl and the ring C=C stretching modes,and thus on the strength of the coupling between them.

In contrast to this, the 1285 cm^{-1} band in DOO^+ is readily assigned to the vinyl CH deformation mode due to its shift observed upon deuterium substitution of both vinyl hydrogens ($D12^+$, DAA^+). In most of the other spectra this is the only band of reasonable intensity between 1100 and 1500 cm^{-1}. Obviously, in the cation radical spectra neither the ring CH deformation mode nor the C-ring stretching modes appear.

The assignment of ring deformation modes in the cation radical spectra is only possible for a few compounds due to their low intensity and the partially low signal/noise ratio. The frequencies of these bands are much closer to those of the neutral compounds,as in the case of the anion radicals.

4. Conclusions

This series of spectra allows to get reliable assignments for the RR bands of TS^- and at least for the vinyl CH deformation mode of TS^+. Some previously made assignments by us and other authors have turned out to be wrong. It is especially helpful for the evaluation of the structures of the radical ions by a normal mode analysis /3/ to know the assignments of the vinyl CH deformation modes, as they are sensitive to the twisting angle around the vinyl bond. The work on isotopically labelled trans-stilbene ion radicals will be continued.

5. References

1 W. Hub, S. Schneider, F. Dörr, J. D. Oxman, F. D. Lewis: J. Am. Chem. Soc. 106, 708 (1984)

2 W. Hub, U. Klüter, S. Schneider, F. Dörr, J. D. Oxman, F. D. Lewis: J. Phys. Chem. 88, 2308 (1984)

3 H. Hamaguchi, T. Urano, M. Tasumi, C. Richter, W. Hub: to be published

Application of Time-Resolved Resonance Raman Spectroscopy to Some Organic Photochemical Processes in Solution

H. Hamaguchi

Department of Chemistry, Faculty of Science, The University of Tokyo, Bunkyo-ku, Tokyo 113, and
Institute for Molecular Science, Okazaki 444, Japan

The knowledge of the structure and kinetics of electronically and/or vibrationally excited molecules is essential for a proper understanding of various photochemical processes which involve these species as the intermediates. Time-resolved resonance Raman spectroscopy can afford information on both of these properties. Vibrational Raman frequencies reflect the molecular structure on the one hand, while intensities and band shapes should be related to the dynamics, on the other. The present paper reviews our recent research activities in the application of this technique to some typical photochemical reactions in solution.

1. Photoisomerization of Stilbene (SB) and 2-styrylanthracene (SA)

Determination of the structure of S_1 stilbene in solution has been an issue for these few years. We established the vibrational assignments of S_1 SB on the basis of the observed frequency-shifts in several deuterated analogues [1]. The unusually low frequency of the olefinic CH ip-bend and its characteristic behavior upon deuteration were interpreted as due to a structural change between S_1 and S_0 [2]. Recently, Raman spectra of the cation and anion radicals of SB were measured for nine deuterated compounds, and the frequencies of the CH ip-bends were determined [3]. The frequency and isotopic shift of the anion were very close to the S_1 values whereas those of the cation were located in between S_1 and S_0 (Table 1). These results suggest that the structure of the anion is more similar to S_1 rather than to S_0. It is known that SB takes a planar structure in crystal at room-temperature but is twisted to a non-planar structure in the gaseous phase [6,7]. Our model calculation for S_0 SB showed that the decrease of the frequency and isotopic shift in going from crystal to solution could be reproduced by twisting the two phenyl rings [5]. It is therefore highly likely that S_0 SB in solution also takes a twisted structure, as was already suggested [8]. Further decrease of the frequency and isotopic shift in cation, anion, and S_1 can be regarded as the consequence of further deviation from the planar structure and the change in the force constants upon the electronic excitation.

As was pointed out by GUSTAFSON et al. [9] the olefinic C=C stretch of S_1 SB gives a broad asymmetric Raman band. They ascribed the asymmetric shape to the existence of more than one S_1 rotational isomers around the C-phenyl single bonds.

Table 1. Observed frequencies and isotopic shifts[a] of the olefinic CH ip-bends of stilbenes in solution (cm^{-1})

state	S_0(crystal)[b]	S_0[c]	cation[d]	anion[d]	S_1[e]
frequency	1327	1318	1285	1251	1241
isotopic shift	309	303	267	261	259

[a]frequency difference between SB and SB-d_{12}. [b][4]. [c][5]. [d][3]. [e][1].

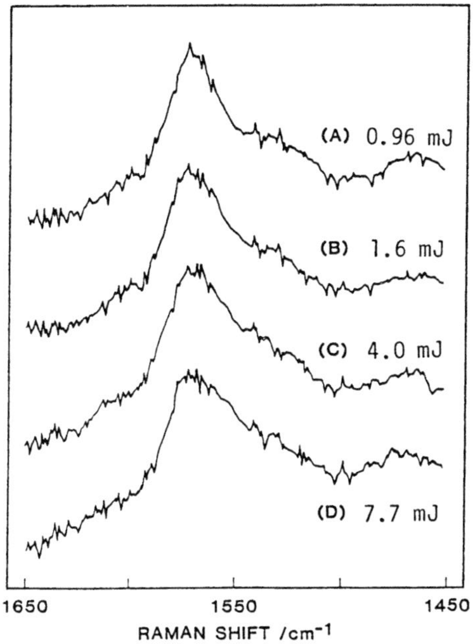

Fig.1 Probe-laser energy-dependence of the bandshape of the C=C stretch band of S_1 SB in n-hexane

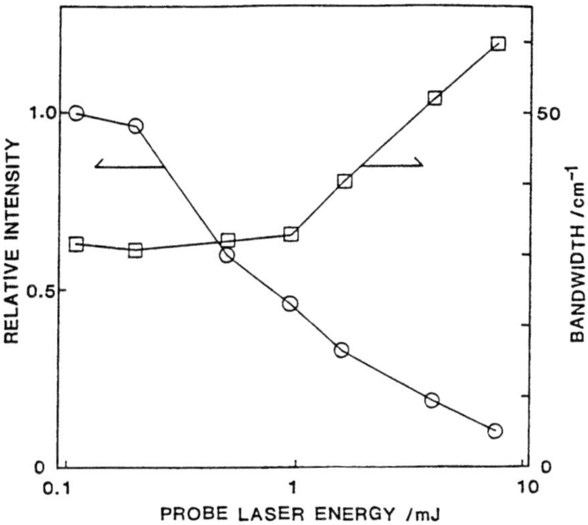

Fig.2 Probe-laser energy-dependence of the area intensity (O) and the half-width (□) of the C=C stretch band of S_1 SB in n-hexane

We found that this bandshape changed significantly with increasing the probe laser power, which was in resonance with the $S_n \leftarrow S_1$ transition (Figs. 1 and 2). As the laser power was increased, the area intensity of the C=C stretch band decreased relatively to the solvent band; at 7.7 mJ/pulse, it was reduced to 1/10 of the value at 0.11 mJ/pulse. This means that the S_1 state of SB was depopulated by the probe laser pulse and that the produced S_n SB did not relax back to the S_1 state, which we detected by resonance Raman scattering. A simple rate equation requires that the depopulation by 1/10 occurs only when the rate of $S_n \leftarrow S_1$ excitation is 10 times faster than that of the relaxation of S_1, the lifetime of which is about 70 ps in solution [10]. In other words, at 7.7 mJ/pulse, the S_1 SB produced by the pump laser pulse was further converted to S_n in less than 10 ps of the excitation[1]. This indicates that the ratio of the vibrationally unrelaxed S_1 molecules to the relaxed ones increased as the probe power was increased; at 0.11 mJ/pulse the ratio was negligibly small, but at 7.7 mJ/pulse it should have been considerably high. Therefore, the bandshape at higher laser energies should reflect a significant contribution from vibrationally unrelaxed S_1 molecules. (In a sense, we effectively attained a time-resolution as short as 10 ps.) The broad bandshape and the slight shift of the band centre seem to suggest that low-frequency large-amplitude vibrations (which can affect the C=C stretching frequency) remains unrelaxed in the experiment at higher energies. The continuous bandshape change with increasing laser power is more reasonably explained by the excitation of multiple low-frequency modes rather than the population change between two or three rotational isomers.

If one of the two phenyl rings of SB is replaced by an anthryl group, 2-styrylanthracene (SA) is obtained. The photochemistry of this molecule was recently

[1] An order of magnitude estimate (10 mJ/pulse energy, 1 mmϕ focusing spot, 2×10^5 molar extinction coefficient for $S_n \leftarrow S_1$ [11]) gives the rate of $S_n \leftarrow S_1$ excitation greater than 10^{11}.

studied thoroughly [12]. The excitation of the cis isomer resulted in a 100 % isomerization to trans, while the reverse process from trans to cis was not observed (termed "one-way" cis to trans photoisomerization).

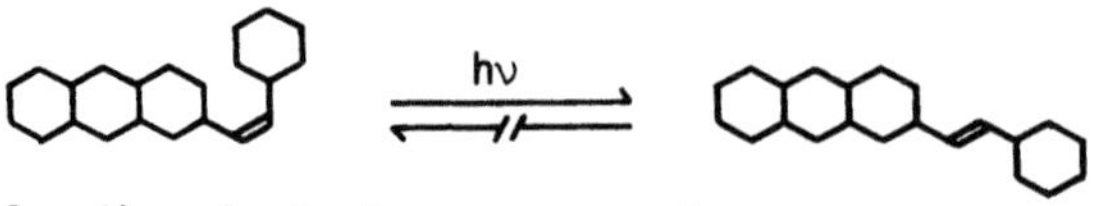

The quantum yield for the cis to trans conversion was much higher than unity, implying that a quantum chain process involving a long-lived excited species plays a role in the process of isomerization [12]. It is therefore important to characterize this long-lived excited species, which is most probably the lowest triplet state of SA. We measured the transient resonance Raman spectra of T_1 SA produced by the photoexcitation of both of the cis and trans isomers [13]. Obtained spectra are given in Fig. 3 together with the corresponding S_0 spectra in the crystalline state. The Raman spectra of S_0 cis-SA was markedly different from that of S_0 trans-SA, whereas the T_1 spectra measured from cis- and trans-SA were identical with each other. This finding, which is similar to the case of retinal [14], indicates that the photoexcitation of the cis and trans isomers gives the same T_1 species which exclusively relaxes to S_0 trans-SA (otherwise the trans to cis isomerization should be observed). In other words, a 100 % cis to trans conversion is complete in the T_1 manifold, before the molecule relaxes to the ground state.

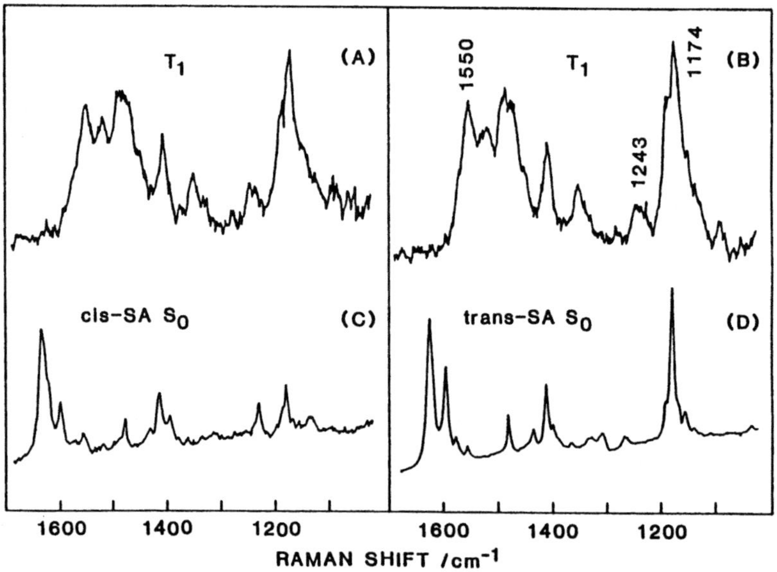

Fig.3 Transient T_1 Raman spectra obtained from cis-SA (A) and trans-SA (B) in benzene, and S_0 Raman spectra of crystalline cis-SA (C) and trans-SA (D)

This novel scheme of photoisomerization, which was previously proposed from the photochemical data [15], is now confirmed by transient Raman studies of two different molecules, retinal and SA. Thus, the conventional picture of the photoisomerization of olefins needs to be altered; there are two types of isomerization pathways, one through the non-adiabatic transition from S_1 (or T_1) to S_0 at the so called "perpendicular" configuration,and the other along the T_1 potential surface which energetically favours the trans conformation,and which allows a very fast adiabatic conversion from cis to trans.

2. Photoionization of Biphenyl (DP)

The photoexcitation of biphenyl in n-hexane generated a strong transient absorption at 650 nm which we assigned to the $S_n \leftarrow S_1$ transition [16]. Time-resolved resonance Raman spectra of DP in various solutions are given in Fig. 4. In n-hexane, four Raman bands, 1555, 1194, 1007, and 313 cm^{-1}, were observed at 0 ns (4-A) and they disappeared after 30 ns of the photoexcitation (4-B). This temporal behavior is in harmony with the lifetime of the S_1 state (5 ns) and hence the four Raman bands were reasonably assigned to S_1 DP. In an acetonitrile solution, on the other hand, the spectral pattern changed greatly. Five extra bands, 1680, 1616, 1502, 1343, and 1013 cm^{-1} were observed (4-C) and remained even after 50 ns (4-D). By adding ammonia to the acetonitrile solution, these extra bands were completely quenched out in the 50 ns spectrum. From these results we concluded that the cation radical of DP was formed photolytically in acetonitrile, and that it did not occur in n-hexane. A preliminary study on the laser power-dependence of the S_1 and the cation band intensities suggested a step-wise two-photon (266 nm x 2) process for the mechanism of the photoionization.

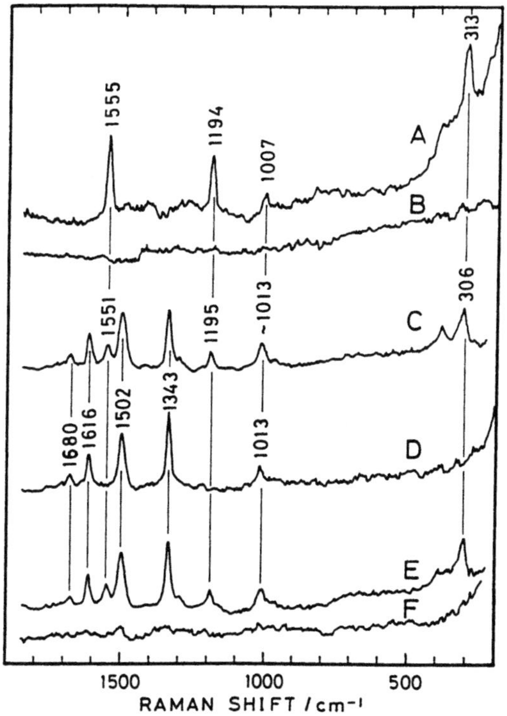

Fig.4 Transient Raman spectra of DP in solution
(A) in n-hexane at 0 ns,
(B) in n-hexane at 30 ns,
(C) in acetonitrile at 0 ns,
(D) in acetonitrile at 50 ns,
(E) in acetonitrile + ammonia at 0 ns,
(F) in acetonitrile + ammonia at 50 ns.

3. Photoreduction of Benzophenone (BP)

It is well known that the photoreduction of benzophenone, the most popular organic photochemical reaction in solution, is initiated by an electron-transfer to the T_1 state. Fig. 5 compares the transient resonance Raman spectra of T_1 and S_0 BP in a carbontetrachloride solution [17]. In the S_0 spectrum, the C=O stretch and the phenyl C=C stretch bands were observed at 1668 and 1603 cm^{-1}, respectively. These two bands disappeared in the T_1 spectrum and, instead, two broad bands appeared at 1540 and around 1100 cm^{-1}. The former band is assignable to the phenyl C=C stretch; the corresponding frequency in S_1 stilbene is 1533 cm^{-1}. The substitution of the carbonyl carbon by ^{13}C did not shift the 1185 and 1147 cm^{-1} band, but slightly changed the shape of the broad feature near 1100 cm^{-1}. Therefore, we tentatively

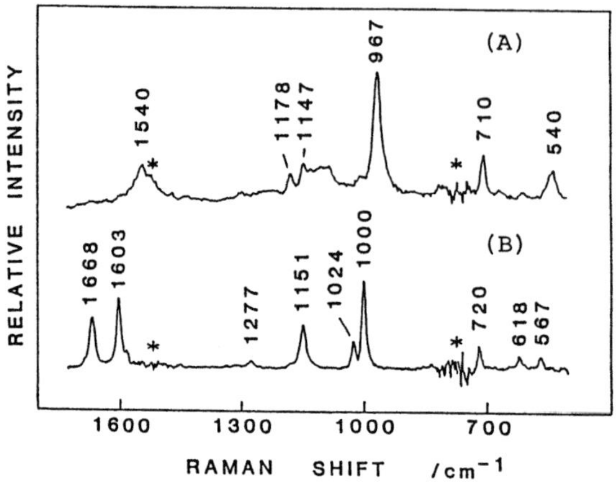

Fig.5 Transient resonance Raman spectrum of T_1 BP (A) and Raman spectrum of S_0 BP (B) in carbontetrachloride. Asterisks mark the region overlapped by the solvent bands which are already subtracted.

assign this broad band to the carbonyl stretch. If it is so, the T_1 carbonyl stretching frequency in solution is significantly lower than the value 1218 cm^{-1} in crystal at 4.2 K [18]. In addition, the unusual bandshape suggests a strong interaction between the vibrational level of the carbonyl stretch and some other vibrational or electronic levels. We note the possibility of an interaction including the T_2 state which is nearly degenerate with T_1.

Acknowledgement

The author is grateful to the collaborators; Prof. Mitsuo Tasumi, Mr. Chihiro Kato, Miss Taeko Urano, Mr. Tahei Tahara of the University of Tokyo, and Mr. Takashi Karatsu, Dr. Tatsuo Arai, Prof. Katsumi Tokumaru of University of Tsukuba.

References

1. H. Hamaguchi, T. Urano, and M. Tasumi: Chem. Phys. Lett. 106, 153 (1984).
2. H. Hamaguchi, J. Mol. Structure 126, 125 (1985).
3. W. Hub: private communication.
4. Z. Meić and H. Güsten: Spectrochim. Acta 34A, 101 (1978).
5. H. Hamaguchi, T. Urano, and M. Tasumi: to be published.
6. C. J. Finder, N. G. Newton, and N. L. Allinger: Acta Cryst. B30, 411 (1974).
7. A. Traetterberg, E. B. Frantsen, F. C. Mijlhoff, and A. Hoekstra: J. Mol. Structure 26, 57 (1975).
8. A. Bree and M. Edelson: Chem. Phys. 51, 77 (1980).
9. T. L. Gustafson, D. M. Roberts, and D. A. Chernoff: J. Chem. Phys. 79, 1559 (1983); ibid. 81, 3438 (1984).
10. R. M. Hochstrasser: Pure Appl. Chem. 52, 2683 (1980).
11. K. Yoshihara: private communication.
12. T. Karatsu, Y. Kuriyama, T. Arai, H. Sakuragi, and K. Tokumaru: J. Am. Chem. Soc. submitted.
13. H. Hamaguchi, M. Tasumi, T. Karatsu, T. Arai, and K. Tokumaru: J. Am. Chem. Soc. submitted.
14. H. Hamaguchi, H. Okamoto, M. Tasumi, Y. Mukai, and Y. Koyama: Chem. Phys. Lett. 107, 355 (1984).
15. T. Arai, T. Karatsu, H. Sakuragi, and K. Tokumaru: Tetrahedron Lett. 24, 2873 (1983).
16. C. Kato, H. Hamaguchi, and M. Tasumi: Chem. Phys. Lett. submitted.
17. H. Hamaguchi, T. Tahara, and M. Tasumi: to be published.
18. S. Dym, R. M. Hochstrasser, and M. Schafer: J. Chem. Phys. 48, 646 (1968); Y. Udagawa, T. Azumi, M. Ito, and S. Nagakura: ibid. 49, 3764 (1968).

Time-Resolved Resonance Raman Studies of Photoenolization of ortho-Alkyl Substituted Benzophenones

H. Takahashi, H. Ohkusa, H. Isaka, S. Suzuki, and S. Hirukawa

Department of Chemistry, School of Science and Engineering, Waseda University, Tokyo 160, Japan

1. Introduction

Although it is now well established that 2-alkylbenzophenones undergo reversible photoenolization by α-hydrogen abstraction from the internal alkyl group [1], the identification of the transient species in the photoenolization is still controversial.

Detection of the ketone triplet state was first claimed by ZWICKER et al. [2] in the flash photolysis of 2-benzylbenzophenone. This assignment was subsequently discounted by PORTER and TCHIR [3] who observed five transients in the flash photolysis of 2,4-dimethylbenzophenone. They assigned the transients to the ketone triplet state, trans and cis dienols, and dihydroanthrone structure [4]. The assignment of one transient was left open to question, but was suggested to be the dienol triplet state or 1,4-biradical. Similar results were obtained for 2-isopropyl- and 2-benzylbenzophenones.

The assignment of Porter and Tchir has been questioned by SAMMES [5] on the basis of chemical trapping experiments. HAAG et al. [6] proposed a different assignment based on nanosecond flash photolysis of 2,4-dimethylbenzophenone. Their assignment is consistent with the assignments by UJI-IE et al. [7] and NAKAYAMA et al. [8] for 2-methylbenzophenone which are based on picosecond and nanosecond laser photolysis experiments.

Our objective in this study is to obtain further information for the identification of the transient species in the photoenolization of 2-methyl- and 2-benzylbenzophenones using time-resolved resonance Raman spectroscopy, and so resolve the above-mentioned discrepancy.

2. Experimental

Our experimental setup for time-resolved resonance Raman spectroscopy will be described elsewhere [9]. The UV pulses from a nitrogen laser were used for photo-excitation. A tunable flash lamp pumped dye laser was used as a source for Raman excitation. Raman signals were detected on a photodiode array detector attached to a 50 cm single monochromator with appropriate filters. The delay time between the firing of the dye laser and that of the nitrogen laser was varied by a delay circuit constructed in our laboratory.

2-Benzylbenzophenone was synthesized by the Grignard reaction of phenylmagnesium bromide with 2-benzylbenzoyl chloride. The crude product was purified by distillation and thin layer chromatography. 2-Methylbenzophenone was purchased from Aldrich Chemical Co. Inc.

3. Results and Discussion

Two transients have been observed by time-resolved resonance Raman spectroscopy of 2-methylbenzophenone. These transients were assigned to the cis and trans dienols. On the other hand, only one transient species corresponding to the cis dienol has been detected for 2-benzylbenzophenone. The assignments are based on the experimental data given below and shall be justified in the discussion.

2-Methylbenzophenone

Figure 1 shows the time-resolved resonance Raman spectra of 2-methylbenzophenone in methanol-d_3 excited by 475 nm light. At 200 ns after the UV irradiation a doublet at 1541 and 1522 cm^{-1} and a weak band at 1275 cm^{-1} were observed. At 5 µs the band at 1522 cm^{-1} decreased in intensity and at 10 µs it almost disappeared. On the other hand, the band at 1541 cm^{-1} faded slowly but still remained at 2 ms. The time-dependence of the band at 1275 cm^{-1} appeared to be similar to that of the 1522 cm^{-1} band. The decay time as well as the intensity of these transient bands was not affected by the presence of oxygen.

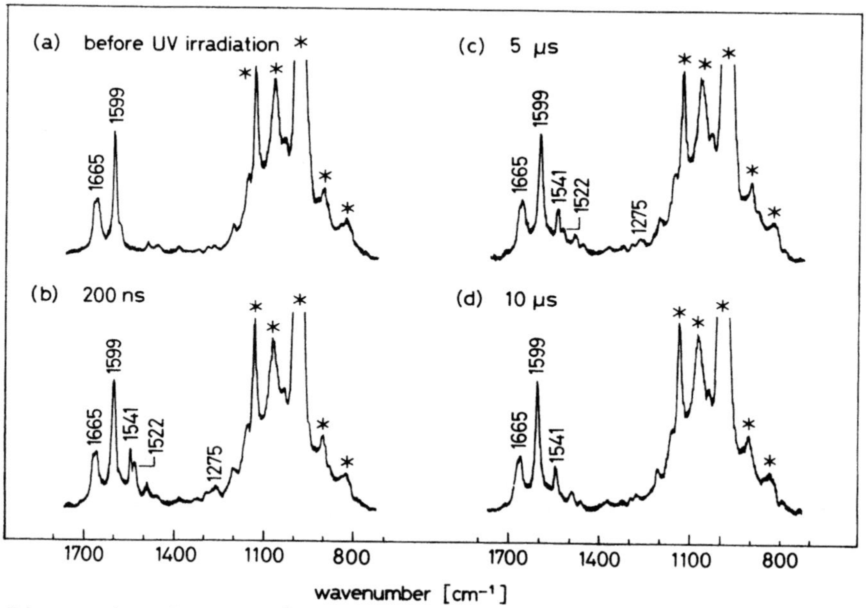

Figure 1 Time-resolved resonance Raman spectra of 2-methylbenzophenone in Methanol-d_3 excited by 475 nm light. * due to solvent

The Raman bands at 1522 and 1541 cm^{-1} can be assigned to the C=C stretching vibrations of the cis and trans dienols, respectively, and the band at 1275 cm^{-1} to the CH_2 in-plane bending vibration. Since in the cis dienol the O-H and $C=CH_2$ groups are situated close to each other, the thermal back reaction (reketonization) should be faster in the cis dienol than in the trans dienol. Thus, it is reasonable to assign the shorter-lived transient to the cis dienol and the longer-lived transient to the trans dienol. The lack of oxygen quenching and the relatively long lifetimes of these transients tend to rule out the triplet state of the ketone or dienol.

The above assignment is consistent with the recent picosecond and nanosecond laser photolysis experiment by NAKAYAMA et al. [8] who detected four transient absorptions and assigned the absorption at 520 nm having the decay time of 2.8 ns to the ketone triplet state, the absorption near 530 nm having the lifetime of 26 ns to the dienol triplet state and the absorption near 400 nm which decayed with two components with lifetimes of 8.1 μs and 8 ms to the cis and trans dienols, respectively. We believe that the Raman band at 1522 cm^{-1} arises from the transient with the lifetime of 8.1 μs and the band at 1541 cm^{-1} from the transient with the 8 ms lifetime.

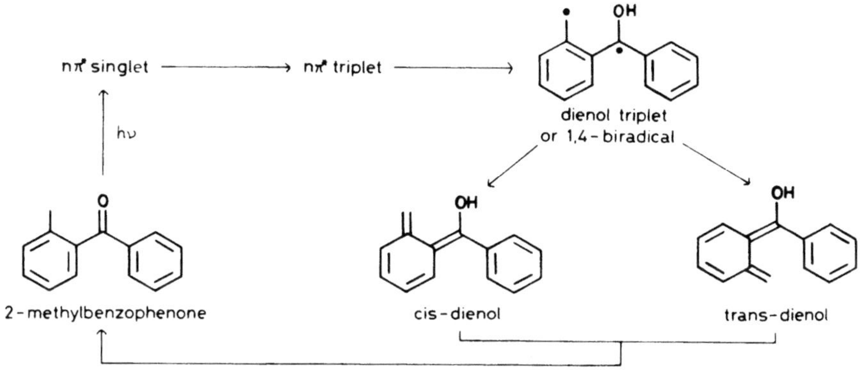

NERDEL and BRODOWSKI [10] showed that trapping of the dienol with maleic anhydride gave only one Diels-Alder adduct having the all-cis configuration. SAMMES [5] pointed out that the trapped dienol must have a trans configuration on the basis of stereo-specificity of Diels-Alder reaction, viz. endo-addition. This fact threw some doubt on the existence of the cis dienol. However, the elusiveness of the cis dienol in the trapping experiment may be explained as due to its very rapid reketonization; the lifetime of ca. 5 μs is too short for the cis dienol to be trapped chemically.

2-Benzylbenzophenone

Figure 2 shows the time-resolved resonance Raman spectra of 2-benzylbenzophenone in methanol-d_3 excited by 475 nm light. At 200 ns after UV irradiation two bands appeared at 1521 and 1295 cm^{-1} and at 2 μs these bands disappeared almost completely.

The transient which gave rise to these bands is considered to be identical to the transient having the lifetime of 1.1 μs reported by PORTER and TCHIR [3]. Although they suggested the assignment of this transient to the dienol triplet state or 1,4-biradical, we attribute it to the cis dienol on the basis of spectral similarity to the cis dienol of 2-methylbenzophenone and the lack of oxygen influence on the Raman intensity and lifetime.

For the dienols of 2-benzylbenzophenone, four stereoisomeric configurations are possible with respect to the C=C(OH)Ph and C=CHPh bonds. However, due to severe steric hindrance, cis configuration around the C=CHPh bond may actually not exist. Hence, like 2-methylbenzophenone, two isomeric dienols having trans and cis configurations around the C=C(OH)Ph bond would be expected.

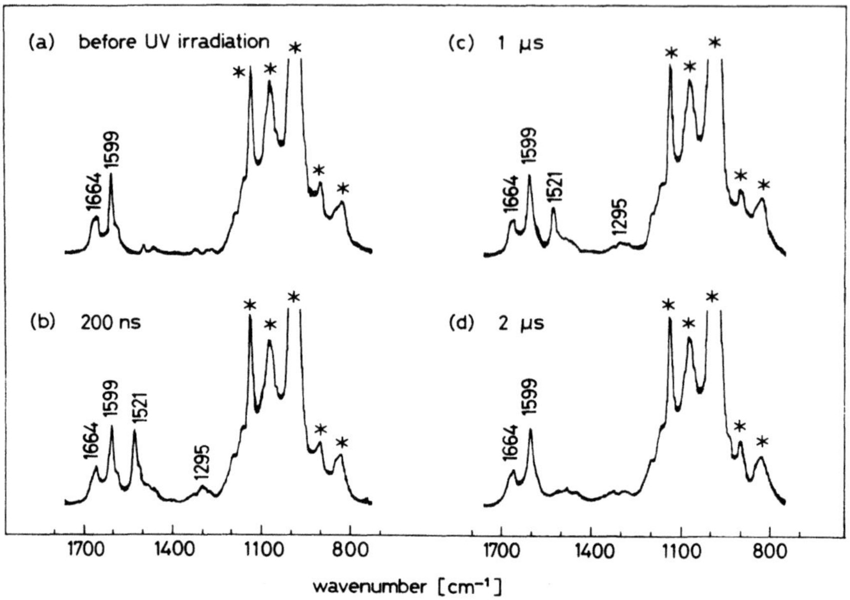

Figure 2 Time-resolved resonance Raman spectra of 2-benzylbenzophenone in methanol-d_3 excited by 475 nm light. * due to solvent

The failure to detect Raman bands of the longer-lived trans dienol may be due to its very low quantum yield, although it is not wholly impossible that the lifetime of the cis dienol is too short to be detected by our system, and the observed transient Raman bands are attributed to the trans dienol.

Solvent Effect on the Lifetimes of the Dienols

The lifetimes of the cis dienols increased with increasing hydrogen-atom accepting ability of the solvent. The time-resolved resonance Raman spectra of 2-methylbenzophenone in dimethyl sulfoxide-d_6 and in acetone-d_6 are compared in Fig. 3 for an illustration. It is seen that the band at 1521 cm^{-1} of the cis dienol is very intense in dimethyl sulfoxide-d_6 and still survived at 50 μs after the UV irradiation, while in acetone-d_6 it disappeared already at 500 ns.

The lifetime of the cis dienol of 2-methylbenzophenone increased in the following order: less than 200 ns in acetonitrile, 200 ~ 500 ns in acetone, 5 ~ 10 μs in methanol and 50 ~ 100 μs in dimethyl sulfoxide. This increasing order of the lifetime corresponds well to the increasing order of hydrogen-bond-acceptor basicity of the solvent obtained by MINESINGER et al. [11]: acetonitrile ca. 0.4, acetone 0.50 methanol 0.62 and dimethyl sulfoxide 0.75. Similar results were obtained for 2-benzylbenzophenone.

The increase of the lifetimes of the cis dienols with increasing hydrogen-bond-acceptor strength of the solvent implies that the back reaction is intramolecular reketonization and is retarded by the hydrogen-bonding between the enol proton and the solvent, because the hydrogen bond obviously provides an additional barrier to the intramolecular hydrogen shift from the O-H to C=CHR groups.

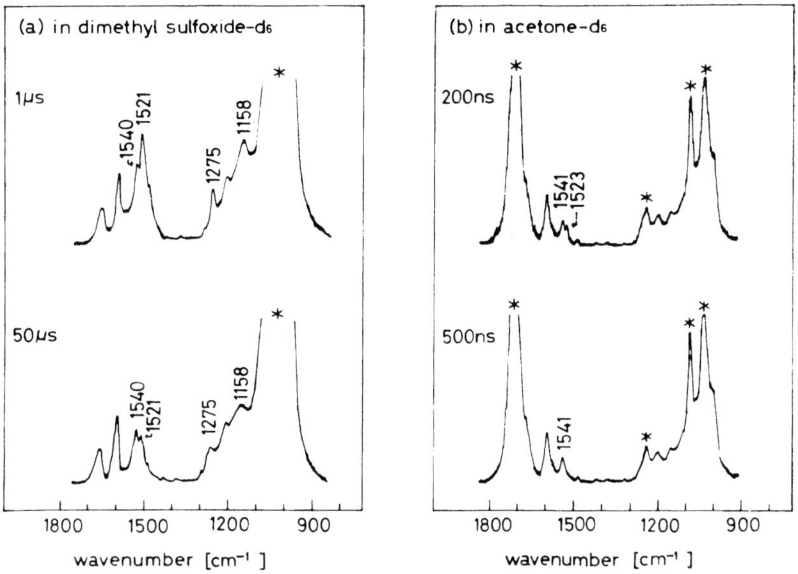

Figure 3 Solvent effect on the lifetime of the cis dienol of 2-methylbenzophenone: (a) in dimethyl sulfoxide-d_6 and (b) in acetone-d_6. * due to solvent

That the lifetime of the cis dienol of 2-benzylbenzophenone is considerably shorter than that of 2-methylbenzophenone in the same solvent may be attributed to less efficient hydrogen-bond formation of the former with the solvent due to steric effect.

On the other hand, the lifetimes of the trans dienols varied considerably from bottle to bottle of the solvent. This might indicate that the decay of the trans dienols to starting ketones could be due to acid or base impurities,or a trace of water in the solvent, and suggest that the trans dienols reketonize by intermolecular proton exchange: protonation on the C=CHR group and deprotonation from the O-H group by solvent or impurities.

References

1. N. C. Yang and C. Rivas, J. Am. Chem. Soc. 83, 2213 (1961).
2. E. F. Zwicker, L. I. Grossweiner and N. C. Yang, J. Am. Chem. Soc. 85, 2671 (1963).
3. G. Porter and M. F. Tchir, Chem. Commun. 1372 (1970); J. Chem. Soc.(A) 3772 (1971).
4. E. F. Ullman and K. R. Hoffman, Tetrahedron Lett. 1863 (1965).
5. P. G. Sammes, Tetrahedron 32, 405 (1976).
6. R. Haag, J. Wirz and P. J. Wagner, Helv. Chim. Acta 60, 2595 (1977).
7. K. Uji-ie, K. Kikuchi and H. Kokubun, Chem. Lett. 499 (1977); J. Photochem. 10, 145 (1979).
8. T. Nakayama, K. Hamanoue, T. Hidaka, M. Okamoto and H. Teranishi, J. Photochem. 24, 71 (1984).
9. S. Suzuki, S. Hirukawa and H. Takahashi, to be published.
10. F. Nerdel and W. Brodowski, Chem. Ber. 101, 1398 (1968).
11. R. R. Minesinger, M. E. Jones, R. W. Taft, M. J. Kamlet, J. Org. Chem. 42, 1929 (1977).

Resonance CARS Spectra of Metastable Merocyanines Produced by UV-Photolysis of Spirooxazines

U. Klüter, W. Hub, and S. Schneider

Institut für Physikalische und Theoretische Chemie,
Technische Universität München, D-8046 Garching, F.R.G.

Introduction:

UV-photolysis of spirooxazines leads to the reversible formation of "photo-merocyanines" (e.g. P1, P71).

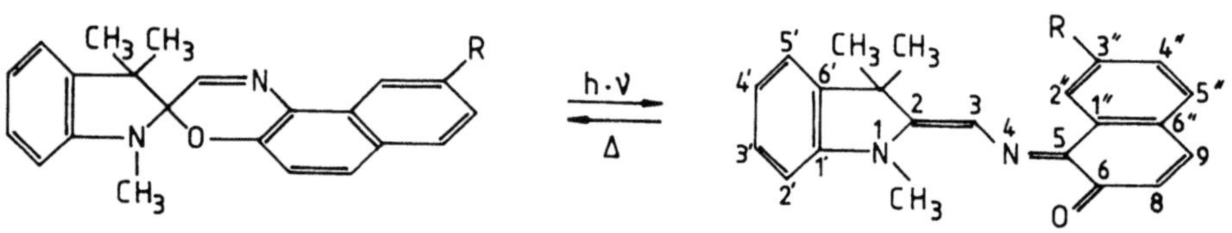

P1: R = H
P71: R = O-CO-CH_3

The absorption maximum of the metastable products at about 600 nm varies with solvent polarity (fig. 1). A much larger shift of the absorption band is observed for the photoproducts of spiropyrans in the same solvents /1/.

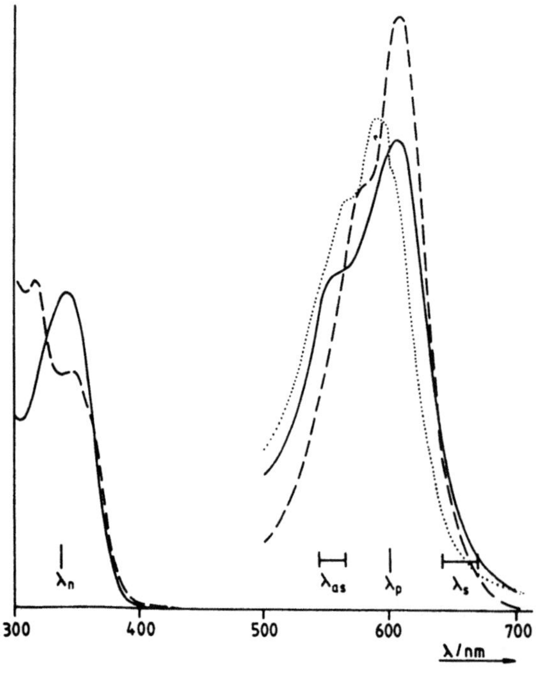

Figure 1
Absorption spectrum of the sprooxazines and their transient photo-products

dashed R= H in MeOH
dotted R= O-CO-CH_3 in CH_3CN
solid R= O-CO-CH_3 in MeOH

It could be shown for stable merocyanines that each isomeric form shows a solvatochromic effect related to the size of the ground-state dipole moment. In connection with "photo-merocyanines", the transformation between isomeric forms is discussed as a source for the solvent shift. From a resonance CARS investigation of TAKEDA et al. /2/ on photocolored solutions of a spiropyran compound (6-nitro-BIPS) in different solvents, it was concluded that a mixture of different isomers exists in each solvent. We have measured the time-resolved CARS spectra of the photoproducts of two spirooxazine compounds in methanol and acetonitrile solution, in order to study the proposed existence of the photomerocyanines' different isomeres.

Experimental:

The solutions ($c \approx 10^{-3}$ m) flowing through a quartz cell (d = 2mm) were photolyzed by a nitrogen laser pulse (337 nm, 5ns, 1mJ). After a delay of 150 ns the CARS signal was generated by two excimer pumped dye laser pulses (600 and 641..669 nm) in a scanning arrangement. The signal was discriminated from the straylight by a Jobin Yvon HRS2 monochromator. The integrated output of the attached R928 photomultiplier was recorded by a MINC minicomputer.

Discussion:

MO-calculations verify that the merocyaninic part of the photoproduct is responsible for the absorption around 600 nm. Typically for merocyanines, the bond order alternation is strong in the electronic ground-state and almost vanishes in the first excited singlet state.

PI - BOND ORDERS

BOND	S_0	S_1	BOND	S_0	S_1
1 - 2	0.55	0.57	5 - 1″	0.38	0.45
2 - 3	0.72	0.62	6 - 8	0.36	0.41
3 - 4	0.53	0.56	8 - 9	0.87	0.82
4 - 5	0.72	0.58	9 - 6″	0.39	0.42
5 - 6	0.36	0.39	1″- 6″	0.59	0.54
6 - 7	0.75	0.70	1 - 1′	0.41	0.46

Therefore, only vibrations related to the atomic centers 1 - 7 should appear in the resonance CARS spectrum.

The spectra of the two photomerocyanines are shown in fig. 2 and fig. 3. For each compound, several pairs of bands exist, which change their relative intensities in different solvents while maintaining their frequencies:

P1: 1135/1152 1194/1210 1400/1423 1533/1557 1605/1620

P71: 1159/1170 1401/1423

The remaining bands in the CARS spectrum of P1 (fig. 2) show similar frequencies and intensities in both solvents. For P71 (fig. 3) however, most of the bands change significantly.

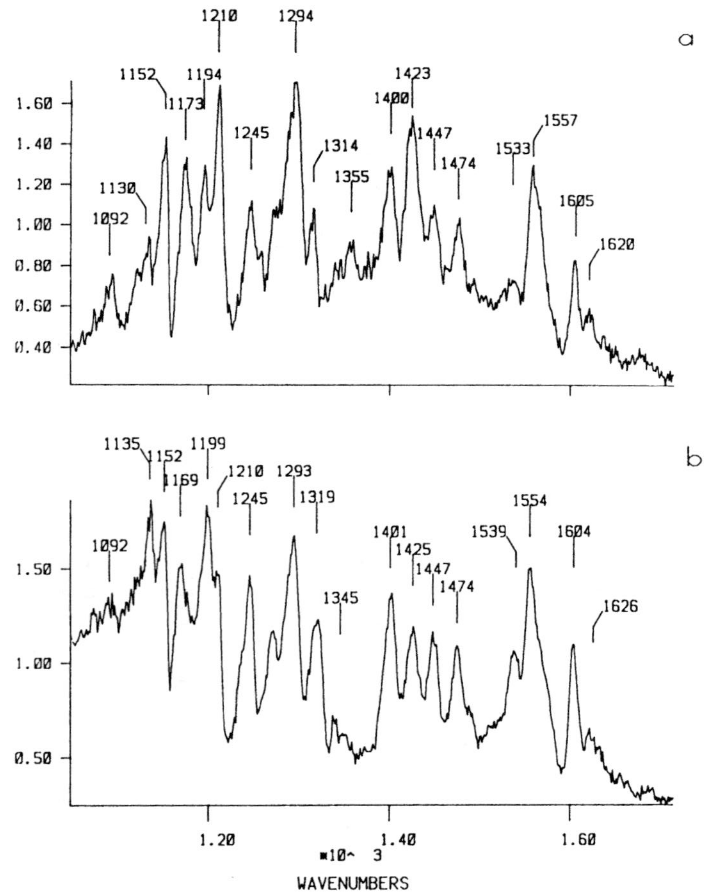

Figure 2: CARS spectra of P1
(R = H) in a) CH_3CN and b) MeOH

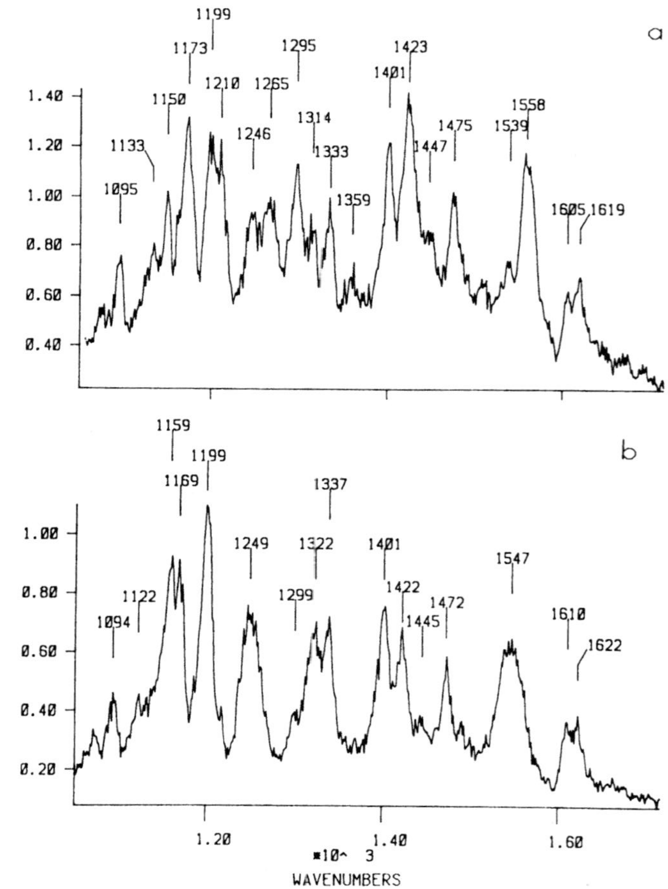

Figure 3: CARS spectra of P71
(R = $O-CO-CH_3$) in a) CH_3CN and b) MeOH

In acetonitrile, both species show nearly the same frequencies in the CARS spectra, although there are modest differences of their relative intensities.

Conclusion:

1 In both solvents, the photomerocyanines P1 and P71 are present as an equilibrium of different isomers.

2 In acetonitrile, P1 and P71 mainly exist in two isomeric forms. The different substituents R only influence the equilibrium constant.

3 In methanol however, P71 predominantly adopts a third configuration, which may be explained by the interaction of the acetylic substituent with this solvent.

/1/ R.C. Bertelson in Photochromism G.H. Brown Ed. (Wiley, New York 1971) p.118

/2/ S. Takeda, K. Kuroyanagi, S. Matsubara, H. Takahashi in Proceedings of the IXth International Conference on Raman Spectroscopy, Tokyo 1984, p.322

Photoisomerization of Stilbene and 1,4-Diphenylbutadiene in Compressed Gases and Liquids – Density-Dependent Solvent Shift and Transport Contributions

G. Maneke, J. Schroeder, J. Troe, and F. Voß

Institut für Physikalische Chemie der Universität Göttingen,
Tammannstraße 6, D-3400 Göttingen, F.R.G.

1. Introduction

Reactions in the liquid phase are strongly influenced by reactant-solvent interactions, which one usually investigates by simply changing the solvent. This approach entails the simultaneous variation of a whole set of parameters that may affect the reaction, so that it appears desirable to employ methods, where only a single parameter is varied continuously over a wide range [1]. One such parameter is the bath gas or solvent pressure, which we have chosen to study the density dependence of halogen photolysis and atom recombination from the low-pressure gas right through the gas-liquid transition region into the dense liquid under several kilobars of pressure [2-6]. We wish to report here on similar experiments concerning the photoisomerization of trans-stilbene and 1,4-diphenylbutadiene (DPB), which have been studied in detail with picosecond time-resolution under collision-free conditions, in dense gases, and a variety of liquid solvents, see, e. g. refs. [7-16], and references therein. The gas phase reaction can be represented very well by RRKM-theory for both stilbene [17] and DPB [18]. If one calculates the limiting high-pressure rate-constant k_∞ from the specific rate-constants k(E) of the isolated molecule reaction, one obtains a value that is more than one order of magnitude below the rate coefficients measured in low viscosity liquid solvents for the case of stilbene, while for DBP it is only slightly above the experimental liquid phase value. Recently, we have suggested a simple model involving a density-dependent lowering of the isomerization barrier in the first excited singlet state, which could be used to describe the observed behaviour [19]. Another question arises as a result of the approximate $\eta^{-0.5}$-dependence of the rate coefficient in high viscosity liquid solvents [15,16,20-22], which is not in accord with the predictions of Kramers theory [23]. A solvent -induced barrier shift could also serve as an explanation for this phenomenon [19], as an alternative to interpretations considering frequency-dependent friction coefficients or non-Markovian effects (see discussion in [19] and references therein). The aim of our experiments was to gain more insight into the way the solvent density and viscosity influence the photoisomerization rate by changing them continuously from the gas to the compressed liquid phase.

2. Experimental Technique

We use picosecond transient absorption spectroscopy to monitor the decay of the first excited singlet state of the trans-conformation of both stilbene [7] and DBP [24]. The wavelength for excitation is 308 nm, for probe 616 nm or 500 nm. The time-resolution in the pump-probe arrangement is ca. 2 ps. Details of the combined dye-exiplex-laser amplifier system can be found in [23]. The sample lengths are 20 mm in the liquid and 200 mm in the gas high-pressure cells. Experiments were carried out in gaseous and liquid ethane and propane, up to pressures of 6 kbar in the liquid.

3. Results and Discussion

From the measured first order decay of the transient absorption at 616 nm we obtained the isomerization rate coefficient k by subtracting the liquid phase radiative rate; i.e. $k_r = 6 \times 10^8\ s^{-1}$ for stilbene and $8 \times 10^8\ s^{-1}$ for DPB, and assuming that under our experimental conditions the nonradiative decay is dominated by the isomerization. Table 1 summarizes our rate coefficients in liquid and gaseous ethane. Whereas measurements in liquid phase were done at 297 K, we had to use somewhat higher temperatures in the gas phase, which are indicated in brackets. At the highest density, the pressure in the liquids is ca. 6 kbar.

Table 1 Isomerization Rate Coefficients for t-Stilbene and DPB in Gaseous and Liquid Ethane

t-Stilbene		DPB	
[M] $[10^{-2}$ mol $cm^{-3}]$	k $[10^9\ s^{-1}]$	[M] $[10^{-2}$ mol $cm^{-3}]$	k $[10^9\ s^{-1}]$
1.105	43 ± 5	1.102	5.9 ± 0.3
1.23	33 ± 2	1.28	3.9 ± 0.1
1.37	31 ± 2	1.42	2.8 ± 0.2
1.68	21 ± 1	1.61	2.2 ± 0.3
1.79	18 ± 4	1.78	1.2 ± 0.1
1.89	17 ± 2	1.96	1.2 ± 0.3
2.09	10 ± 1	2.07	0.8 ± 0.1
2.21	9 ± 1	2.29	0.5 ± 0.1
0.131 (311 K)	50 ± 20	0.146 (388 K)	52 ± 2
0.133 (329 K)	70 ± 7	0.126 (388 K)	42 ± 10
0.088 (366 K)	60 ± 10		
0.093 (368 K)	78 ± 10		
0.211 (370 K)	100 ± 30		

We represent our data according to the models outlined previously [5,19] as a function of the inverse of the diffusion coefficient in figs. 1 and 2. In the low-pressure regime D^{-1} represents a density scale , whereas in the dense liquid it is proportional to viscosity. We have therefore a continuous scale in terms of collision frequency for the whole density range.

The discrepancy between the high-pressure limiting rate-constant and the liquid phase isomerization rate for t-stilbene is clearly visible in fig. 1. There curve 1 depicts the falloff curve for the rate coefficient as calculated from the RRKM-fit, k_{th}, multiplied by the Kramers-term [23] , F_{kr}, which describes the influence of barrier-recrossings due to frictional forces. Curve 3 represents the rotational relaxation limit of the reaction one would encounter in the absence of any barrier for isomerization in the S_1-state. The data are represented very well by curve 2, which is a fit of the solvent-induced barrier shift model 19 to the rate coefficients measured in ethane; it consists of

(i) a cluster equilibrium $K_{Cl} = [St\text{-}M]/[St][M]$,

(ii) a lowering of the barrier in the cluster by ΔE_{Cl}, and

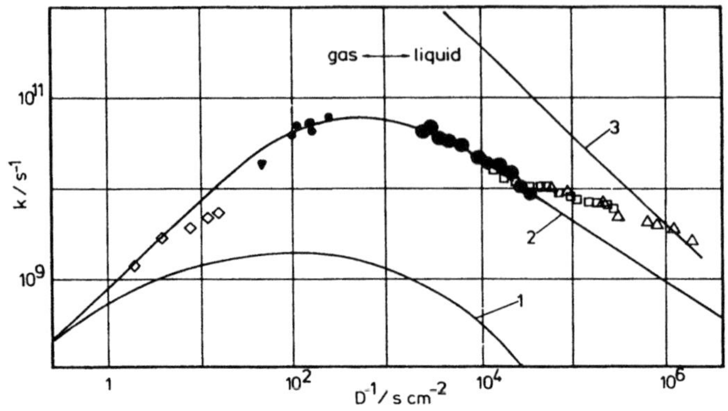

Fig. 1 Isomerization rate coefficients k for t-Stilbene in various solvents: ● ethane (l), this work; • ethane (g), temperature-scaled, this work; ▼ propane (g) [23]; ◇ methane (g) [15]; □ linear alkanes (l) [20]; △ n-hexane (l) [25] Curves see text

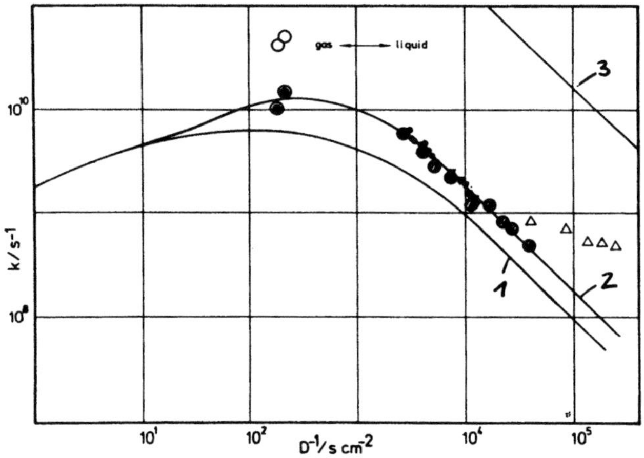

Fig. 2 Isomerization rate coefficients k for DPB in various solvents: ● ethane (l), this work; ○ ethane (g), this work; ◉ ethane (g), temperature-scaled, this work; • ethane (l), [22]; ▼ propane (l) [22]; △ n-octane (l), [21].

Curves as in fig. 1, see text

(iii) a lowering of the barrier due to solvent compression ΔE_{cage}, which is proportional to density.

This leads to the following expression for the rate-constant

$$k = \frac{k_{th} F_{Kr}}{1 + K_{Cl}[M]} \left(1 + K_{Cl}[M]\exp(-\Delta E_{Cl}/kT)\right) \exp(-\Delta E_{cage}/kT)$$

The same treatment is applied to the DPB-data (fig. 2). Table 2 lists the values of the fit parameters for both curves, using a value of $D\beta / 2\omega_B = 3.28 \times 10^{-4}$ cm^2 s^{-1} in the Kramers-term

$F_{Kr} = \sqrt{(\beta/2\omega_B)^2 + 1} - \beta/2\omega_B$.

Table 2

	t-Stilbene	DPB
ΔE_{Cl} [cm^{-1}]	-813	-245
K_{Cl} [cm^3 mol^{-1}]	1×10^3	1×10^3
$\frac{\Delta E_{cage}}{[M] - [M]_o}$ [cm^2 mol^{-1}]	-160	0

There is a striking difference between the isomerization rates of t-stilbene and DPB, which is also reflected in the shift parameters. In stilbene the photoisomerization is extremely susceptible to solvent density, we have a large cluster shift of the barrier, $k \gg k_\infty$, at low viscosities and in dense liquid ethane the barrier is continuously decreasing with increasing pressure. DPB, on the contrary, shows only a small cluster shift, at low viscosities $k \approx k_\infty$, and in dense liquid ethane there is no further shift, i.e. the rate decreases according to Kramers-Smoluchowski theory.

This difference in behaviour is probably a consequence of the ordering of A and B excited states in the gas phase, together with a strong shift of the B-state with respect to the A-states with increasing solvent density [23,24]. Without detailed knowledge of the excited state potential surfaces, however, it is not possible to give definite answers.

Acknowledgements: Financial support by the Deutsche Forschungsgemeinschaft (Sonderforschungsbereich 93 "Photochemie mit Lasern") is gratefully acknowledged.

1. J. Troe, in "High-Pressure Chemistry" (ed. H. Kelm; D. Reidel Publ., Dordrecht, Boston, and London, 1978), p. 489; Ann. Rev. Phys. Chem. 29, 223 (1978)
2. H. Hippler, K. Luther, and J. Troe, Ber. Bunsenges. Phys. Chem. 77, 1104 (1973)
3. K. Luther and J. Troe, Chem. Phys. Lett. 24, 85 (1974)
4. K. Luther, J. Schroeder, J. Troe, and U. Unterberg, J. Phys. Chem. 84, 3072 (1980)
5. B. Ottó, J. Schroeder, and J. Troe, J. Chem. Phys. 81, 202 (1984)
6. H. Hippler, V. Schubert, and J. Troe, J. Chem. Phys. 81, 3931 (1984)
7. B.I. Greene, R.M. Hochstrasser, and R.B. Weisman, J. Chem. Phys. 71, 544 (1979); Chem. Phys. 48, 289 (1980)
8. B.I. Greene and R.C. Farrow, J. Chem. Phys. 78, 3336 (1983)
9. T.J. Majors, U. Even, and J. Jortner, J. Chem. Phys. 81, 2330 (1984)
10. J.A. Syage, P.M. Felker, and A.H. Zewail, J. Chem. Phys. 81, 4685 (1984); J. Chem. Phys. 81, 4706 (1984)
11. M.F. Scherer, J.W. Perry, F.E. Doany, and A.H. Zewail, J. Phys. Chem. 89, 894 (1985)
12. P.M. Felker, W.R. Lambert, and A.H. Zewail, J. Chem. Phys. 82, 3003 (1985)
13. J.F. Shepanski, M.F. Scherer, and A.H. Zewail, Chem. Phys. Lett. 103, 9 (1983)
14. A. Amirav, M. Sonnenschein, and J. Jortner, submitted for publication
15. S.H. Courtney and G.R. Fleming, submitted for publication
16. V. Sundström and T. Gillbro, Ber. Bunsenges. Phys. Chem. 89, 222 (1985)
17. J. Troe, Chem. Phys. Lett. 114, 242 (1985)
18. J. Troe, A. Amirav, and J. Jortner, Chem. Phys. Lett. 115, 245 (1985)
19. J. Schroeder and J. Troe, Chem. Phys. Lett. 116, 453 (1985)
20. G. Rothenberger, D.K. Negus, and R.M. Hochstrasser, J. Chem. Phys. 80, 201 (1984)
21. S.P. Velsko and G.R. Fleming, J. Chem. Phys. 76, 3553 (1982)
22. S.H. Courtney and G.R. Fleming, Chem. Phys. Lett. 103, 443 (1984)
23. G. Maneke, J. Schroeder, J. Troe, and F. Voß, Ber. Bunsenges. Phys. Chem. 89, 000 (1985)
24. G. Maneke, J. Schroeder, J. Troe, and F. Voß, to be published
25. L.A. Brey, G.B. Schuster, and H.G. Drickamer, J. Am. Chem. Soc. 101, 129 (1979)

Part V

Transient Species

Advantages of Resonance CARS for the Study of Short-Living Radicals and Photoisomers

A. Lau, W. Werncke, M. Pfeiffer, H.-J. Weigmann, and J.T. Tschö

Central Institute of Optics and Spectroscopy of the
Academy of Sciences of the GDR, Rudower Chaussee, DDR-1116 Berlin, GDR

One important advantage of resonance CARS in comparison to spontaneous Raman methods results from the fact, that all relevant molecular information is inherent in the line shapes of CARS spectra accessible from measurements of relative intensities. But only three line shape parameters, I_{max}, I_{mid} and I_{min} [1], or even only two, I_{mid}/I_{max} and I_{min}/I_{max} are available from a CARS line in one spectra for the calculation of the five parameters of the third order susceptibility, appearing in the resonance CARS equation

$$I(\delta) = \left| \chi_{nr}^{(3)} + b' - ib'' + \frac{R - iI}{\delta + i\gamma} \right|^2 \qquad (1)$$

$\chi_{nr}^{(3)}$ is the electronic background contribution of the solvent

b', b'' are real and imaginary parts of the non-Raman resonant susceptibility of the dye under investigation

R/γ , I/γ are real and imaginary parts of the resonance enhancement Raman susceptibility of the same dye, normalized to the Raman line width γ

$\delta = \omega_{vib} - \omega_p + \omega_s$ is the Raman frequency mismatch and ω_p, ω_s are the frequencies of the pumping lasers; ω_{vib} is the frequency of the vibration under investigation

In principle, it is necessary to evaluate I and R only in order to determine the Raman cross-section, or even only I/R if only parameters of the excited state potential curve are wanted. But since b' and b" also influence the CARS line shape, a neglection of b' and b" in a line shape analysis results in incorrect I and R values.

Because under resonance conditions these non-Raman resonant terms are acting more or less strongly, we have to take them into account. To overcome the lack of experimental information, we combine the line shape data, obtained from a series with different sample concentrations, which but usually would cause a complicated analysis.

We propose here a practical and very simple procedure in using a linear relation between two auxiliary quantities μ and ν and relative susceptibility parameters [2].

$$\mu = \frac{I}{R} - \frac{b''}{R/\gamma}\nu \qquad \mu = \mu\,(I_{max}, I_{mid}, I_{min}, \gamma)$$

$$\nu = \nu\,(I_{max}, I_{mid}, I_{min}, \gamma)$$

It is possible to conctruct special nomograms from which the I/R and $b''/R/\gamma$ values can be obtained using directly the experimental data I_{mid}/I_{max} and I_{min}/I_{max} for three different concentrations. For each concentration there are two points in the nomogram. The physical meaningful solution is selected, by best fitting the points to a straight line. This is demonstrated for the 614 cm^{-1} and 1365 cm^{-1} vibration of Rh 6G in ethanol excited with a pump radiation at 580 nm in Fig 1 and gives the following values $^{I}/_{R} = 0.15 \pm 0.07$ and $^{b''}/R/ = -0.18 \pm 0.02$ and $^{I}/_{R} = -2.1 \pm 0.1$ and $^{b''}/R/ = 1.31 \pm 0.03$ respectively.

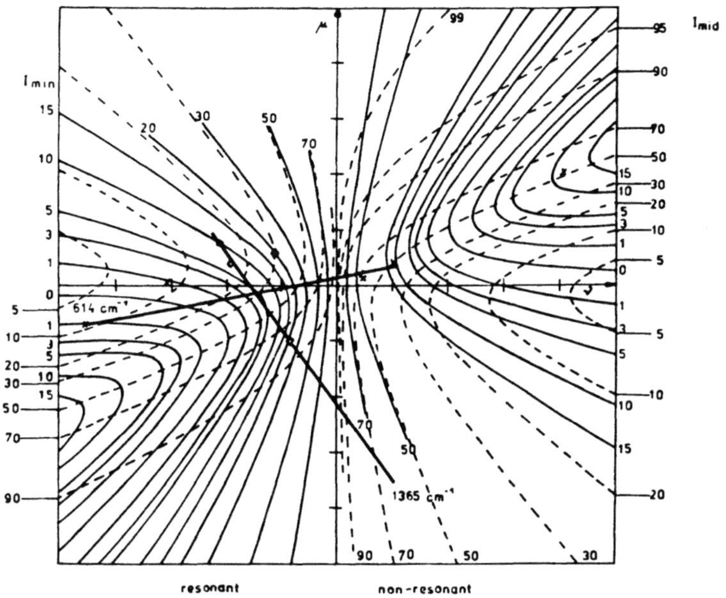

Fig 1 μ, γ -nomogram for the estimation of $^{I}/_{R}$ and $^{b''}/R/_{\gamma}$.

From these values for different excitation wavelengths one can evaluate the potential curve parameters on the basis of an appropriate molecular model [3], in analogy to the procedure usually applied in Raman excitation profile analysis.

Spectra of photoisomers and ion radicals are used to demonstrate the possibilities of CARS to study photophysically and photochemically-induced structure changes.

In both cases the shifted absorptions of the photoinduced species are used in selecting proper CARS excitation wavelengths for the assigment of the CARS spectra to the special short-living species.

If the absorption of the photoinduced species are strongly shifted ($>$ 50 nm) as in the case of the photochemically generated dimethylterephthalate anion radical (530 nm) and the N-ethylcarbazole cation radical (780 nm) they can be distinguished by selective resonance enhancement as in spontaneous Raman spectroscopy. Table 1 compares the vibrational frequencies of dimethylterephthalate (DMTP) and its short-living anion radical.

Table 1

Frequencies		Assignment
DMTP [cm^{-1}]	DMTP anion radical [cm^{-1}]	
1718	1656	(C=O)
1608	1612	(CC_{ring})
1433	-	(CC_{ring})
1279	-	(C-O)
1193	1178	
1105	-	

From the frequency decrease of the carbonyl group in the DMTP radical in comparison to the parent molecule (1718 → 1656 cm^{-1}), it can be concluded that the corresponding bond order in the radical is lowered appreciably in agreement with expections of chemical considerations.

If the absorption shift between different species are rather small (some few 10 nm) as for photoisomers of cyanine dyes, the spectra of the stable and photoinduced species are both resonance-enhanced by the same pump frequency. The origin of the observed CARS lines can but be distinguished by line shape analysis even if the vibrational frequencies of both species are not or nearly not shifted, as is demonstrated in Figure 2 for bisdimethylamino-heptamethine

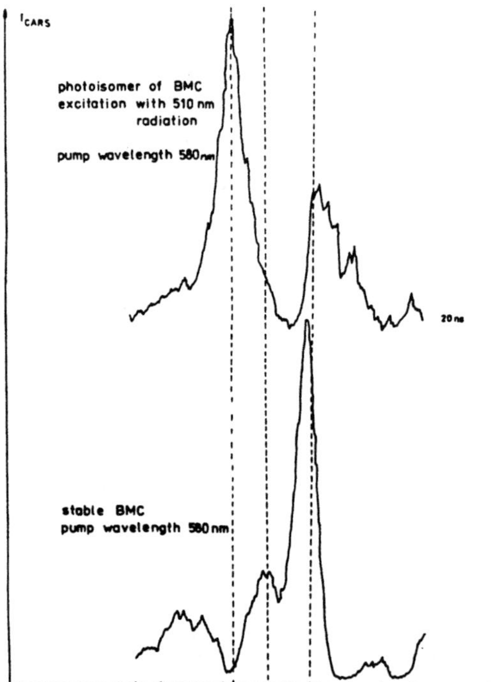

Fig 2
CARS spectra of BMC (bottom) and BMC photoisomer (top) excited at 510 nm. The line at 1115 cm^{-1} changes its asymmetric feature

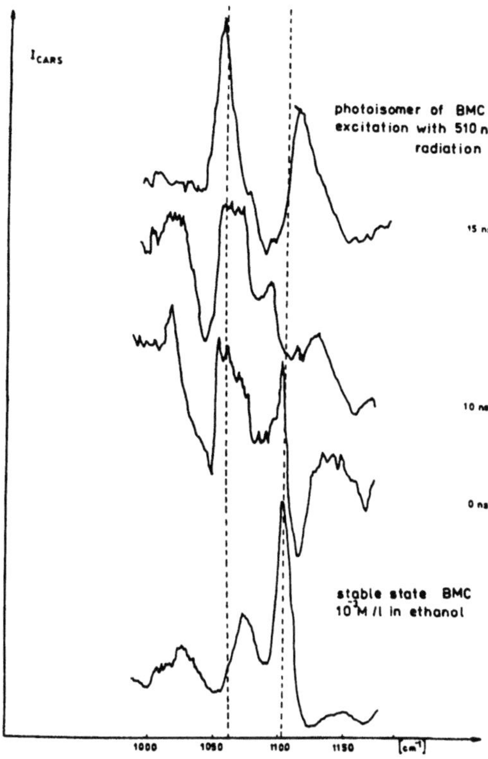

Fig 3
CARS spectra of photoisomeric states generated at 510 nm for different delay times of the CARS probing

perchlorate (BMC). Figure 3 shows the time-dependent transformation of the stable molecule into a longer living (> 5 μs) photoisomer during photoexcitation in a ns time-scale.

Many lines are only unessentially shifted, but a few new lines arise for the photoisomer. The result shows that in the ns time-scale only one very short living species and one or only few longer living photoisomers are existent, in contrast to some chemical considerations expecting many photoisomeric states.

[1] J.W. Fleming, S. Johnson: J. Raman Spectr. 8, 284 (1979)
[2] M. Pfeiffer, A. Lau, W. Werncke to be published
[3] M. Pfeiffer, A. Lau, W. Werncke: J. Raman Spectr. 15, 20 (1984)

Resonance CARS Study of α,ω-Diphenylpolyenes in the Electronic Excited States

A. Kasama, M. Taya, T. Kamisuki, Y. Adachi, and S. Maeda

Research Laboratory of Resources Utilization, Tokyo Institute of Technology, Nagatsutacho, Midori-ku, Yokohama 227, Japan

1. INTRODUCTION

There has been renewed interest in the spectroscopic and quantum-chemical investigation of linear polyenes, concerning the low-lying singlet excited state of forbidden symmetry (A_g in C_{2h}). The last state is now approved to be lower than the first allowed level (^1B_u) in most polyenes[1], as illustrated in Fig.1 for the systems investigated in the present paper. It is also well known that this state has large contribution from a doubly excited electron configuration.

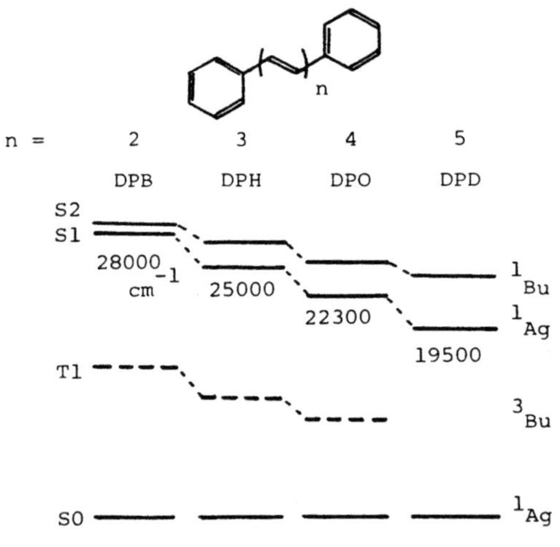

Fig.1 Energy-level diagram of α,ω-diphenylpolyenes:
DPB; diphenylbutadiene
DPH; diphenylhexatriene
DPO; diphenyloctatetraene
DPD; diphenyldecapentaene

Vibrational frequencies in the first excited singlet(S_1) state have been obtained in several polyenes from the vibronic structures of one- and two-photon excitation (1PE and 2PE) spectra at low temperatures. Recently, the time-resolved resonance Raman(TR^3) technique was applied to the S_1 state of diphenylbutadiene(DPB) in room-temperature solutions[2,3]. In the present paper, we report the Raman spectra of a series of α,ω-diphenylpolyenes in the S_1 and lowest triplet(T_1) states as observed by the resonance CARS technique. The results are discussed with primary interest in the unusually high frequency of the chain stretching mode in the S_1 state of longer polyenes.

2. EXPERIMENTAL

The CARS measurement was carried out on the 10^{-3}~10^{-4}Mdm^{-3} solutions in various solvents at room-temperature. The polyene molecules were pumped up to the S_1 level by 337 nm N_2 laser pulses of 10 ns duration, and the CARS generation of the excited population was effected by two dye lasers pumped by the same N_2 laser pulses. For measuring the triplet state, a third harmonic Nd:YAG laser was used for the photoexcitation, and the CARS measurement was made by appropriately delayed probe pulses. The dye laser linewidth was 1 cm^{-1} or less.

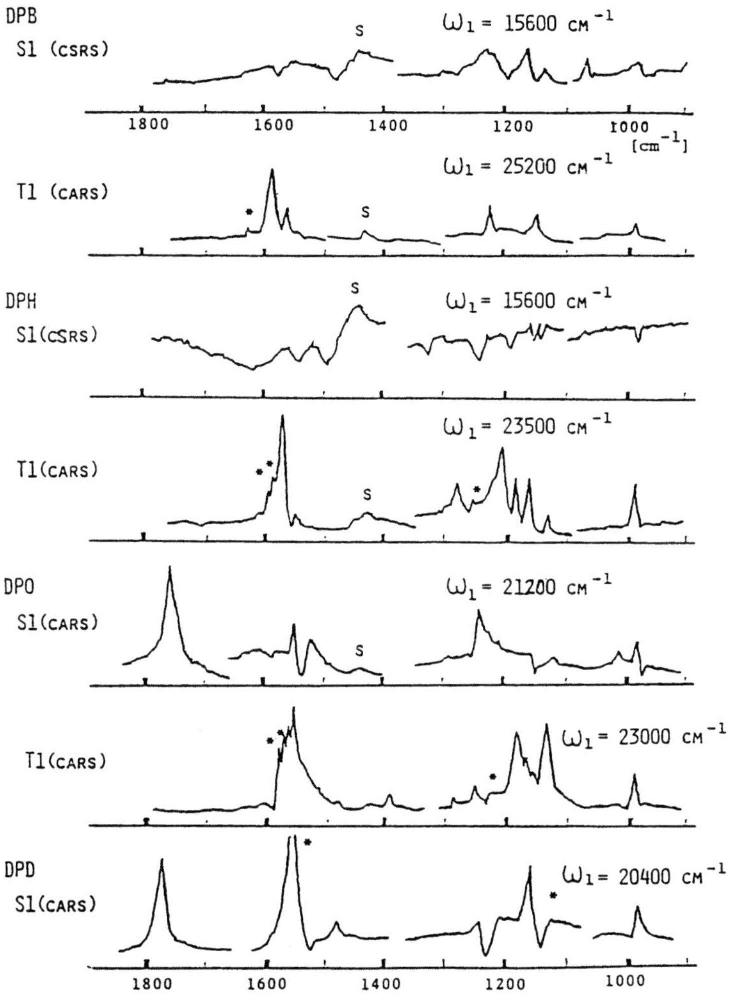

Fig.2 Resonance CARS or CSRS spectra of α,ω-diphenylpolyenes in the S_1 and T_1 states (*: S_0 state line, s: solvent line)

Table 1 Vibrational frequencies[cm^{-1}] of α,ω-diphenylpolyenes in the S_o, S_1 and T_1 states

DPB			DPH			DPO			DPD	
S0	S1	T1	S0	S1	T1	S0	S1	T1	S0	S1
1625	1575	1585	1608	1620	1570	1580	1755	1560	1610	1775
1598			1596	1543			1545	1555	1552	
					1550	1574				1536
	1480	1560	1589	1490			1525	1482	1483	
										1485
				1325				1395	1440	
					1270	1230				
				1285			1243	1285	1311	1245
1285	1230		1255							
		1220		1240	1200	1158		1248	1282	
1177										1236
	1164			1186	1185			1185	1218	
						1149				
1138				1164				1170	1205	
	1135	1145	1144		1164	1144				1162
				1148				1143	1143	
					1130		1008			
1067	1070			1135						
995	990	980	998	977	987	998	973	990	995	998

3. RESULTS and DISCUSSION

Resonance CARS spectra recorded by the closely resonant conditions with the transient S_N-S_1 or T_N-T_1 transitions are shown in Fig.2, and the obtained Raman frequencies are listed in Table 1. The frequencies are in good agreement with the TR^3 results on the S_1[2,3] and T_1[4] states of DPB and the T_1 state of DPO[5], and are in consistency with the 2PE structures in DPB[6], DPH[7] and DPO[8]. Most striking feature is the extremely intense Raman signals at 1755 cm^{-1} in DPO(n-heptane solution) and at 1775 cm^{-1} in DPD(THF solution) both in the S_1 state. The former frequency corresponds to the broad structure observed in the 2PE spectrum of DPO[8], and similar vibrational frequencies have been reported for octatetraene(1PE and 2PE [9,10]), 2,10-undecapentaene(1PE[11]) and 2,12-dimethyltridecahexaene(1PE[12]). It seems little doubtful that the frequency is assigned to a fundamental stretching vibration of polyene chain on every evidence[1], but no corresponding signal appears in the present S_1 spectra of DPH and DPB and in all the triplet spectra.

The frequencies noted above are remarkably higher than the ground state $\nu_{C=C}$ ($\lesssim$ 1600 cm^{-1}), exhibiting a shift contrary to the ordinary expectation for the excited state. The calculated π-bond order[13] shows that the bond strength in the 2^1A_g state is more or less levelled off along the chain from the distinctly alternate distribution in the ground state. It is shown by calculation that such a levelling-off always results in a decrease of the highest stretching frequency. Therefore, the change in the diagonal stretching force constants(K_{ii}) by the excitation cannot possibly cause the observed higher frequency shift, and the observed high frequency must be attributed to the cross terms($(f_{ij})_k(\Delta r_i)(\Delta r_j)$) between the CC stretching coordinates in the potential function:

$$E_k = \frac{1}{2}\sum_i (K_{ii})_k(\Delta r_i)^2 + \sum_{i<j}\sum (f_{ij})_k(\Delta r_i)(\Delta r_j).$$

The binding nature in the 2^1A_g state of polyenes is known to be essentially covalent from the theoretical and experimental evidences. Such a state can be properly described in the MO representation only by taking account of the configuration interaction with multiply excited electron configurations. Alternatively, the covalent nature of the state may be suitably treated by the valence bond(VB) approach. So, we employed a simple VB formalism for estimating the cross term force constants.

Simplifying the problem by considering unsubstituted polyene chains, the VB eigenfunction was obtained from the appropriate number(four in hexatriene and seven in octatetraene) of bond eigenfunctions. Then, the cross term force constants(f_{ij}) for each pair of bonds were calculated by a standard second order perturbation treatment. The resulting f_{ij} for the ground(1^1A_g) and 2^1A_g states of octatetraene is given in Fig.3, where marked contrast is observed between those two states.

In the same figure is shown the effect of the cross terms(f_{ij}) on the CC stretching frequency pattern of octatetraene in the 1^1A_g and 2^1A_g states, where the diagonal force constants(K_{ii}) were determined on the basis of π-bond order[13] and the C-H bending motions were neglected. The remarkable increase of the highest frequency (bold lines) by increasing f_{ij} values indicates that the observed high ν_{CC} values can be accounted for as due to the notable contribution of the cross terms in the 2^1A_g potential surface. The normal modes including the C-H bending motions are shown in Fig.4. It is noted that the f_{ij} values required to give the observed large shift are no more than several percent of the diagonal values. The results indicate the importance of the vibronic effect in the excited states,where the bond delocalization is remarkable in general. On the other hand, the nonappearance of similar high-frequency modes in DPB and DPH may perhaps be understood as due to the increasingly greater electronic and vibrational perturbations from the phenyl substituents in shorter polyenes.

Finally, we observed an appreciable frequency shift of the above-mentioned ν_{CC} (2^1A_g) of DPO by solvents: $\Delta\nu = -25$ cm^{-1} from n-heptane to dioxane, and some solvent effects were also observed in the ~1600 cm^{-1} region of the 2^1A_g spectrum of DPH. Those effects are being investigated more extensively.

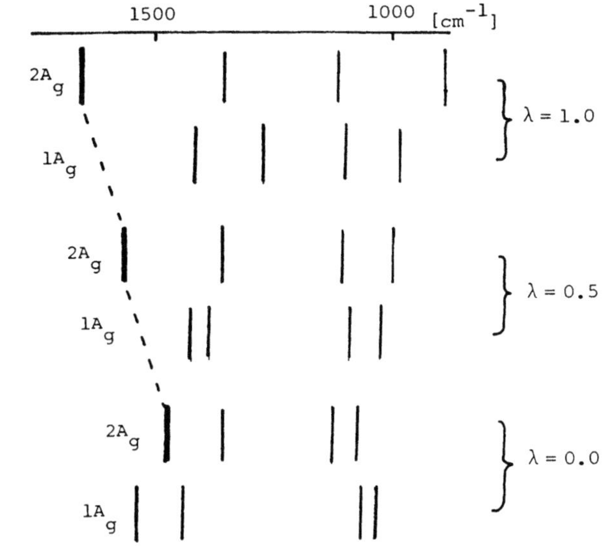

$f_{ij} = \lambda \times f'_{ij}$

[md/Å]	$1A_g$	$2A_g$
f'_{12}	0.190	0.062
f'_{13}	-0.122	0.211
f'_{14}	0.089	0.217
f'_{15}	-0.122	0.211
f'_{16}	0.190	0.062
f'_{17}	-0.113	-0.382
f'_{23}	0.212	-0.112
f'_{24}	-0.152	0.238
f'_{25}	0.212	-0.112
f'_{26}	-0.326	-0.069
f'_{34}	0.402	-0.440
f'_{35}	-0.291	0.120

(i,j: CC bond number)

Fig.3 Change of ν_{CC} frequency pattern of octatetraene by f_{ij} values

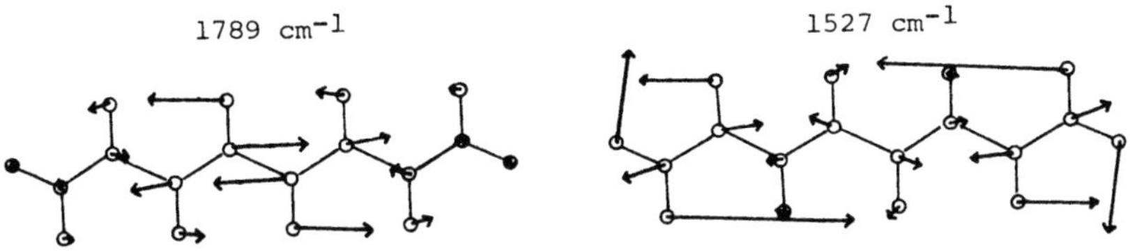

Fig.4 Normal modes of 2^1A_g octatetraene

References

1. B. Hudson, B. Kohler and K. Schulten: Excited States, Vol.6(E. Lim ed.), p.1, Academic Press, New York (1982)
2. R. Wilbrandt, N.H. Jensen and F.W. Langkilde: Chem. Phys. Lett. 111, 123 (1984)
3. T.L. Gustafson, D.A. Chernoff, J.F. Palmer and D.M. Roberts: Proc. IX Internat. Conf. Raman Spectrosc., 126 (1984)
4. R. Wilbrandt, W.E.L. Grossman, P.M. Killough, J.E. Bennett and R.E. Hester: J. Phys. Chem. 88, 5964 (1984)
5. G.N.R. Tripathi: Proc. IX Internat. Conf. Raman Spectrosc., 328 (1984)
6. J.A. Bennett and R.R. Birge: J. Chem. Phys. 73, 4234 (1980)
7. H.L.B. Fang, R.J. Thrash and G.E. Leroi: Chem. Phys. Lett. 57, 59 (1978)
8. H.L.B. Fang, R.J. Thrash and G.E. Leroi: J. Chem. Phys. 67, 3389 (1977)
9. R.M. Gavin, Jr. and C. Weisman: J. Chem. Phys. 68, 522 (1978)
10. M.F. Granville, G.R. Holtom and B.E. Kohler: J. Chem. Phys. 72, 4671 (1980)
11. R.L. Christensen and B.E. Kohler: J. Chem. Phys. 63, 1837 (1975)
12. R.A. Auerbach, R.L. Christensen, M.F. Granville and B.E. Kohler: J. Chem. Phys. 74, 4 (1981)
13. K. Schulten, I. Ohmine and M. Karplus: J. Chem. Phys. 64, 4422 (1976)

Ultraviolet Resonance Raman Studies of Electronic Excitations

B.S. Hudson

Chemical Physics Institute and Department of Chemistry, University of Oregon, Eugene, OR 97403, USA

1 Introduction

Electronic excitation of molecules often results in significant changes in chemical bonding and may initiate chemical reaction. This is particularly true for small molecules, because excitation of a single electron can be a major influence on the total bonding of the system. In such cases the bond lengths and the shape of a molecule may be very different in the excited state than in the ground state, or, indeed, the excited state potential may be dissociative or pre-dissociative. These bonding changes may cause the resulting absorption spectra to be very difficult to interpret because of spectral diffuseness, congestion due to the excitation of many degrees of freedom, or very different vibrational frequencies in the excited state from those of the ground state.

Resonance Raman spectroscopy has proven to be a useful way to study such electronic excitations. One reason for this is that the Raman process essentially projects the excited electronic state potential surface onto the ground state surface, permitting utilization of the extensive knowledge of ground state molecular vibrations to be used to understand the bonding of the excited state. The availability of ultraviolet radiation sources has permitted the application of this resonance light scattering technique to electronic excitations of small molecules, where really dramatic geometry changes are more the rule than the exception [1-13]. This brief review discusses this aspect of resonance Raman scattering with emphasis on the use of the spectra obtained with this method to probe kinetic processes of three types: photochemical transformations, vibrational relaxation from highly excited states and conventional ground state isomerization reactions. These three applications depend on distinct aspects of the Raman scattering process. The pattern of the spectrum observed under resonance conditions reflects the geometry of the excited electronic state. This initially excited region of the upper state potential surface acts as the precursor for subsequent photochemical transformations. Large geometry changes along some normal mode in a particular electronic state and excitation with radiation well into the electronic absorption band leads to production of long series of overtone and combination bands. Stimulated excitation of these bands can be used to populate vibrational levels not accessible by other methods. Finally, excitation within electronic transitions results in considerable resonance enhancement of the Raman process. This fact, in combination with the general observation that vibrational frequencies are quite sensitive to molecular geometry, permits the use of resonance Raman scattering as a kinetic probe of conformational changes. Ultraviolet excitation makes a greater variety of species amenable to this resonance enhancement effect.

2 Experimental Procedures

The experimental methods used to perform ultraviolet resonance Raman experiments have been described in detail elsewhere [1,9]. The basic laser source in our experiments is a Q-switched, flash lamp pumped Nd:YAG laser producing radiation at 1064 nm and operating with a repetition rate of 30 Hz. This fundamental radiation is converted to higher harmonics using non-linear optical crystals. The resulting radiation is at 213, 266, 355 or 532 nm. The 266, 355 or 532 nm beams can then be used to generate stimulated Raman scattering in hydrogen gas. This produces a series of Stokes and anti-Stokes shifted lines with a spacing of 4155 cm^{-1} from the excitation radiation. The primary limiting factor determining the shortest wavelengths usable with this type of device is oxygen absorption from air. The shortest wavelength we have used to date is 184 nm. A solar-blind photomultiplier and a box-car integrator are used as the detector.

3 Photochemical Transformations: Ammonia and Alkenes

Very rapid processes following photoexcitation may result in diffuse spectra. On the other hand, if these processes are only "moderately fast", with rates on the order of 10^{11} to 10^{14} sec^{-1}, then the absorption spectra will have some residual structure. Our emphasis has been on this case of moderately fast reactions. The very rapid limit, corresponding to completely diffuse absorption spectra, has been emphasized by other workers [15-17].

Excitation of ammonia to its "A" singlet state involves promotion of a lone pair electron to a 3s level. The A state is planar with three equal N-H bond lengths somewhat longer than the ground state values. Thus, the upper state geometry differs from that of the ground state due to displacement along the pyramidalization and symmetric N-H stretching coordinates. The A state of ammonia dissociates very rapidly to form NH_2 and H atoms both in their ground electronic states. This reaction coordinate is therefore predominantly along the asymmetric N-H stretching direction. The rate of this dissociation is sufficiently rapid in NH_3 that the individual rovibronic lines are broadened to the point where rotational structure is lost. The resulting residual vibrational structure consists of a series of bands with widths on the order of 200 wavenumbers due to the overlapping rotational bands.

Resonance Raman spectra of ammonia obtained with ultraviolet radiation resonant with the transition to the A state have several interesting features [4,6,14]. The series of lines corresponding to excitation of the pyramidalization and symmetric N-H stretching motions is observed as expected. The most interesting feature of the spectra, however, is that excitation in an odd (anti-symmetric) vibrational level of the upper state pyramidalization series results exclusively in transitions to odd inversion split levels of the ground state [4,6]; the complementary results are observed for even levels [14]. This means that only a single vibrational level is contributing to the Raman scattering process. We may then ask how many rotational levels contribute to the Raman process. The answer to this question is revealed in the intensity distribution of the rotational side bands observed with each vibrational transition. Analysis of this intensity distribution for 212.8 nm excitation shows that levels that are within about 30 cm^{-1} of the excitation radiation contribute to the intensity. This means that the intrinsic linewidth of the rovibronic lines of the A state of ammonia have widths on this order, corresponding to a dissociation rate of about 0.15 ps at this energy-level.

The famous N->V electronic transition of ethylene has been the subject of considerable argument for many years. Here the upper V state is apparently twisted to a perpendicular configuration. Presumably there is also a change in the C-C bond length. Resonance Raman spectra of ethylene [2] exhibit the expected enhancement in intensity of the even quanta overtones of the torsional mode and the C-C stretch motion, as expected. The demonstration of the activity of the torsional motion is the first direct demonstration that ethylene twists significantly in its excited state. There is also, however, considerable resonance enhancement of activity in the HCH in-plane bending mode, a motion not expected to be active in this transition.

Very recent studies of the spectra of methyl- and chloro-substituted ethylenes have led to the unexpected observation that many of these species do not appear to twist in their excited "V" states. Specifically, **trans** dichloro- and dimethyl-ethylene (**trans**-2-butene) and the 1,1-disubstituted isobutylene do not show activity in the torsional vibrational mode with excitation anywhere in the spectral region associated with the pi-pi* transition. Apparently hyperconjugation effects cause the excitation to spread into the substituent groups and reduce the anti-bonding nature of the orbital involved in the excitation. The **cis** 1,2-disubstituted species show strong excitation of the torsional mode with the greatest intensity being in the even overtones. Tetrachloroethylene shows excitation of the fundamental transition of the torsional mode. Since this transition is forbidden for a planar centrosymmetric geometry, we must conclude that this molecule is at least slightly non-planar.

Similar studies of butadiene [12] have shown that this simple polyene twists slightly at its terminal CH_2 groups in its low-lying excited states. Activity in certain overtone bands has solved a long standing problem [19,20] in molecular electronic spectroscopy concerning the location of a forbidden electronic transition in this molecule.

4 Highly Excited Vibrational Levels: Benzene and Carbondisulfide

The resonance Raman spectra of benzene [1,3,14] and carbondisulfide [7] vapor (Figs 1 and 2) are of interest for a number of reasons [18] but for the present purposes the major reason for discussion of these spectra is that they exhibit very strong intensity in high-order overtones and combinations. This is relevant to many current studies aimed at the determination of relaxation rates for highly excited vibrational levels of molecules. Specifically, resonance excitation of spontaneous Raman scattering can be used as a screening method for finding electronic excitations giving rise to strong intensity in high-energy vibrational levels. These can then be pumped by stimulated emission pumping. This is an efficient method for preparing molecules in excited levels not accessible with other methods.

5 Specific Group Resonance Enhancement: Proline Isomerization

Ultraviolet resonance Raman scattering is potentially a useful method for the study of the structure and conformational rearrangement reactions of biopolymers [5,8,9,11]. We have recently been particularly interested in one specific reaction of proteins relating to their folding kinetics. The mechanism of folding of proteins to make their "native" structure is a significant unsolved problem in molecular biology [21]. One recent advance in this area has been the recognition [22] that some "slow" kinetic phases in

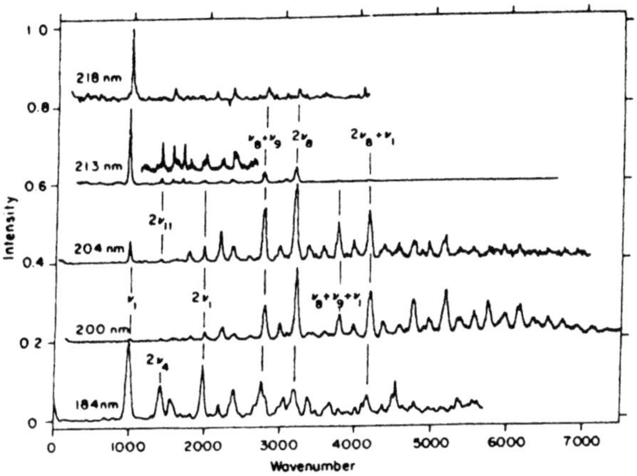

Figure 1. Ultraviolet Resonance Raman spectra of benzene vapor. The excitation wavelengths are indicated. From [13].

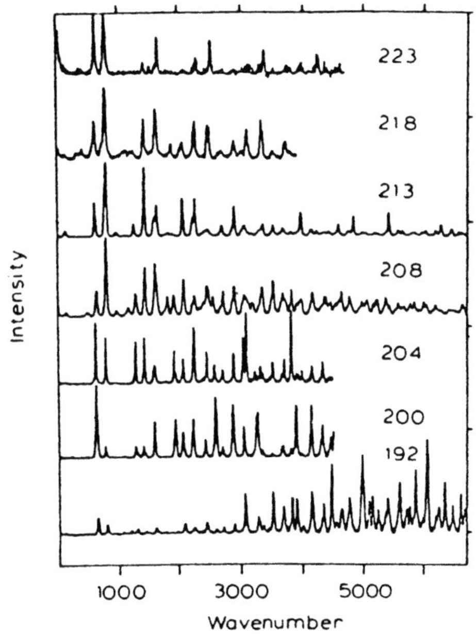

Figure 2. Ultraviolet resonance Raman spectra of carbondisulfide. From [7]. The excitation wavelengths are indicated.

protein refolding may be associated with the isomerization of the peptide bond linkage between a proline residue in the structure and the residue immediately to its N-terminal side, i.e., the X-proline peptide bond. These slow kinetic phases have time-constants that vary from seconds to hours depending on conditions.

There is, however, no direct experimental evidence that these slow phases are due to proline isomerization. There are few methods for determining the relative populations of the **cis** and **trans** conformations of this bond that are applicable to the unfolded form of large proteins. We have therefore explored the use of ultraviolet resonance Raman scattering to assess its utility in this regard. Studies of simple peptides [8,9] show that there is strong resonance enhancement of the amide II C-N stretch vibration with excitation near 200 nm. Also, studies using carbon-13 and nitrogen-15 labeling have shown that the enhanced bands are strongly shifted [8,9]. This permits examination of specific peptide linkage types since specific amino acids can be isotopically labeled and biosynthetically incorporated. Similar studies of constrained cyclic dipeptides indicate that there is a difference in the frequency of this enhanced band for the **cis** and **trans** isomers of the X-proline bond. The problem of detecting this particular peptide bond vibration in the presence of all of the other vibrations of a protein has been approached by utilizing the fact that the X-proline linkage, being a tertiary peptide bond, has an absorption spectrum that is slightly red-shifted relative to that of all the other secondary peptide linkages. Recent experiments using excitation in the region near 230 - 240 nm have shown that with this excitation wavelength the X-proline vibrations are enhanced relative to those of other peptide bonds by a factor of the order of 100. This should permit the determination of the conformation state and rate of interconversion of X-proline bonds in proteins even when there are only a few such bonds in a protein. Studies of the folding of ribonuclease A are now in progress.

6 Conclusions

The examples given here illustrate the application of ultraviolet resonance Raman scattering to kinetic processes with time-scales ranging from 0.15 ps to hours and to molecules ranging from triatomic species to proteins with molecular weights of tens of thousands. The apparatus needed to perform these experiments is simple and reliable and data collection is rapid. The contributions of this technique to kinetic processes will clearly be considerable and varied.

7 Acknowledgements

The author wishes to acknowledge his co-workers who performed the ultraviolet resonance Raman experiments described in this paper: R.R. Chadwick, R.A. Desiderio, D.P. Gerrity, W. Hess, P.B. Kelly, L. Mayne, and, especially, L.D. Ziegler.

8 References

1. L.D. Ziegler and B. Hudson: J. Chem. Phys. **74**, 982 (1981)
2. L.D. Ziegler and B. Hudson: J. Chem. Phys. **79**, 1197 (1983)
3. L.D. Ziegler and B. Hudson: J. Chem. Phys. **79**, 1134 (1983)
4. L.D. Ziegler and B. Hudson: J. Phys. Chem. **88**, 1110 (1984)
5. L.D. Ziegler, B. Hudson, D.P. Strommen and W.L. Peticolas: Biopolymers **23**, 2067 (1984)
6. L.D. Ziegler, P.B. Kelly and B. Hudson: J. Chem. Phys. **81**, 6399 (1984)
7. R.A. Desiderio, D.P. Gerrity and B. Hudson: Chem. Phys. Lett. **115**, 29 (1985)
8. L.C. Mayne, L.D. Ziegler and B. Hudson: J. Phys. Chem., in press
9. B. Hudson and L. Mayne: Methods in Enzymology, in press
10. P.B. Kelly and B. Hudson: Chem. Phys. Lett. **114**, 451 (1985)
11. W.L. Kubasek, B. Hudson and W.L. Peticolas: Proc. Natl. Acad. Sci. USA, **82**, 2369 (1985)
12. R.R. Chadwick, D.P. Gerrity and B. Hudson: J. Chem. Phys. **115**, 24 (1985)
13. D.P. Gerrity, L.D. Ziegler, P.B. Kelly, R.A. Desiderio and B. Hudson: J. Chem. Phys, in press
14. W. Hess, D.P. Gerrity, L.D. Ziegler and B. Hudson: unpublished results
15. D. Imre, J.L. Kinsey, A. Sinha and J. Krenos: J. Phys. Chem. **88**, 3956 (1984)
16. D. Imre, J. Kinsey, R. Field and D. Katayama: J. Phys. Chem. **86**, 2564 (1982)
17. E. Heller: Acc. Chem. Res. **14**, 368 (1981)
18. L.D. Ziegler and B. Hudson: Excited States, Vol. **5**, E.C. Lim, editor (New York: Academic Press 1982), pages 41-139
19. B. Hudson, B.E. Kohler and K. Schulten: in Excited States, Vol. **6**, E. C. Lim, editor (New York: Academic, 1982) pgs 1-95
20. B. Hudson and B.E. Kohler: Synthetic Metals **9**, 241-252 (1984)
21. C. Ghelis and J. Yon: Protein Folding (Academic, New York 1982)
22. J.F. Brandts, H.R. Halvorson and M. Brennan: Biochemistry 14, **4953** (1975)

Excited States of Polyenes Studied by Time-Resolved Resonance Raman Spectroscopy

F.W. Langkilde, N.-H. Jensen, and R. Wilbrandt

Chemistry Department, Risø National Laboratory, DK-4000 Roskilde, Denmark

After the first publication of a triplet state resonance Raman spectrum (1) we have during the past years studied the spectra of polyenes excited into their short-lived lowest triplet (2-9, 11,12) and recently lowest excited singlet (10) state. Among the systems studied are various carotenoids (2-5), retinal isomers (6-8), diphenylbutadiene (9,10) and recently trienes (11,12). In the present paper we present ground state Raman spectra of all-trans-1,3,5-heptatriene (heptatriene) and all-trans-2,4,6-octatriene (octatriene), and time-resolved resonance Raman spectra of the lowest excited triplet states of trans-1,3,5-hexatriene (hexatriene), heptatriene, and octatriene.

Various calculations exist for excited states of hexatriene. Lasaga et. al. (13) have calculated energies, equilibrium geometries, and vibrational frequencies for the two lowest excited singlet states of hexatriene under the assumption of planar excited state geometries. Bonačić-Koutecký and Shingo-Ishimaru (14), and Ohmine and Morokuma (15), have calculated energies and equilibrium geometries for the lowest triplet state of hexatriene. Ohmine and Morukoma found that a geometry with twisting around the central C=C bond is 5.8 kcal/mole lower in energy than a planar geometry. A geometry with twisting around a terminal C=C bond was found to be 1.1 kcal/mole lower than the planar geometry. For twisting around the central bond, the C_1-C_2 bond length was found to be 1.373Å, the C_2-C_3 bond length 1.399Å, and the C_3-C_4 bond length 1.454Å.

The lowest triplet states of the polyenes studied in this paper were investigated with a pump-probe technique using two pulsed lasers. The triplet states were populated in the following way: A 13 ns pulse from an excimer laser excites acetone at 308 nm, acetone converts to the triplet manifold with an intersystem crossing efficiency close to unity (16), and the triene is excited to its lowest triplet state by energy-transfer (17). The $T_n \leftarrow T_1$ absorption spectrum is known for neo-alloocimene (18) and heptatriene (11). For both compounds the spectrum shows a maximum around 315 nm. The resonance Raman spectra of the T_1 states of hexatriene, heptatriene, and octatriene were excited by 10 ns pulses at 315 nm obtained from the second harmonic of a Nd:YAG pumped dye laser, this wavelength being in resonance with the $T_n \leftarrow T_1$ transition of the trienes.

Ground state Raman spectra using 514.5 nm excitation, a horizontal scattering plane, and 90° scattering, of neat heptatriene and of 1.7 M octatriene in CS_2 are shown in figs. 1A and 1B, respectively. Only the spectra obtained in I_{VV} configuration are shown, wavenumbers are listed in table 1. The spectra shown have been smoothed by 4 cm^{-1}.

Transient Raman spectra of Ar-saturated solutions of 15.0 mM hexatriene, of 6.5 mM heptatriene, and of 4.9 mM octatriene, in acetonitrile with 0.54 M acetone as sensitizer are shown in figs. 1C, 1D, and 1E, respectively. The spectra shown are subtraction spectra: Spectra obtained using the probe laser alone were multiplied by scaling factors of the order of 0.20 and subtracted from spectra obtained

This work was partially supported by The Danish Natural Science Research Council.

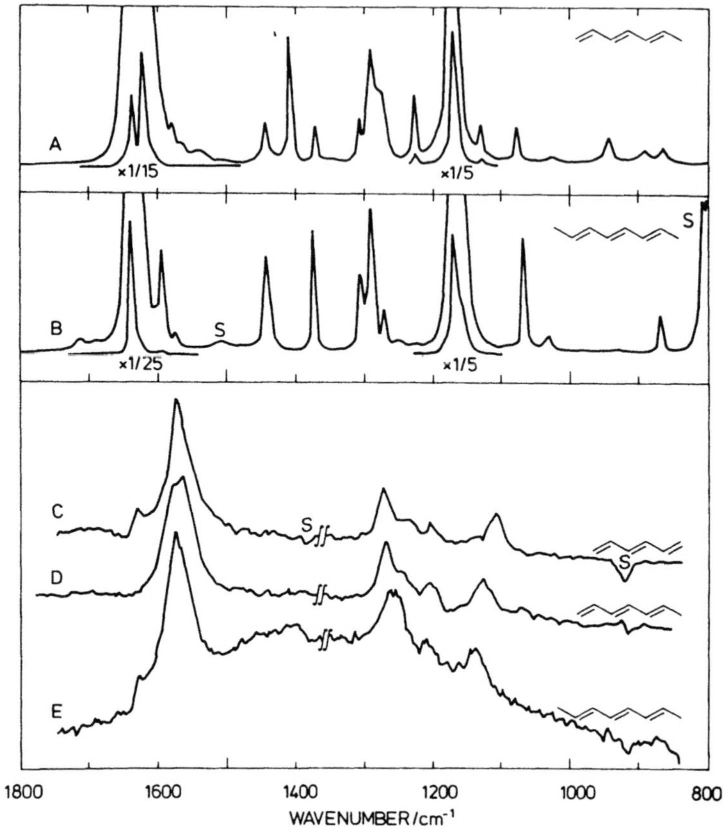

Fig. 1 A: Ground state Raman spectrum of neat heptatriene. B: Ground state Raman spectrum of 1.7 M octatriene in CS_2. C: Time-resolved resonance Raman spectrum of 15.0 mM hexatriene, 0.54 M acetone in acetonitrile. Subtraction spectrum at 50 ns delay between pump and probe pulses, using scaling factors 0.44 (left) and 0.42 (right) in the subtraction. D: Subtraction spectrum of 6.5 mM heptatriene at 70 ns delay, using scaling factors 0.33 (left) and 0.29 (right). E: Subtraction spectrum of 4.9 mM octatriene at 90 ns delay, using scaling factors 0.20 (left) and 0.16 (right). S denotes solvent

using pump and probe lasers together. This had the effect of removing the contribution of ground state Raman bands of solvent, sensitizer, and triene from the transient spectra. The scaling factors reflect the decreased intensity of the spectra due to transient absorption when triene triplets are present. As this transient absorption is stronger at 800 than at 1800 cm^{-1}, different scaling factors had to be used when constructing the low- and high-wavenumber regions of the spectra. The subtraction procedure has been discussed in detail previously (11,12). Wavenumbers for the transient bands are listed in table 1. The results for hexatriene are preliminary.

The spectra of the lowest triplet states of the three trienes studied in this paper have strong similarities. Each shows a strong band around 1570 cm^{-1}, three bands in the region 1170-1320 cm^{-1}, and one band at 1100-1140 cm^{-1}. But there are clear differences as well. The exact position of each band differs between the three compounds, the strong band around 1570 cm^{-1} has a larger bandwidth for heptatriene than for the two other compounds, and the relative intensity of the band around 1250 cm^{-1} is largest for octatriene, smallest for hexatriene.

Table 1. Observed vibrational wavenumbers (cm^{-1}) of the S_0 states of all-trans-1,3,5-heptatriene and all-trans-2,4,6-octatriene, and of the T_1 states of trans-1,3,5-hexatriene, all-trans-1,3,5-heptatriene and all-trans-2,4,6-octatriene. For the S_0 states wavenumbers are only tabulated for the region 800-1800 cm^{-1}

Heptatriene S_0		Octatriene S_0		Hexatriene T_1	Heptatriene T_1	Octatriene T_1
1642vs	1293s	1716w	1271m	1570	1567	1574
1625vs	1279m sh	1638vs	1252w	1269	1268	1264
1584w	1230m	1594s	1169s	1236	1245	1252
1571w	1174s	1576w	1157s sh	1201	1201	1205
1545w	1132m	1443s	1068s	1107	1125	1136
1509w	1079m	1373s	1030w			
1447m	1029w	1305m	931w			
1412s	945m	1289s	868m			
1373m	894m					
1348w	867m					
1309m						

For all three compounds the band around 1570 cm^{-1} is likely to represent the C=C double bond streching mode shifted to lower wavenumbers in the excited triplet state because of weakening of the double bonds with the lower degree of alternation in the excited state (13,15). An assignment of the remaining bands is very difficult. As seen in the calculations described,reversal of bond orders may take place in going from the ground state to the lowest triplet state. Another factor which complicates the assignment is the strong coupling of vibrational modes in the excited state, e.g. between C-C single bond stretching modes and CCH bending modes (19).

As the lowest triplet state is an intermediate in the sensitized pnoto-isomerization of many polyenes (20-22), we used GC to analyse the isomeric purity of the samples before and after laser irradiation. This showed conversion from the original all-trans isomers to isomers with cis conformation around one or more double bonds. This information was in agreement with changes in Raman spectra upon laser irradiation.

In addition to the trienes studied in this paper, we have studied various other short polyenes. From comparisons of the ground state and triplet spectra of these compounds,a picture of the lowest excited triplet state of short polyenes is developing.

References

1. R. Wilbrandt, N.-H. Jensen, P. Pagsberg, A.H. Sillesen and K.B. Hansen, Nature 276, 167 (1978)

2. N.-H. Jensen, R. Wilbrandt, P. Pagsberg, A.H. Sillesen and K.B. Hansen, J.Am.Chem.Soc. 102, 7441 (1980)

3. N.-H. Jensen and R. Wilbrandt, in "Photosynthesis I. Photophysical processes - Membrane energetization", G. Akoyunoglou, Ed., Balaban Int. Science Services, Philadelphia, pp. 97-114 (1981)

4. R. Wilbrandt and N.-H. Jensen, Ber.Bunsenges.Phys.Chem. 85, 508 (1981)

5. R. Wilbrandt and N.-H. Jensen, in "Time-resolved vibrational spectroscopy", G.H. Atkinson, Ed., Academic Press, New York, pp. 273-285 (1983)

6. R. Wilbrandt and N.-H. Jensen, J.Am.Chem.Soc. 103, 1036 (1981)

7. G.H. Atkinson, J.B. Pallix, T.B. Freedman, D.A. Gilmore and R. Wilbrandt, J.Am.Chem.Soc. 103, 5069 (1981)

8. R. Wilbrandt, N.-H. Jensen and C. Houeé-Levin, Photochem.Photobiol. 41, 175 (1985)

9. R. Wilbrandt, W.E.L. Grossman, P.M. Killough, J.E. Bennett and R.E. Hester, J.Phys.Chem. 88, 5964 (1984)

10. R. Wilbrandt, N.-H. Jensen and F.W. Langkilde, Chem.Phys. Letters 111, 123 (1984)

11. F.W. Langkilde, R. Wilbrandt and N.-H. Jensen, Chem.Phys. Letters 111, 372 (1984)

12. F.W. Langkilde, N.-H. Jensen and R. Wilbrandt, Chem.Phys. Letters, to be published

13. A.C. Lasaga, R.J. Aerni and M. Karplus, J.Chem.Phys. 73, 5230 (1980)

14. V. Bonačič - Koutecký and Shingo - Ishimaru, J.Am.Chem.Soc. 99, 8134 (1977)

15. I. Ohmine and K. Morokuma, J.Chem.Phys. 73, 1907 (1980)

16. R.F. Borkman and D.R. Kearns, J.Chem.Phys. 44, 945 (1966)

17. N.J. Turro and Y. Tanimoto, J.Photochem. 14, 199 (1980)

18. A.A. Gorman and I. Hamblett, Chem.Phys. Letters 97, 422 (1983)

19. U. Dinur, R.J. Hemley and M. Karplus, J.Phys.Chem. 87, 924 (1983)

20. W.H. Waddell, R. Crouch, K. Nakanishi and N.J. Turro, J.Am.Chem.Soc. 98, 4189 (1976)

21. Y.C.C. Butt, A.K. Singh, B.H. Baretz and R.S.H. Liu, J.Phys.Chem. 85, 2091 (1981)

22. N.-H. Jensen, A.B. Nielsen and R. Wilbrandt, J.Am.Chem.Soc. 104, 6117 (1982)

Resonance Raman Spectra of the Triplet States and Anion Radicals of Trans-Nitrostilbenes

C. Richter and W. Hub

Institut für Physikalische und Theoretische Chemie, Technische Universität München, Lichtenbergstr. 4, D-8046 Garching, F. R. G.

1. INTRODUCTION:

The direct isomerization of stilbene proceeds via a singlet pathway. The TRRR spectrum of the excited singlet state of trans-stilbene has been published by Hamaguchi et al. /1/ and by Gustafson et al. /2/.

Isomerization of stilbene also occurs in the triplet state, which may be populated by sensitized excitation or in the course of photolytically initiated electron transfer-reactions. Görner et al. /3/ postulated an equilibrium between a planar (λ_{max}=480nm) and a twisted (λ_{max}=350nm) configuration of the triplet state of trans-nitrostilbenes.

We have measured the RR spectra of the triplet state and of the anion radical of trans-4-nitrostilbene (NS) and of trans-4,4'-dinitrostilbene (DNS) with 450nm/480nm excitation. The comparison of these spectra gives us the possibility to get some information about the structure of these species. The RR spectra of the nitrostilbene anion radicals are compared with those of the trans-stilbene (TS) cation and anion radical.

2. EXPERIMENTAL:

Photolysis : 337nm
Probe : 480nm/450nm
Δt : 50ns
Concentration: $3*10^{-4}$ - $5*10^{-4}$ M
Solvent : Acetonitrile, Methanol, Ar satured
Mechanism of photoreduction of DNS:

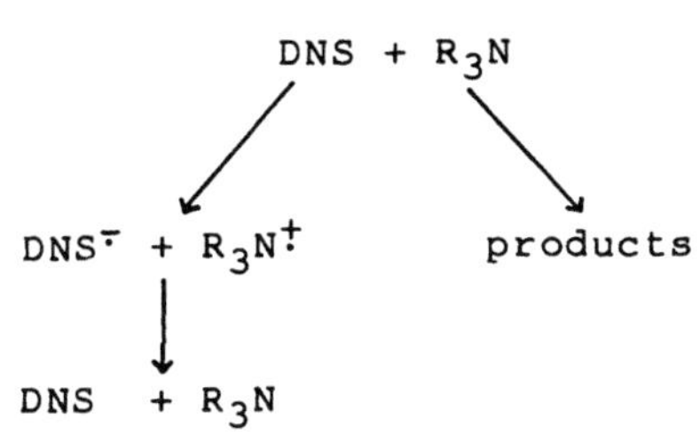

3. RESULTS:

The TRRR spectra obtained after photolysis (Δt=50ns) of a NS/DNS solution are shown in figures 1b and 2b. The triplet state of NS and DNS was identified by its lifetime (τ_{NS}=85ns, τ_{DNS}=170ns) and its absorption band at 450nm/480nm.

In the presence of tert. amine the spectra in figures 1c and 2c are observed, which belong to the anion radicals (τ=500ns). TRRR bands from the triplet state can not be found here.

All bands in the RR spectra of the anion radical and triplet state of NS/DNS are shifted to lower frequencies compared to the neutral groundstate molecules (table 1). The shifts of both C=C stretching frequencies of DNS are similar: -68 cm^{-1} (triplet) and -52 cm^{-1} (anion).

The NO stretching frequency is the same for both transient species (triplet: 1310 cm^{-1}; anion radical: 1306 cm^{-1}), whereas the C=C stretching and CH bending modes are found at higher frequencies in the DNS anion radical than in the triplet state.

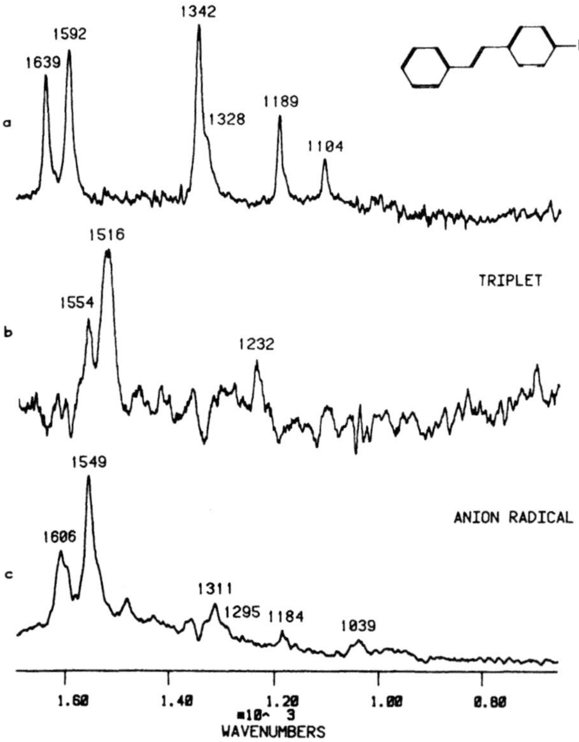

Figure 1. (a) RR spectrum of NS in acetonitrile. (b) RR spectrum of the triplet state of NS in methanol. (c) RR spectrum of the radical anion of NS in acetonitrile.

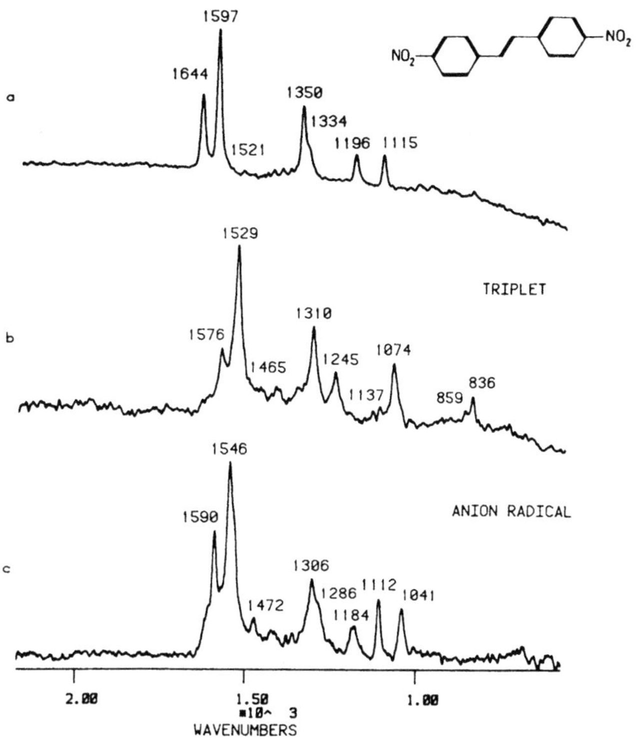

Figure 2. (a) RR spectrum of DNS in acetonitrile. (b) RR spectrum of the triplet state of DNS in acetonitrile. (c) RR spectrum of the radical anion of DNS in acetonitrile.

Table 1: Resonance Raman Frequencies of TS, NS, DNS

	TS		NS			DNS		
	neutral	anion	neutral	anion	triplet	neutral	anion	triplet
C=C stretch (e)	1642	1553	1639	1606	1554	1644	1590	1576
C=C stretch phenyl	1598	1577	1592	1549	1516	1597	1546	1529
N=O stretch asym			1524	1485		1521	1472	1465
N=O stretch sym			1342	1311		1350	1306	1310
CH def (e)	1325	1251	1328	1295	1232	1334	1286	1245
CH def phenyl	1187	1180	1189	1184		1196	1184	1137

4. CONCLUSIONS:

The large shift of the NO stretching frequency by about -40 cm^{-1} shows a strong delocalization of π-electron density into the NO_2 groups of the triplet state and the anion radical. Accordingly the C=C stretching frequencies are lowered. We conclude, that the anion radical and the triplet state of nitro-stilbenes absorbing around 480nm are essentially planar. This is in accord with the results of Goerner et al. /3/.

The C=C stretching and CH bending modes are found at higher frequencies in the anion radical than in the triplet state spectra of DNS and NS. This can be due to the higher π-electron density in the anion radical.

The comparison of the RR spectra of DNS and TS show the influence of the nitro groups. The groundstate frequencies of TS are not shifted by nitro-substitution. In contrast to this the C=C stretching frequencies in the anion radical spectra of DNS and TS are significantly shifted (table 1). It seems as if π-electron density is increased in the olefinic part of $DNS^{\cdot-}$ compared to $TS^{\cdot-}$. This points to a more planar structure of $DNS^{\cdot-}$ than of $TS^{\cdot-}$. The similarity of the vibrational frequencies of the olefinic part of $DNS^{\cdot-}$ and $TS^{\cdot+}$ ($\tilde{\nu}_{C=C}$=1590/1605 cm^{-1}, $\tilde{\nu}_{CH(e)}$=1286/1285 cm^{-1}) also suggests a planar configuration of $TS^{\cdot+}$.

/1/ H. Hamaguchi, T. Urano, and M. Tasumi: Chem. Phys. Letters 106, 153 (1984)

/2/ T.L. Gustafson, D.M. Roberts, and D.A. Chernoff: J. Chem. Phys. 79, 1559 (1983)

/3/ H. Görner and D. Schulte-Frohlinde: Ber. Bunsenges. Phys. Chem. 88, 1208 (1984)

Resonance Raman Spectroscopy of Transient Species in the Pulse Radiolytic Oxidation of Aniline†

G.N.R. Tripathi, R.H. Schuler, and L. Qin

Radiation Laboratory and Department of Chemistry, University of Notre Dame. Notre Dame, IN 46556, USA

1. Introduction

Recent advances in time-resolved resonance Raman spectroscopy make it practical to examine the vibrational spectra of transient intermediates produced in radiation and photochemical processes. Appropriate studies provide structural and kinetic information on these transients, and also mechanistic information on their mode of formation and decay. Recently, we have applied this technique, combined with pulse radiolysis methods, to examine in detail the vibrational spectra and structure of several transient species, including the prototype aromatic oxy radicals such as semiquinone [1] and phenoxyl [2]. In the latter case the decay of phenoxyl radical at $\sim 10^{-5}$ [M] concentrations has been shown to proceed entirely by second order processes ($2k = 2.6 \times 10^9$ [$M^{-1}s^{-1}$]). In the study reported here we have examined, by Raman spectroscopy, the oxidation of aniline in aqueous solution by azide radical ($N_3\cdot$). In acidic solution the aniline radical cation, which is isoelectronic with phenoxyl radical, is produced by electron-transfer to the azide radical

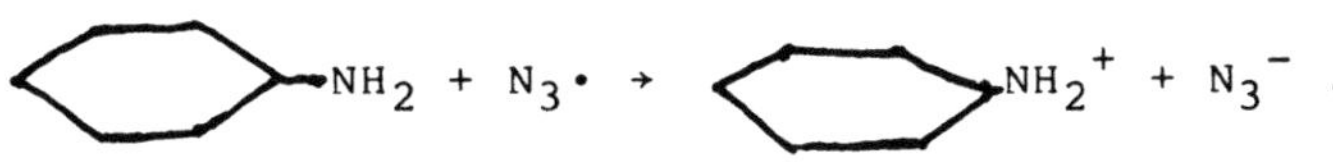

(1)

The Raman studies on this radical are particularly important for structural elucidation, since it has not yet been possible to observe its ESR spectrum in solution. We have shown that the decay of the aniline radical cation results in the formation of benzidine radical cation, presumably because of secondary oxidation of benzidine formed by combination of the initial radicals. In basic solution anilino radical, which is also isoelectronic with phenoxyl, is formed. The Raman experiments conclusively show that the radical cation is the initial intermediate, which subsequently deprotonates by reaction with base. We briefly describe here some structural and kinetic features of the important transients produced in reaction.

2. Experimental

Time-resolved resonance Raman apparatus used in this study has been described previously [3]. Pulse irradiation was with ~ 100 [ns] pulses of 2 [MeV] electrons from a Van de Graaff accelerator, operated at a repetition rate of 7.5 pulses per second. The electron beam was tightly focused (1 [mm^2]) and doses ($\sim 2 \times 10^{18}$ [ev/g/pulse]) were sufficient to produce $\sim 2 \times 10^{-4}$ [M] radicals/pulse. The aniline was vacuum-distilled prior to use. Solutions, 10 [mM] in aniline and 0.1 [M] in NaN_3 were prepared with water from a Millipore Q system and saturated with N_2O to purge O_2 and convert e^-_{aq} to $OH\cdot$. The solution was flowed continuously (rate, ~ 2 [cm^3/s]) through the Raman cell (~ 0.05 [cm^3]), to replenish the sample between pulses. The Raman scattering was excited by ~ 10 [ns] laser pulses obtained from a XeCl excimer pumped dye laser system. For time resolution, the laser pulse was electronically delayed with respect to the electron pulse. A 0.85 [m] spectrograph with 110 [mm], 1800 [g/mm] holographic grating was used as the dispersive element. The Raman signals were detected by a synchronously gated (~ 20 [ns]) intensified silicon photodiode array detector having ~ 700 active pixels (Reticon 1420-3) and averaged and processed by PAR OMAII optical multichannel analyser system. Further averaging and processing was done off-line using a Vax 11/780 computer. The Raman frequencies of the transient species were measured by reference to the spectra of ethanol, cyclohexane and benzene recorded at the same laser and spectrometer setting. Frequency measurements were accurate to ± 1 [cm^{-1}].

3. Results and Discussion

Pulse radiolytic studies on aqueous solutions containing 1 [mM] aniline and 0.1 [M] NaN_3 were performed by optical absorption technique in the pH range 5 to 11 to optimize the chemical conditions for Raman measurements [4]. The OH radicals produced by pulse radiolysis of water react rapidly with N_3^- ($k = 1.2 \times 10^{10}$ [$M^{-1}s^{-1}$]), so that in a 0.1 [M] NaN_3 solution the $N_3\cdot$ radicals are produced in 10^{-9} [s]. Rate-constant for the reaction of $N_3\cdot$ with aniline has been estimated as ~ 4.4×10^9 [$M^{-1}s^{-1}$]. Fig. 1 depicts the optical absorption spectra of aniline radical cation and anilino radical observed 2 [μs] after the electron pulse at, respectively, pH 5.2 and 10.5. The radical cation exhibits two absorption peaks at 406 [nm] (ext. coeff. 3310 [$M^{-1}cm^{-1}$]) and 423 [nm] (ext. coeff. 4100 [$M^{-1}cm^{-1}$]) and is similar to phenoxyl in spectral features. The absorption spectrum of anilino radical is broad and considerably weaker with λ_{max} at 401 [nm] (ext. coeff. 1250 [$M^{-1}cm^{-1}$]).

Resonance Raman Spectrum of Aniline Radical Cation: From the rate-constant mentioned above the half period for oxidation of 10 [mM] aniline by $N_3\cdot$ is estimated as 23 [ns]. Fig. 2 presents the Raman spectrum, excited at 425 [nm], 100 [ns] after the electron

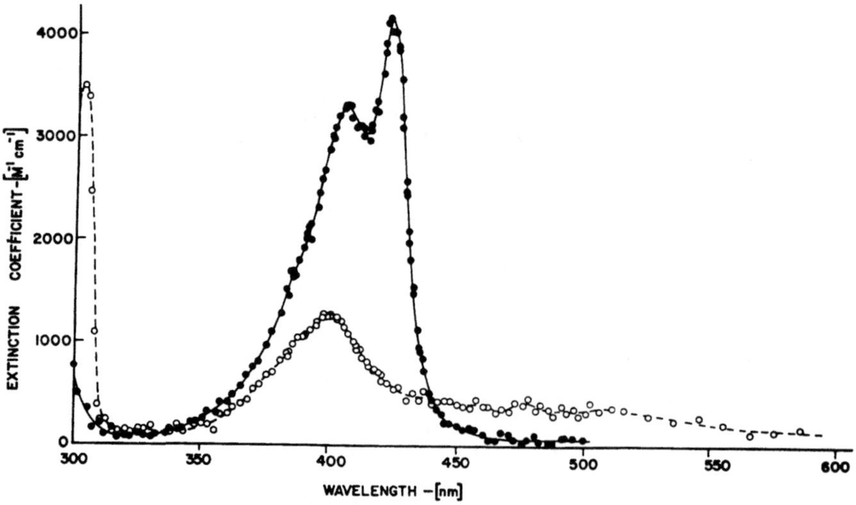

Fig. 1. Absorption spectra of the aniline cation (•) and anilino radical (o) observed 2 [μs] after the electron pulse.

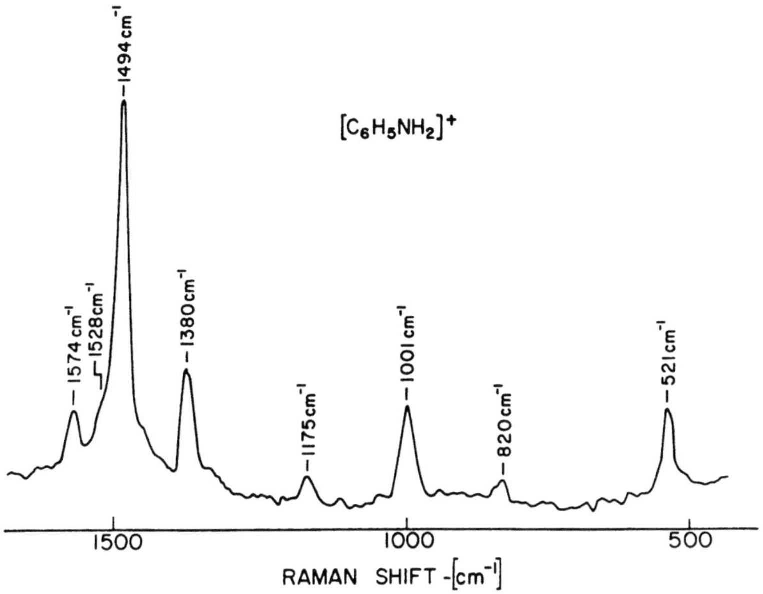

Fig. 2. Resonance Raman spectrum of the aniline radical cation observed 100 [ns] after the electron pulse, at pH 5.5.

pulse at pH 5.5. The spectral features are similar to those of phenoxyl radical. From resonance Raman conditions, spectral pattern, and effects of deuteration on the spectrum the species is positively identified as the aniline radical cation. Eight fundamentals at 1574, 1528, 1494, 1380, 1175, 1001, 820 and 521 [cm^{-1}] and several combinations have been observed. The 1494 [cm^{-1}] band is most intense in the spectrum, and exhibits an upward shift of 12 [cm^{-1}] on deuteration of the amino group. This band has been assigned due to predominantly CN stretching mode, although significant contributions from NH_2 and ring CH bending motions are also involved. The corresponding vibration occurs at 1278 [cm^{-1}] in aniline. An increase of 216 [cm^{-1}] in the frequency indicates that the CN bond acquires considerable double bond character on the formation of the radical. The weaker bands at 1574 and 1528 [cm^{-1}] are assigned to Wilson modes 8a (CC stretch) and 19a (CC stretch). These modes are not resonance-enhanced in phenoxyl. The former is reduced by 28 [cm^{-1}] and the latter increased by 27 [cm^{-1}] in frequency in the radical cation as compared to aniline. This implies that the ring structure is considerably distorted, with relatively weaker C_2-C_3 and C_5-C_6 and stronger C_1-C_2 and C_1-C_6 bonds with respect to aniline. The Raman bands at 1175, 1001, 820 and 521 [cm^{-1}] are assigned to Wilson modes 9a (CH bend), 18a (CH bend), 12 (CCC bend) and 6a (CCC bend) with fair degree of certainty. The assignment of the prominent band at 1380 [cm^{-1}] is somewhat uncertain at this stage [5].

Second Order Reaction of Cation: It is difficult to determine the second order rate-constant for reaction between the aniline cation radicals accurately by optical absorption techniques because the products absorb appreciately in the same region as the radical. One finds, for example, that in acidic solutions the absorption at 423 [nm] changes very little over the first several hundred microseconds after the pulse, even though there is an appreciable increase at longer wavelengths. It is clear from this latter observation that the radical cations, in fact, combine on the hundred microsecond time-scale. Since the radical concentration is only micromolar in optical absorption studies, the second order rate-constant must be of the magnitude of 10^9 $[M^{-1}]$ but lack of optical resolution between the individual components makes more detailed interpretation questionable. Resonance Raman spectroscopy provides the selectivity required to examine the individual components. Time-resolved resonance Raman studies of the decay of the radical ($\sim 10^{-4}$ [M]) has been monitored by the 1494 $[cm^{-1}]$ band. The decay pattern is very similar to that of phenoxyl radical. Initially, the radical decays at a rate corresponding to a second order rate-constant for reaction between like radicals ($2k = 2.8 \times 10^9$ $[M^{-1}s^{-1}]$). However, as products build up, decay is more rapid than expected solely from second order processes, because of a superimposed reaction of the radical cations with the products. It was noted that as the Raman signals of aniline cation in the 1300-1800 $[cm^{-1}]$ region decreased, three rather intense lines appeared at 1688, 1614 and 1540 $[cm^{-1}]$. Raman signals, observed 100 [μs] after oxidizing aniline with $N_3\cdot$, with excitation at 450 [nm] are reported in Fig. 3. It is seen in this figure that the features are identical to that of the cation radical produced directly from benzidine. It is clear from this result that coupling of the radical cation at the para position represents a major channel

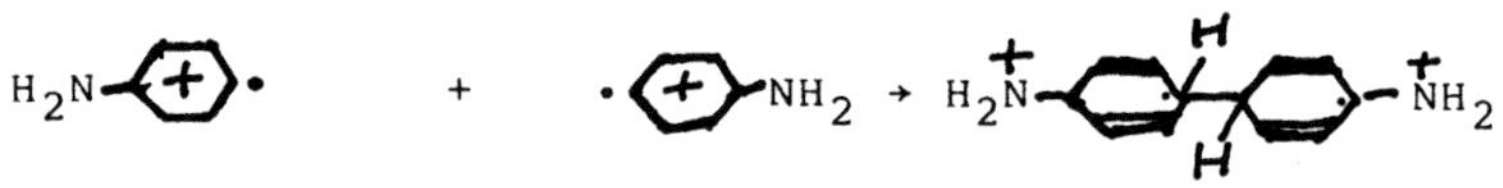

(2)

for the decay of the radicals. Presumably reaction (2) is followed by deprotonation to give benzidine,which is subsequently oxidized

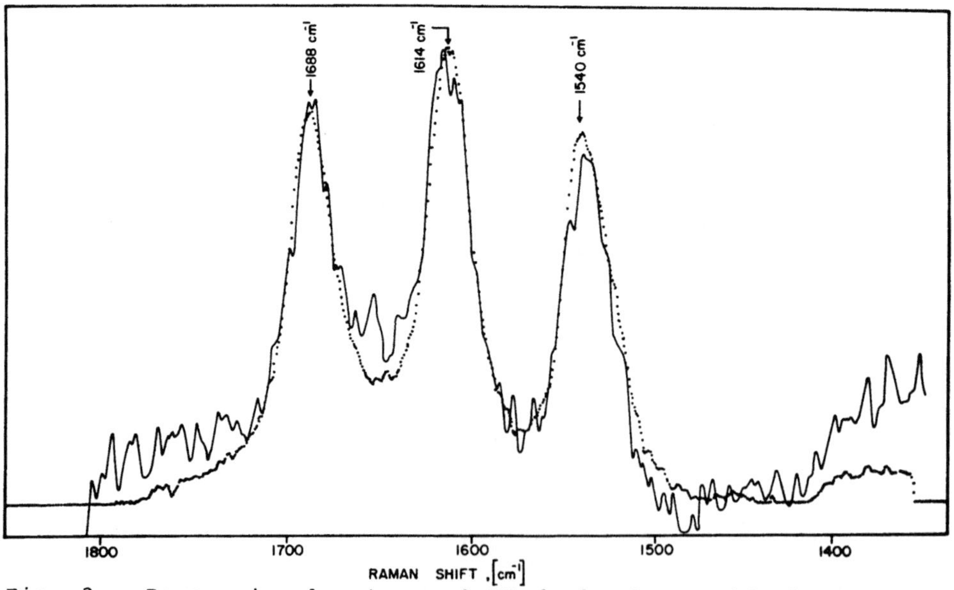

Fig. 3. Raman signals observed 50 [μs] after oxidation of aniline (solid spectrum) and benzidine (dotted spectrum); at a relative sensitivity of 0.15 by $N_3\cdot$. Excitation is in the 450 [nm] absorption band of the benzidine radical cation.

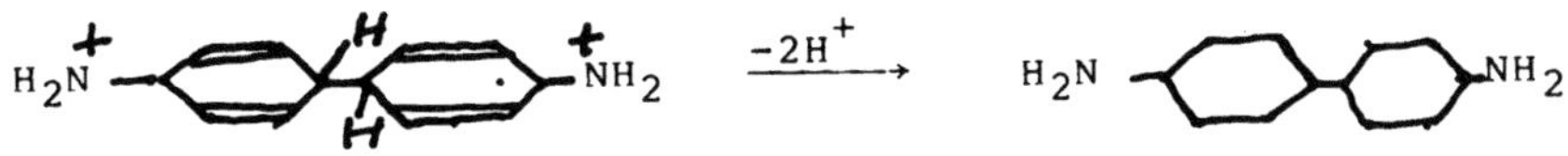

(3)

by electron-transfer to the aniline radical cation still remaining in the system

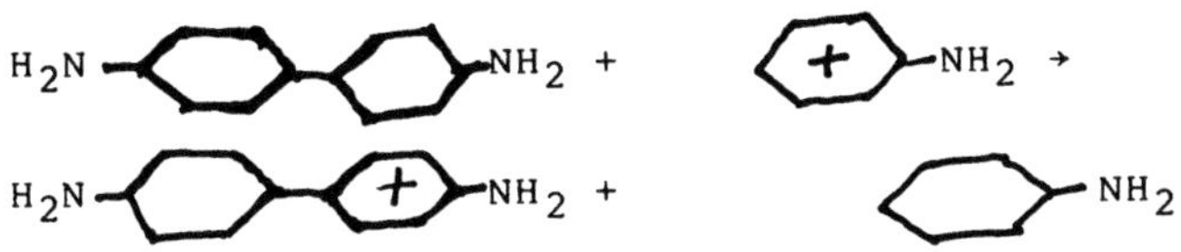

(4)

Because the product concentration is ~ 10^{-4} [M] in these Raman studies, the rapid decay of aniline radical cation on microsecond time-scale in readily explained. Since production of benzidine cation radicals via this mechanism requires three aniline cation radicals we estimate from relative intensities in Fig. 3 that 30% of the second order processes involve coupling at 4 position of two radicals.

Raman Spectra in Basic Solutions: As indicated in Fig. 1., the optical absorption spectrum obtained 2 [μs] after the electron pulse at pH 10.5 is that of anilino radical. We have positively established from the Raman studies that precursor of this radical is aniline radical cation. At pH ~ 10.4, the Raman spectrum, observed

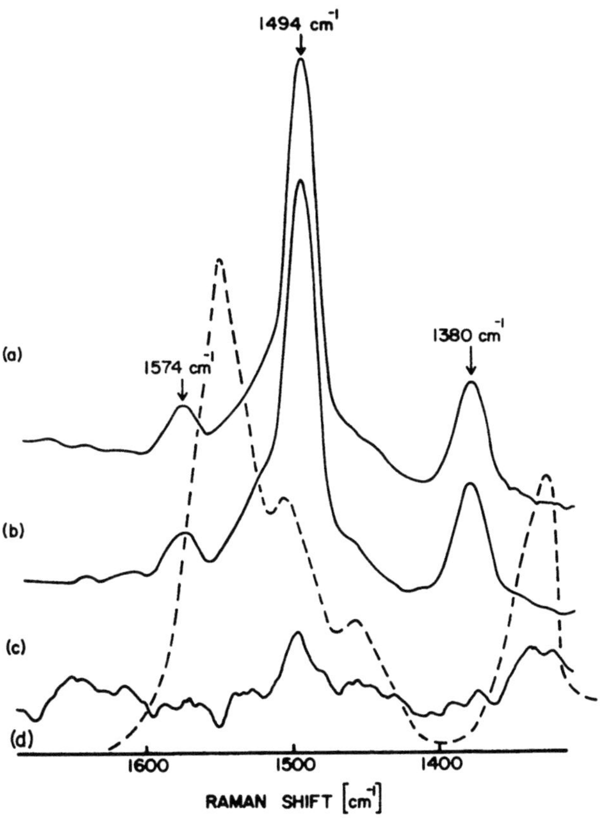

Figure 4. Raman spectrum characteristic of the aniline radical cation observed 0.1 [μs] after pulse at (a) pH 6.23 (scaled by a factor of 0.13) and (b) at pH 10.4. Excitation was in the absorption band of the radical cation at 425 [nm]. The 1494 [cm^{-1}] signal decays by a factor of 6 at 0.3 [μs], as indicated in (c). (d) Spectrum (---) of anilino radical (pH ∿ 11) excited at 400 [nm], 0.5 [μs] after the pulse.

by excitation at 425 [nm], 100 [ns] after the pulse exhibits characteristic features of the aniline radical cation (Fig. 4). The signals decay by a factor of 6 at 300 [ns] and are not observable at 500 [ns]. We estimate a decay half-period of 80 [ns] from Raman studies, in good agreement with the period expected at this pH from absorption measurements. By excitation at 400 [nm], 500 [ns] after the electron pulse, the Raman spectrum of the anilino radical is observed. Because of low extinction coefficient of the radical, the Raman signals are at least an order of magnitude weaker than those of the radical cation. By extensive signal averaging (12000 pulses) however, a reasonably good Raman spectrum of the anilino radical has been obtained. A portion of the spectrum is shown in Fig. 4d. The spectrum of this radical [6] is quite different from that of the aniline cation or phenoxyl radical, and suggests that the electronic structure is significantly different in one or both of the resonant electronic states. A detailed analysis of the vibrational structure of this radical will be presented elsewhere.

4. References

† The research described herein was supported by the Office of Basic Energy Sciences of the Department of Energy. This is Document No. NDRL-2718 from the Notre Dame Radiation Laboratory.

[1] G.N.R. Tripathi and R.H. Schuler, J. Chem. Phys. 76, 2139 (1982); also see, G.N.R. Tripathi, J. Chem. Phys. 74, 6044 (1981); R.H. Schuler, G.N.R. Tripathi, M.F. Prabenda and D.M. Chipman, J. Phys. Chem. 87, 5357 (1983); G.N.R. Tripathi and R.H. Schuler, J. Phys. Chem. 87, 3101 (1983).

[2] G.N.R. Tripathi and R.H. Schuler, Chem. Phys. Lett. 88, 253 (1982); J. Chem. Phys. 81, 113 (1984).

[3] G.N.R. Tripathi and R.H. Schuler, J. Phys. Chem. 88, 1706 (1984); G.N.R. Tripathi, "Time-resolved Resonance Raman Spectroscopy of Radiation-chemical Processes" in Multichannel Image Detectors, Vol II (ed. Y. Talmi), ACS Symposium Series, No. 236, 171 (1983).

[4] L. Qin, G.N.R. Tripathi and R.H. Schuler, Z. Naturforsch. 40a, (1985), in press.

[5] G.N.R. Tripathi and R.H. Schuler, Chem. Phys. Lett. 110, 542 (1984).

[6] G.N.R. Tripathi and R.H. Schuler, Proceedings of the Ninth International Conference on Raman Spectroscopy, The Chemical Society of Japan, Tokyo, Japan, 1984, page 326.

Intramolecular Kinetics of Vibrationally Excited Singlet and Triplet States in Some Cyclic Hydrocarbons

H.G. Löhmannsröben and K. Luther

Institute of Physical Chemistry, University of Göttingen,
Tammannstraße 6, D-3400 Göttingen, F.R.G.

1. Introduction

In recent years multiphoton ionization (MPI) has developed into a versatile tool to investigate the dynamics of different molecular processes, as unimolecular reactions [1], radiationless transitions [2] and collisional energy-transfer [3]. For radiationless processes MPI became particularly useful, as it allows to study the dynamics of vibrationally excited species in the gas phase at low pressures under collision-free conditions, where for various reasons other spectroscopic techniques, like absorption spectroscopy or laser-induced fluorescence, are often not well suited.

The simplest experimental MPI approach involves a one-color two-photon ionization via an intermediate molecular state. If the photoionization from this excited state competes with other molecular transitions, simple ion yield measurements allow to determine lifetimes of the resonant state down into the femtosecond range. We have used this method to determine the extremely fast rates of internal conversion (IC) in several alkyl-substituted cycloheptatrienes (R-CHT). More general experimental possibilities are given by two-color-ionization techniques. In this case, a first laser with a variable wavelength excites molecules into a selected vibronic level or - at high densities of states - into a manifold of states with a particular total vibrational-rotational energy. Ionization with a second laser after variable delay times allows to probe the time-evolution of these photoexcited states, which are formed from the initially populated ones by a fast intramolecular process like intersystem crossing. The wavelength of the second laser is often conveniently fixed as to reach high into the ionization continuum, and thus to achieve nearly energy-independent ionization cross-sections within the range of interest. By varying the wavelength of the first laser the energy-dependence of excited state lifetimes can thus be studied. This two-color photoionization scheme has been successfully employed in supersonic beams [4]. In a somewhat different approach, we studied the energy-dependence of triplet lifetimes of alkyl-substituted benzenes in a low-pressure static time-of-flight mass spectrometer (TOF-MS) arrangement.

2. Experimental

In one-color experiments total ion yields were measured in a quartz cell with parallel metal plates as ion collector. Integrated signals were fed into a computer for data storage and processing. Ionization of cycloheptatriene and the methyl, ethyl, and isopropyl-substituted compounds (CHT, MeCHT, EtCHT, iPrCHT respectively) was carried out at a pressure of 100 mTorr with 10 ns laser pulses at 266 nm. For reference two-photon ionization of toluene was performed with a frequency-doubled dye laser at 266.77 nm using the $S_o(v=0) \rightarrow S_1(v=0)$ transition as a first step. We worked under low intensity conditions in the 1-10 kW cm^{-2} range where only parent ion formation occurs. Parallel beam geometry was used. The laser intensity was permanently monitored and shot-to-shot normalization of the ion signal was to account for residual employed energy fluctuations. Careful tests were performed to avoid saturation, recombination, surface catalysed ionization etc. Further details are given elsewhere [5].

Two-color experiments were carried out in a TOF-MS which has been described [6]. Briefly, photoexcitation was carried out by means of an excimer pumped dye laser in the range λ_1=264-275 nm. An excimer laser at λ_2= 193 nm was used for the ionization step. Both beams were directed antiparallel through the reaction chamber of the TOF-MS, with laser intensities of 1-50 kW cm^{-2} for the excitation and about 0.5 MW cm^{-2} for the ionisation. Pressures were usually 1-3 µTorr. The laser pulse width was approximately 15 ns and the time-delay between both pulses had a jitter of ± 20ns.

3. Results and Discussions

3.1 One-Color Experiments

Around 250 nm all cycloheptatrienes absorb in a broad structurless ($S_o \rightarrow S_1$) band with a cross-section of α_1=7.7x10^{-18} cm^{-2} at 266 nm. At this wavelength absorption of a second photon is sufficient to ionize the RCHT. Nevertheless, the ionization yields of the RCHT are exceedingly small. Under the excitation conditions of 5.4 kW cm^{-2} the probability for ionization of CHT molecules in the S_1 state by a second photon was found to be 1.2x10^5 times smaller than in its aromatic isomer toluene (at 266.77 nm). The process competing with the ionization in CHT is known to be an efficient internal conversion ($S_1 \rightarrow S_o^*$) with a quantum yield of ≥ 0.97, which therefore practically determines the lifetime of the S_1 state. At low pressures the IC is followed by a rearrangement reaction of hot R-CHT (S_o^*) into the corresponding hot aromatic isomers (S_o^*) [7]. Although the S_o absorption cross-sections α_1(266 nm) are practically constant for the various RCHT, the ion yield increases with the size of the substituted alkyl group, with a nearly 20 times bigger value in iPrCHT than in CHT (table 1). Rate-constants for the internal conversion k_{IC} are derived by kinetic analysis from the measured ion yields, and the pumping rates $\alpha_2 I$ (laser intensity I, S_1 absorption cross-section α_2). As it is shown in table 1, k_{IC} decreases with increasing size of the alkyl group, in accordance with the effect of substitution on the relative ion yield.

Table 1 Relative ion yield and rate-constants of internal conversion k_{IC} ($S_1 \rightarrow S_o$) of cycloheptatrienes

	CHT	MeCHT	EtCHT	iPrCHT
$Y_{ION}/Y_{ION,CHT}$	1	5.4	11.4	18.7
k_{IC}/s^{-1} ($\alpha_1 = \alpha_2$)	9.2x10^{12}	1.9x10^{12}	7.2x10^{11}	3.8x10^{11}

This influence of the molecular size on the rate of internal conversion can be rationalized e.g. on the basis of statistical type of theories [8] which show a considerable degree of analogy with established concepts in the fields of unimolecular reactions like the statistic adiabatic channel model [9].

The assumption $\alpha_1 = \alpha_2$ appears to be a reasonable choice for the RCHT; as direct data on α_2 are not yet available. For further details see Ref. [5].

3.2 Two-Color Experiments

Ion signals at the masses 106 and 91 were observed when the xylenes or ethylbenzene were irradiated only by the second laser at λ_2=193 nm. The parent ion peak is much enhanced, when excitation by the tunable first laser at λ_1 preceeds the ionization pulse at 193 nm. This enhancement is due to efficient ionization of the excited molecules in the singlet (S_1) and triplet (T_1) states. By tuning λ_1 the average vibrational-rotational energy of the molecules in S_1 and T_1 is varied. Therefore, the time-dependent enhancement of the ion signal measures the S_1 and T_1 lifetimes as functions of excess energies in these electronic states. Fig. 1 shows schematically relevant energy-levels and the ionization paths for o-xylene before and after intersystem crossing (ISC). As an example, the time-dependence of the transient ion signal after excitation of p-xylene is shown in Fig. 2. It displays a clearly biexponential shape corresponding to the decay kinetics of the S_1 and the T_1 state with distinctly separated half-life times. This work concentrates on the kinetics of the triplet state. A summary of the results is given in Table 2. The rate-constants k_T of the decay of triplets with an excess energy E_{VT} above the zero point energy are received from the measured apparent rate-constants by a correction for losses of molecules out of the sampling volume during the time until the ionization laser pulse occurs. The details of the procedure are given elsewhere [10].

From the results of our measurements, some interesting conclusions can be drawn. The triplet lifetime of toluene has been subject of two recent MPI studies in cold beams [12, 13] which differed in their results on k_T by a factor of two. It has been suggested that this discrepancy originated from long-range interactions, which cause collisional relaxation and increased triplet lifetimes. Our data confirm the short

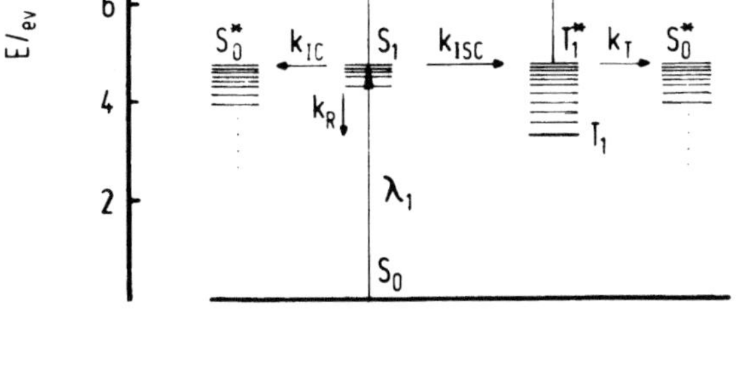

Fig. 1.
Photoinduced processes in the 2-color MPI of o-xylene

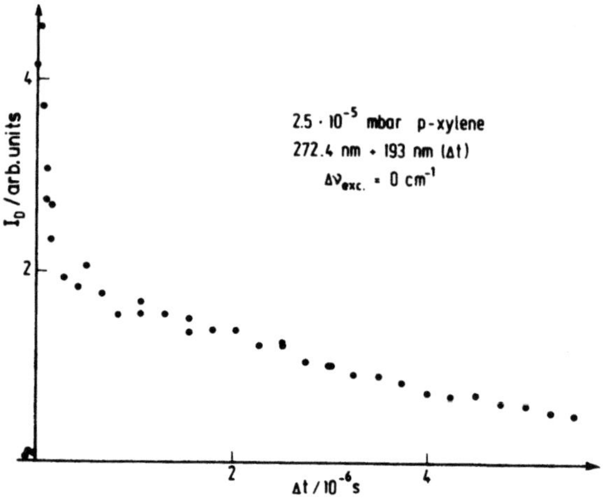

Fig. 2.

Time-dependence of the transient ion signal in p-xylene after excitation into $S_1(v=0)$.

Table 2 Triplet decay rates at various vibrational energies

Substance	Vibrational energy in S_1 (cm^{-1})	Vibrational energy in T_1, E_{VT} (cm^{-1})	$k_T/10^5$ (s^{-1})
p-xylene	0	8300	1.6 ± 0.2
	770	9070	2.1 ± 0.3
	1125	9425	2.9 ± 0.8
o-xylene	0	8600	2.4 ± 0.3
	200	8800	3.2 ± 0.4
	530	9130	3.3 ± 0.8
ethylbenzene	0	8400	4.0 ± 0.5
toluene (p=2µTorr)	0	8400	3.3 ± 0.5
(p=40µTorr)	0	8400	3.4 ± 0.3

lifetimes of [12] even at much higher pressures, and show no pressure-dependence in the range of 2-40 µTorr. This indicates that at least in a static system, long-range interactions are of much less importance (more than a factor of 10^3 in the cross-section) than it has been proposed for molecular beam conditions.

The triplet decay rates of p-xylene, o-xylene and ethylbenzene at high energies E_{VT} have been determined for the first time. They are found to be two orders of magnitude higher than those reported for vibrationally relaxed systems [11] . This behaviour is in accordance with the energy-dependence of triplet lifetimes on other aromatic systems, and seems to represent a general scheme: From low-temperature matrices ($k_T \simeq 10^0 - 10^1\ s^{-1}$) over room-temperature conditions ($k_T \simeq 10^3 - 10^4\ s^{-1}$) a steep increase of the rate-constant is found up to energies $E_{VT} \simeq 8500\ cm^{-1}$ ($k_T \simeq 3 \times 10^5\ s^{-1}$) where the energy-dependence slows down very much towards higher energies. This energy-dependence is not explained by the standard theories of radiationless transitions, but in its general pattern it resembles the typical energy-dependence of a different type of intramolecular process , that of unimolecular reactions. New theoretical approaches along such lines as mentioned already above, [8], may lead here to a better understanding.. Further details of this work are given in Ref. [10].

Acknowledgements

We thank Dr. P.M. Borrell for her help in the first period of the experiments, and Dr. M. Stuke for his cooperation and the possibility to use his TOF-MS apparatus in part of this work. Financial support from the Deutsche Forschungsgemeinschaft (Sonderforschungsbereich 93 "Photochemie mit Lasern") is gratefully acknowledged.

References

1. e.g. : U. Boesl, H.J. Neusser, E.W. Schlag, Chem. Phys. Lett. 52, 413 (1977)
2. M.A. Duncan, T.G. Dietz, M.G. Liverman, R.E. Smalley, J. Phys. Chem. 85, 7 (1981)
3. H.G. Löhmannsröben, K. Luther, to be published
4. J.L. Knee, P.M. Johnson, J. Chem. Phys. 89, 948 (1985), and references therein.
5. P.M. Borrell, H.G. Löhmannsröben, and K. Luther, to be published
6. M. Stuke, Appl. Phys. Lett. 45, 1175 (1984)
7. H. Hippler, K. Luther, J. Troe, and H.J. Wendelken, J. Chem. Phys. 79, 239 (1983)
8. H. Hornburger, H. Kono, and S.H. Lin, J. Chem. Phys. 81, 3554 (1984); H. Hornburger, H. Schröder, J. Brand, J. Chem. Phys. 80, 3197 (1984)
9. M. Quack and J. Troe, Ber. Bunsenges. Phys. Chem. 78, 240 (1974); J. Troe, J. Chem. Phys. 79, 6017 (1983)
10. H.G. Löhmannsröben, K. Luther, and M. Stuke, in preparation
11. K.W. Holtzclaw, M.D. Schuh, Chem. Phys. 56, 219 (1981)
12. T.G. Dietz, M.A. Duncan, and R.E. Smalley, J. Chem. Phys. 76, 1227 (1982)
13. C.E. Otis, J.L. Knee, and P.M. Johnson, J. Chem. Phys. 78, 2091 (1983); J. Phys. Chem. 87, 2232 (1983)

In Situ Raman Studies of Homopolymerisation and Copolymerisation Reactions

H.J. Bowley, I.S. Biggin, and D.L. Gerrard

BP Research Centre, Chertsey Road, Sunbury-on-Thames, Middlesex, TW16 7LN, England

A knowledge of the rates of polymerisation of monomers to yield homopolymers and copolymers is of considerable interest, both academically and industrially. Such a knowledge applied to a wide range of monomers, under an extensive variety of temperatures and pressures, and also different atmospheres (such as N_2, CO_2, O_2 etc) is particularly valuable to the industrial chemist. This is because it is often the case that monomers have to be distilled during their preparation, and a knowledge of polymerisation rate can be used to determine the best conditions (such as temperature, pressure, atmosphere) under which this distillation should be carried out. Also, with the increased tendency towards producing speciality copolymers, a knowledge of the polymerisation rates of the constituent monomers proves valuable in predicting their compositions.

Raman spectroscopy is a particularly valuable technique for studying polymerisation reactions, since the design of a cell operating at a wide range of temperatures and pressures, and using a glass sample container presents no intrinsic problems. Having such a cell, the polymerisation reaction can then be followed, to completion if necessary, under a range of conditions and atmospheres. Most monomers used to produce commercially useful polymers contain a - C = C - group via which the polymerisation occurs, and this gives a very strong Raman signal.

As the polymerisation proceeds the intensity of this band decreases, and this decrease can be monitored quantitately. There are two methods that can be used to quantify this; first by using another band as an internal standard or, secondly, by keeping all of the spectroscopic conditions constant and monitoring the change in the integrated area of the -C = C - band. Care has to be taken with the first method, as shown by Gulari et al [1] to ensure that the band being used as the internal standard has the same intensity whether it is in a monomer or a polymer molecule. The carbonyl band of methyl methacrylate was found to have invariant intensity throughout the polymerisation reaction when conditions were maintained constant. The ring "breathing" mode of styrene at 1000 cm^{-1} however, has been shown to exhibit a significant degree of variance.

Conventional Raman spectrometers are too slow to follow rapid reactions, and so to cover all the possibilities which might reasonably be expected in this type of study, a diode array detector was used throughout. This is capable of giving very good quality spectra of the specific region of interest in a maximum accumulation time of 1 to 20 seconds. In this way a wide range of polymerisation reactions can be followed with either the pure monomer or in solution, and with or without a wide range of initiators.

EXPERIMENTAL

An Anaspec 36 spectrometer, fitted with a Tracor Northern Reticon type S intensified diode array, was used throughout. The 514.5nm line of a

Spectra-Physics Model 2000 argon ion laser was used for all of the work. Monomers were flash-distilled immediately before use in order to remove stabiliser. The initiator used was benzoyl peroxide, and this was purified by dissolving in chloroform in a separating funnel, removing the aqueous layer and then precipitating with methanol. Benzoyl peroxide was used at a level of 0.5% $^{w}/v$ throughout.

The cell used was a Ventacon polymerisation cell, capable of temperatures up to 250°C. All of the polymerisations were carried out at 80°C and atmospheric pressure under an atmosphere of nitrogen. The laser was used in its light mode in order to stabilise its output. The diode array was centred at 1650 cm^{-1}, and the intensity of the vinyl group determind from the areas of this band in the case of styrene and both its areas and its ratio to the carbonyl band for methyl methacrylate. Second order rate-plots were used to determine the rate-constants.

In the case of copolymerisation reactions,the overlap of the vinyl groups of the two monomers was so severe that deconvolution was necessary, and this was achieved by Fourier self-deconvolution (FSD)[2]. The relative concentration of the vinyl groups as the reaction proceeded were determind from the heights of the deconvoluted bands.

RESULTS

Second order rate-constant (k) were obtained for the homopolymerisations of styrene and methyl methacrylate (MMA) and the copolymerisation of these two monomers. The results are given in Table 1.

TABLE 1

Monomer	k (mol^{-1} $dm^{3}s^{-1}$)
Styrene	1.82×10^{-5}
MMA (based on band area)	5.13×10^{-5}
MMA (based on internal standard)	5.20×10^{-5}
Styrene/MMA (based on styrene)	7.67×10^{-5}
Styrene/MMA (based on MMA)	3.73×10^{-4}

CONCLUSIONS

The results represent the first use of time-resolved Raman specroscopy to determine rate-constants for polymerisation reactions, and in particular shows the ability of the technique to study copolymerisation reactions. The value of a diode array detector to follow such reactions is clearly demonstrated. The techniques of FSD have proved to be of great value in separating overlapping vinyl bands, and are undoubtedly suited to systems such as those considered in the present work, where a very high signal: noise ratio is available.

1. Erdogan Gulari, K. McKeigue and KYS Ng: Macromolecules, 17, 1822, 1984.
2. J.K. Kauppinen, D.J. Moffat, H.H. Mantsch and D.G. Cameron: Anal. Chem. 53, 1454, (1981).

Time-Resolved Spontaneous Raman Scattering by Chrysene in S_1 and Incoherent CARS for T_2 Measurement

T. Kobayashi, S. Koshihara, and T. Hattori

Department of Physics, University of Tokyo, Hongo, Bunkyo, Tokyo 113, Japan

I. Time-Resolved Resonance Raman (TR^3) Spectra of Chrysene in the Lowest Excited Singlet and Triplet States

I.1. Introduction

Recently the vibrational Raman spectra have been measured for many short-lived species in solution, such as excited states or photochemical or photobiological intermediates. There are two measurement methods of the time-resolved Raman spectra, one is the spontaneous Raman and the other is the nonlinear coherent anti-Stokes Raman scattering (CARS).

The spectra of time-resolved spontaneous and coherent anti-Stokes Raman scattering have been measured for several aromatic molecules such as chrysene [1,2,3]. Chrysene molecule in the lowest triplet state is one of the most extensively studied aromatic molecular species with a short lifetime by both the time-resolved Raman spectroscopies. However, the spontaneous Raman spectrum of chrysene molecules in the excited singlet state has not been reported.

In the present paper, we observed for the first time the spontaneous Raman scattering spectrum of chrysene molecules in the lowest excited singlet state at room temperature, and we observed the decaying kinetics of the TR^3 signal intensity of chrysene in the lowest excited singlet state (S_1) and the growing kinetics of that in the lowest excited triplet state (T_1). The decaying time of S_1 and growing time of T_1 agreed with each other within experimental error.

I.2. Experimental

The block diagram of the instrumentation used for the measurement of TR^3 spectra is shown in Fig. 1. The lowest excited singlet state (S_1) with $^1B_{3u}$ symmetry is populated by the third harmonic (355 nm) of a Nd:YAG laser (laser 1) pulse (8 ns (FWHM), 4 mJ). Resonance Raman scattering from the chrysene molecules in the excited singlet state ($^1B_{3u}$) and triplet state (^3B_u) is generated by a pulse from a dye laser (10 ns (FWHM), 1.8 mJ) tuned to 580 nm which is pumped by the second harmonic of another Nd:YAG laser (laser 2). The electronic excitation and Raman pump lasers are focused collinearly to a beam waist of about 0.5 mm inside a quartz sample cell. Raman scattering light is collected at a right angle to both the incident electronic excitation and Raman pump laser beams. A camera lens of 35 mm focal length is used for the condenser lens to focus the Raman scattered light on the entrance slit of a 0.3 m triple monochromator equipped with a 1800 grooves/mm grating. Raman scattering is detected by a coupled system of an image intensifier (Hamamatsu V-1845) and a vidicon camera (EG&G PAR 1254 SIT 1215/1216). Raman spectrum of about 350 cm^{-1} is recorded on the 500-channel vidicon.

Reagent grade chrysene is heated in boiling maleic anhydride (150 g) with chloranil (8 g) as an oxidizing agent for about 5 hrs. Hot toluene was added onto the reacted mixture, and the solution is boiled further for 2 hrs. After removing

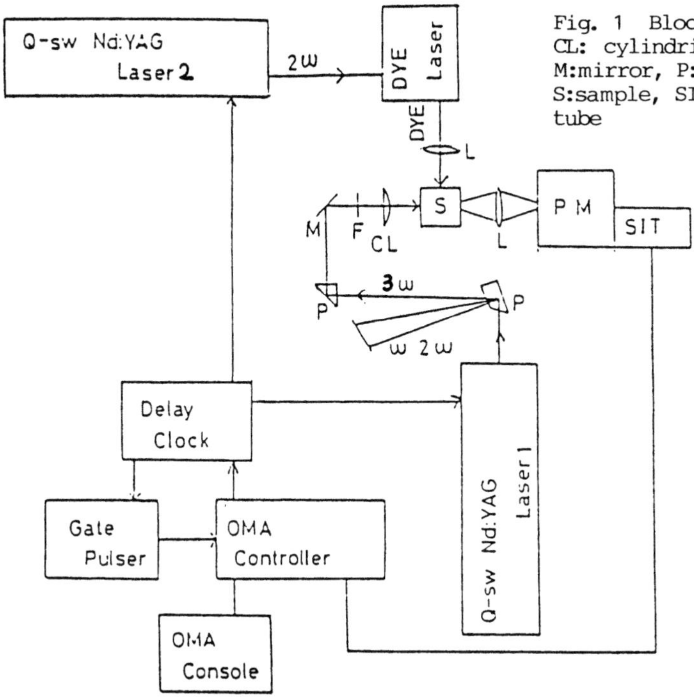

Fig. 1 Block diagram of TR^3 experiment: CL: cylindrical lens, F:filter, L:lens, M:mirror, P:prism, PM: polychromator, S:sample, SIT:silicon intensified target tube

precipitate from the parent solution by filtration, the solid chrysene is obtained by recrystallization. It is then sublimed in vacuo [4]. Fluorescence-grade tetrahydrofuran (THF)(Merck) is used without further purification.

I.3. Results and Discussion

I.3.1 Fluorescence Spectra and Lifetimes of Chrysene in Solution and in Crystalline Phase as Probes of Purity

Fluorescence spectra of aromatic molecules in solution and in crystalline phase can be used as monitors for the purity [5]. Figure 2 shows the fluorescence spectra of chrysene in THF solution. Sample 1 is obtained by the purification process described in the experimental section. Sample 2 is obtained by twice recrystallization of the reagent-grade chrysene (Wako) from benzene solution. The fluorescence spectra of the two samples are quite different from each other.

The fluorescence of sample 2 is contaminated with emission from impurities such as anthracene. The measured fluorescence lifetimes of sample 1 in crystalline phase and in solution are longer than respective values of sample 2.

I.3.2 TR^3 Spectra of Chrysene in S_1 and T_1 States.

Both coherent and spontaneous Raman scattering from the excited states of chrysene have been observed previously [1,2,3]. The coherent Raman scattering in resonance with S_n-S_1 and T_n-T_1 transitions were measured for chrysene embedded into poly-(methylmethacrylate) medium at -100°C [3].

The TR^3 spectra of chrysene between 1250 and 1600 cm^{-1} are shown in Fig. 3. The delay times of the Raman pump laser pulse at 580 nm after the electronic excitation laser pulse at 355 nm are -46, -16, -6, 4, 24, 44, 64, and 84 ns. The peaks between 1460 cm^{-1} and 1490 cm^{-1} are nonresonant Raman scattering

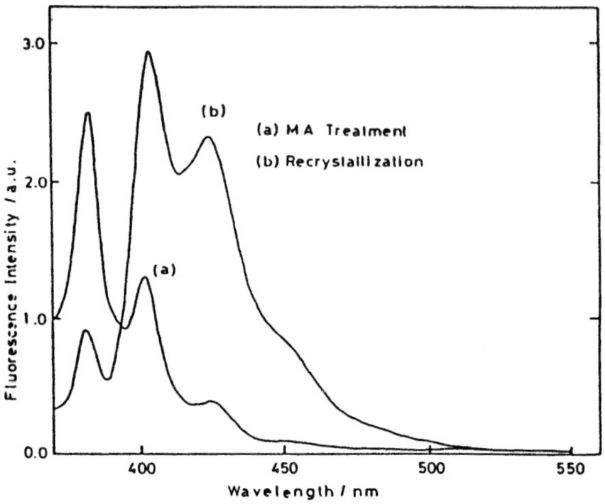

Fig. 2 Fluorescence spectra of (a) highly-pure and (b) twice-recrystallized chrysene in THF solution

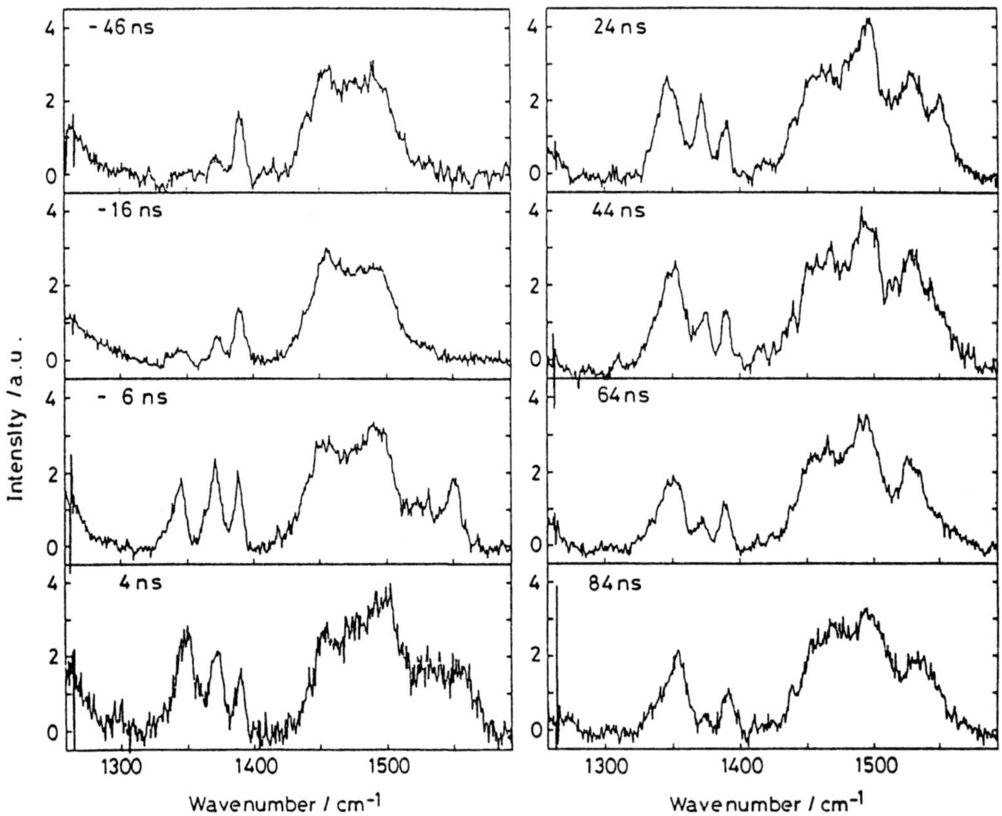

Fig. 3 TR^3 spectra of highly-pure chrysene in THF solution at different delay times

due to THF solvent. Nonresonant Raman scattering peaks from chrysene in the ground state are located at 1373 and 1388 cm^{-1}.

The intensity of the Raman spectrum at 1344 cm^{-1} decreases while that at 1353 cm^{-1} increases with the increase in delay time from 4 ns to 84 ns. The Raman peaks at 1549 cm^{-1} and 1369 cm^{-1} are clearly seen at the delay times at -6 ns and 24 ns while they disappear or become weak at 64 ns and 84 ns. The Raman peak at 1373 cm^{-1} due to chrysene in the ground state is very close to 1369 cm^{-1} scattering from S_1 state. Therefore the disappearance of 1369 cm^{-1} scattering is not as clear as the 1549 cm^{-1} peak at 64 ns and 84 ns. The peaks at 1494 cm^{-1} and 1528 cm^{-1} start to appear at the delay times of 24 and 44 ns and become clear at 64 ns after excitation. Because the accumulated numbers of data are small for delay times of 4 ns and 84 ns, the signal-to-noise ratios of these data are smaller than those of the others. It is the reason why 1549 cm^{-1} and 1528 cm^{-1} lines are not clear in 4 ns and 84 ns spectra, respectively.

From the above-mentioned results, Raman peaks at 1344, 1369 and 1549 cm^{-1} are attributed to S_1 state. The peaks at 1353, 1494, and 1528 cm^{-1} are attributed to T_1 since the frequencies are close to those previously reported TR^3 (1340, 1488, 1520 cm^{-1} Atkinson et al.) of chrysene in resonance with T_n-T_1 transition [1]. The difference in frequencies between the reported values and those measured in the present study is between +13 and +6 cm^{-1}. This may be due to inevitable error in the determination of shift, because the transient Raman peaks are very broad as is shown in Fig. 3. The above-mentioned assignment of the Raman scattering spectrum can also be supported by the intensity-dependence of the signal on the delay time as follows.

Figure 4 shows the dependence of the intensity of Raman signals at 1369 cm^{-1} on the energy of the electronic excitation pulse (355 nm). The Raman intensity is proportional to the electronic excitation laser intensity. All the other lines at 1344 (S_1), 1549 (S_1), 1353 (T_1), 1494 (T_1) and 1528 cm^{-1} (T_1) were also found to be linearly dependent on the excitation laser pulse within experimental error.

I.3.3 Delay Time-Dependence of Raman Scattering Intensities

Atkinson et al. presented in their paper quantitative analyzing method of a transient species from its time-resolved resonance Raman spectra [1]. They obtained the rise and decay times of the TR^3 signal from chrysene (h_{12}) in THF solution in resonance with T_n-T_1 transition to be 100 ns and 1.4 μs, respectively.

Figure 5 shows delay time-dependence of the TR^3 intesities at 1369 cm^{-1} (S_1) and 1528 cm^{-1} (T_1). The Raman signals are integrated over the frequency region of 1363 -- 1375 cm^{-1} for 1369 cm^{-1} shift and 1522 -- 1533 cm^{-1} for 1528 cm^{-1} shift.

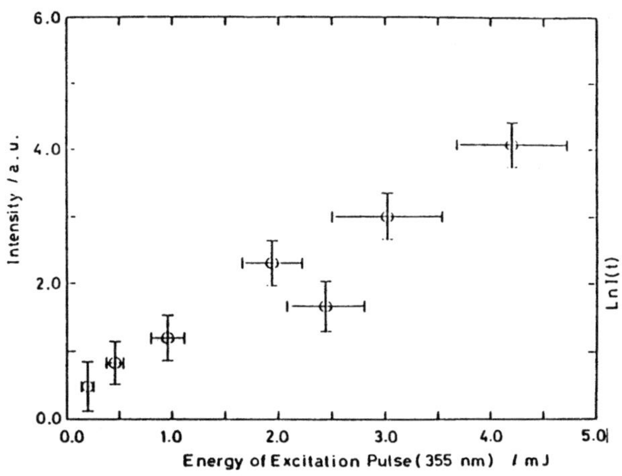

Fig. 4 Dependence of TR^3 signal at 1369 cm^{-1} due to S_1 state on the energy of the excitation pulse at 355 nm

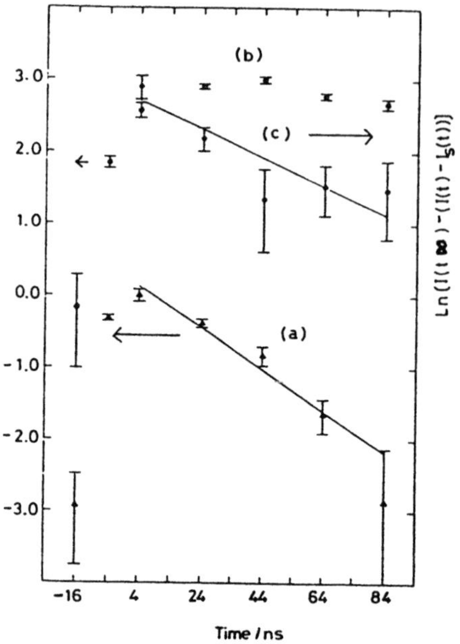

Fig. 5 Time-dependence of TR^3 signal at (a) 1369 cm^{-1} due to S_1 state and at (b) 1528 cm^{-1} due to both T_1 and S_1 states. Subtraction of S_1 component from curve (b) gives curve (c) which shows the growth of T_1.

Since the Raman shift at 1528 cm^{-1} due to T_1 state is very close to 1549 cm^{-1} due to S_1 state, the integrated intensity over the latter frequency region is due both to 1528 cm^{-1} (T_1) and 1549 cm^{-1} (S_1) shifts. Therefore we separated the signal into two, one is due to S_1 (1549 cm^{-1}) and the other to T_1 (1528 cm^{-1}). The separated signal intensity, $I_{T1}(t)$, due to T_1 at different delay times are thus obtained. The difference in the intensities $(I_{T1}(t{>>}80)-(I_{T1+S1}(t)-I_{S1}(t)))$ is plotted against the delay time. The slope of the curve gives the formation time-constant of the T_1 state. The time-constant was obtained to be 51+19−12 ns. It is consistent with the decay time-constant of the scattering intensity at 1369 cm^{-1} which was measured to be 35+6−5 ns.

II. Coherent Raman Spectroscopy with Incoherent Light

We also considered the application of incoherent laser for the measurement of T_2 ($=\gamma^{-1}$) by CARS or CSRS. We call them incoherent CARS or CSRS. Using two light sources ($E(t)$ at ω_1 and $E_3(t)$ at ω_2), one of which is incoherent and split into two beams with delay time τ ($E_1(t) = E_2(t+\tau) = E(t)$), we can obtain the intensity of CARS or CSRS at $2\omega_1 - \omega_2$, $I(\tau)$. If E_3 has finite correlation time Γ_c^{-1} and $< E_3^*(s)E_3(t) > = \exp(-\Gamma_c |s-t|)$, then CARS or CSRS intensity is given as follows as a function of delay time τ between E_1 and E_2.

$$I(\tau) \propto 1+[\exp(-\Gamma_c\tau) + \frac{2\gamma}{\Gamma_c}(1-\exp(-\Gamma_c\tau))]\exp(-2\gamma\tau). \quad (1)$$

In the limiting case of completely incoherent E_3, $I(\tau)$ becomes independent of τ , while on the other hand in the case of completely coherent E_3, $I(\tau)$ is given by

$$I(\tau) \propto 1+(1+2\gamma\tau)\exp(-2\gamma\tau). \quad (2)$$

This method has a great advantage that the resolution time is not limited by pulse widths and time jitter of lasers used,but by correlation time of the radiation field of them.

An experiment is in progress using a mode-locked Ar laser as a coherent light source and an incompletely mode-locked dye laser synchronously pumped by the Ar laser as an incoherent light source.

The authors are indebted to Prof. Iwashima for useful suggestions for the purification of chrysene. This work was partly supported by a Grant-in-Aid for Scientific Research from the Ministry of Education, Science and Culture of Japan and partly by Toray Science and Technology Foundation.

References

1. G.H. Atkinson, D.A. Gilmore, L.R. Dosser and J.B. Pallix: J. Phys. Chem. 86, 2305 (1982)

2. S.M. Beck and L.E. Brus: J. Chem. Phys. 75, 1031 (1981)

3. Ch. Jung, A. Lau, H.J. Weigmann, W. Werncke and M. Pfeiffer: Chem. Phys. 72, 327 (1982)

4. S. Iwashima, T. Kobayashi and J. Aoki: Nippon Kagaku Kaishi 4, 686 (1972)

5. T. Kobayashi, S. Iwashima, S. Nagakura and H. Inokuchi: Mol. Cryst. Liquid Cryst. 18, 117 (1972)

Part VI

Biological Systems

How Protein Structure Controls Oxygen Dissociation in Haemoglobin: Implications of Picosecond Transient Raman Studies

J.M. Friedman

AT & T Bell Laboratories, Murray Hill, NJ 07974, USA

Proteins are macromolecular machines. In cases where the reactive centers are relatively small chromophoric units embedded in a protein matrix, they bear a resemblance to more familiar condensed phase materials such as organic mixed crystals or impurity centers in a lattice. Unlike these systems, the protein matrix has a structure that has been fine tuned over the course of evolution for a specific biological function. A major goal of biological research has been to understand how energy is both stored and utilized within the structure of the protein for the purpose of biological function. Despite the remarkable success of structural chemists and biochemists in determining the equilibrium structures and the functional properties of many proteins, this problem has not been answered on a microscopic level. A major deficiency of this approach is the absence of information of how structural dynamics relates to reactivity. In this paper we examine the implications of recent picosecond Raman studies on the relationship between thermal fluctuations in protein structure and reactivity in haemoglobin.

Haemoglobin is the prototypic cooperative protein. It consists of four subunits (two α and two β subunits) each of which contains a single iron porphyrin or heme to which small ligands such as O_2 or CO can reversibly bind. Cooperativity is manifested in the ligand binding properties of the protein. In presence of low oxygen saturation, ligand binding occurs with low affinity, whereas an increase in the degree of saturation causes a large increase in the ligand binding affinity of the remaining unbound sites. The macroscopic features of this cooperative binding have been explained in terms of a saturation-dependent equilibrium between two distinct well-defined quaternary structures of the protein[1,2]. The quaternary structure is defined and determined largely from the structure of the interface between the α_1 and β_2 subunits and the parallel α_2-β_1 interface[3]. The absence and presence of substantial ligand-binding favors the low affinity T or tense quaternary state and the high affinity R or relaxed quarternary state respectively. The mechanism for how the overall quaternary structure controls the localized events at the binding site is unclear. In this work we focus on one particular contribution to the overall affinity by examining how the quarternary structure influences the local dynamics associated with ligand dissociation.

The ligand off-rate is a macroscopic parameter of ligand reactivity that is strongly influenced by the quaternary state. It is usually measured by following the kinetics associated with either the loss of a bound ligand or its replacement with a faster binding one. The observed rates are proportional to the intrinsic rate at which the iron ligand bond ruptures multiplied by the probability that the dissociated ligand does not geminately rebind. The substantial R-T difference in the off-rate can arise from structural influences on either the intrinsic dissociation step or the geminate rebinding.

When a ligand dissociates from the heme, it either diffuses out through the protein into the solvent or geminately recombines from within the protein. Studies on hemeproteins at cryogenic temperatures have revealed two geminate processes[4]. Rebinding from within the heme pocket and from the bulk protein have been termed process 1 and M respectively[4]. At ambient temperatures, ps(τ<200ps)[5,6] and ns(τ<100ns)[7] geminate processes have been reported which have been tentatively assigned to process 1 and M respectively[6]. In a recent study of process 1 using ps transient absorption techniques, it was found that the yield of process 1 responds to both quarternary and tertiary differences in protein structure[6]. In particular, the geminate yield increased with increasing values of the iron-proximal histidine stretching frequency (ν_{Fe-His}) for the transient at 10 ns after photolysis. The range of values (see Fig. 1) for ν_{Fe-His} include those from T state species at the low end and high affinity R state species at the upper end. The observed R-T difference in the geminate yield does not, however, account for the full difference in the observed off-rates. As noted by other investigators the intrinsic rate of dissociation must also contribute[7e,8].

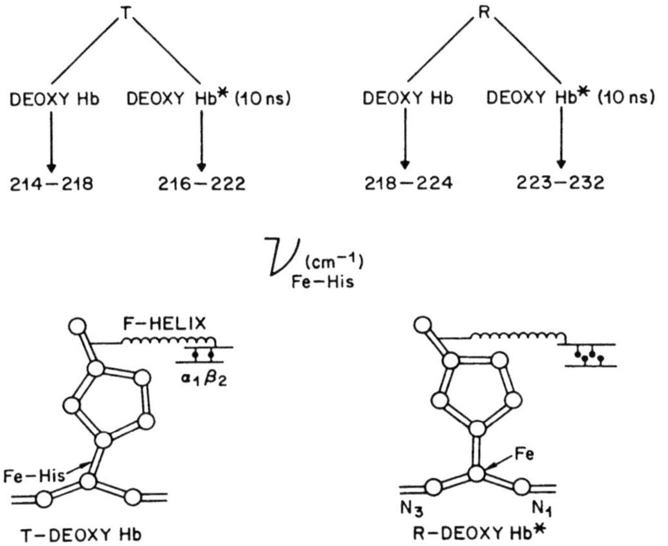

Figure 1. The systematic variations of the frequency of the iron-proximal histidine stretching mode ν_{Fe-His} as a function of protein tertiary structure. In each instance the chromophore is a deoxy heme; however for T and R state Hb* the deoxy chromophore is generated by photodissociating the corresponding R and T liganded species and probing before the R or T liganded <u>protein</u> structure has relaxed. Under these circumstances one has a deoxy heme embedded in a non-equilibrium protein matrix whose tertiary structure is still that of the parent liganded haemoglobin. The bottom of the figure shows the proposed relationship between ν_{Fe-His} and the tilt of the proximal histidine with respect to the heme.

When the coordinate of a fully liganded iron undergoes a thermally-induced fluctuation proximal to the heme plane, the iron-ligand bond weakens and may rupture giving rise to spontaneous dissociation. Differences in the rate at which this occurs can, within the context of transition state theory, be analyzed in terms of differences in the barrier for dissociation. Differences in the barrier can originate from differences in the initial in-plane configuration or the transition state, which is presumably nonplanar. A comparison of the initial states can be made directly using CW Raman and IR spectroscopy, whereas ps time-resolved Raman of the transient species occurring after photodissociation can be used to infer the nature of the difference in the transition state. There have been several spectroscopic studies of the R-T differences in liganded haemoglobins, all of which indicate that the immediate environment, both proximal and distal, about the iron is insensitive to change in quaternary structure for the six coordinate Fe^{+2} system[9]. Similarly a recent EXAFS study on R and T state carboxy carp haemoglobin also reveals minimal differences at the iron[10]. In contrast, extensive studies on the transient deoxy species occurring within 10ns of photodissociation show large R-T differences in ν_{Fe-His}[11]. More recently, using a hydrogen cell to generate 25 ps pulses at 435 nm (10HZ) from the second harmonic of an active passive mode locked Nd:YAG system, it was found that these R-T differences are already fully developed within 25 ps of dissociation[12]. This is significant since for the non-planar five coordinate iron it is the systematic variation in this frequency (see Fig. 1) that reflects the influence upon the heme of the protein structural changes induced by ligand binding[11] and alteration of the quarternary structure[13]. Furthermore, this frequency has been correlated with the contribution to the "inner" potential energy barrier controlling process 1 that responds to proximal perturbations including the R to T transition[6]. It follows that the appearance at the iron of functionally significant R-T induced structural differences subsequent to dissociation is tightly coupled to events that occur at the heme within 25 ps of dissociation. The most obvious such event is the movement of the iron out of the heme plane within a fraction of a picosecond subsequent to dissociation[14,15].

A framework for understanding the structural and functional implication of the above studies can be derived both from what Karplus and coworkers call the allosteric core[16] and from the proposed relationship between the tilt of proximal histidine with respect to the heme plane and the variation in ν_{Fe-His} as shown in Fig. 1[11]. The allosteric core is a segment of the protein that links the ligand binding site with the quaternary structure. It is composed of the heme, the proximal histidine (F8), the α_1-β_2 interface and the segment of the F helix between the proximal histidine and the interface (see Fig. 2). The stability and identity of the quaternary state is determined by the structure of the α_1-β_2 interface[17]. The allosteric core defines a protein tertiary structure about the heme that is responsive to both quaternary structure and events at the heme. There are four spectroscopically identifiable tertiary structures of the protein: deoxy T, liganded T, deoxy R and liganded R[11]. These designations refer to those protein structures within a given quaternary state that are in equilibrium or quasi-equilibrium with either a ligand-free or ligand-bound heme. Use of transient spectroscopy allows one to study any of these structures with either a ligand-bound or ligand-free heme. An induced change in tertiary structure occurs over 100's of ns to 10's of μsec[23] and involves a reorganization of structure within the protein[16]. The diagrams shown in Fig. 2 deal with the very much faster processes associated with ligand dissociation and rebinding within any of these stable or quasistable tertiary structures of the protein. The top of the figure shows the elastic-like response occurring upon dissociation within the liganded T and liganded R protein tertiary structures. The protein tertiary structure determines the "setting" for the "spring" (F helix) that couples the binding site to the α_1-β_2 interface. The frequency (ν) of the iron-proximal histidine stretching mode derived from the ligand-free heme conformer (B) of any stable or unstable protein tertiary structures is a direct indication of this "setting". The strain energies G, G' and G^+ are the result of the repulsive interaction between the heme and the proximal histidine respectively, in the ligand-bound (A), ligand-free (B) and transition state (I) conformation of a given protein structure. Because the repulsive force increases as the iron moves into the plane of the heme[16,18], these strain energies can be ordered as shown in the figure. Protein settings that favor a tilted histidine (low values of ν) have a much larger effect upon G compared to G'. The extent to which G^+ is affected depends upon the displacement of the iron in the transition state[24].

Within the above description, the geometry of the proximal histidine with respect to the heme is determined by two competing influences. A tilted configuration is favored by the pull on the proximal histidine through the F helix by the α_1-β_2 interface. The more T-like the interface the greater the pull (see Fig. 1). At the heme, repulsive forces between the imidazole carbons and the pyrrol nitrogens of the heme favor an untilted configuration. These repulsive forces increase as the iron moves into the heme plane[16,18]. The large but systematic variation in

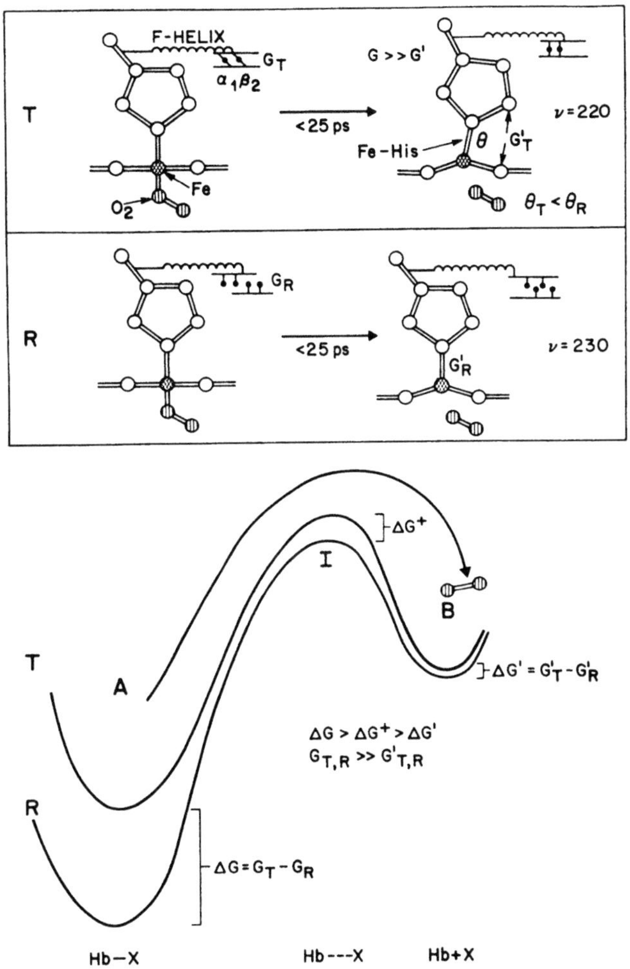

Figure 2. The dynamic relationship between ligand-binding and tertiary structure with respect to the heme-proximal histidine interaction.

ν_{Fe-His} (Fig. 1) for both equilibrium and transient forms of deoxy haemoglobin indicate that with the iron out the heme plane the influence of the interface dominates. The spread in frequencies also indicates that for the nonplanar heme there are only small energy differences due to differences in the heme histidine geometry. When the iron is moved in plane, the CW Raman, IR and X-ray studies taken together point to a situation where the now dominant increased repulsive force maintains an untilted geometry at the heme regardless of the pull from the interface. However, for the T-state it takes an additional ΔG to make the initially tilted non-planar configuration an un tilted planar structure (see Fig. 2). This extra strain energy is then stored in the weakest link in the allosteric core - presumably at the interface in most cases (vide infra). Although the heme does not show evidence of this tilt-induced extra strain in most R-T comparisons of liganded hemoglobins, it is dramatically evident when NO is the ligand. Under these conditions the NO weakened Fe-His bond becomes the weak link[19], thus the competing forces of the T-state interface and the repulsive forces from heme-histidine interaction are sufficient to rupture the Fe-His bond[20]. That this phenomenon is connected to the pull on histidine is reflected in the appearance[21] of five coordinate hemes in the NO derivative of low affinity R state haemoglobin from the green turtle. This low affinity R-state exhibits an anomolously low-frequency ν_{Fe-His} indicating a substantial pull from the interface even though it has an R structure[21]. It follows from the analysis that the energy-difference between two protein structures due to the interaction between the heme and the histidine reaches a maximum and minimum for when the iron is fully in and out of the heme plane respectively. Because the R-T -induced differences in ν_{Fe-His} fully appear within 25ps of dissociation[12], it is likely that the interface and the iron are tightly coupled through an elastic like linkage. Thus movement of the iron will modulate the magnitude of the strain energy, whereas the actual magnitude of the strain energy for a given position of the iron (transient or otherwise) will be a function of the pull from the F helix. Since formation and rupture of the ligand-iron bond is fast compared to protein structure relaxation[23], we can ascribe to any given equilibrium or nonequilibrium protein structure a reaction coordinate diagram involving the movement of the iron. A specific value for ν_{Fe-His} derived from the deoxy heme version of a given structure (see Fig. 2) dictates a specific reaction coordinate diagram.

We are now in a position to reconsider the structural basis for the differences in the rate of dissociation. As shown in the reactions coordinate diagram in Fig. 2, the bottom portion of the T -state potential well for the liganded heme (A) is displaced from that of the R by ΔG. The R and T strain energies (G_R and G_T) are shown to be residing at the interface. Thermal movement of the iron proximal to the heme plane (towards B) weakens the iron-ligand bond and also decreases the strain energies. The R-T difference in strain energy for the transition state (ΔG^+) will be less than for the in plane configuration; consequently, the barrier for dissociation will be lower for the T state by an amount $\Delta G-\Delta G^+$. For the dissociated complex (B), the strain energy (G') is still lower; consequently, $\Delta G^+-\Delta G'$ determines the difference in the barrier for the geminate rebinding. As previously pointed out by Szabo[24] different ligands have the transition state occurring at different displacements. For those ligands having a transition state resembling the in-plane complex, the geminate process will be more responsive to protein structure than the dissociation process ($\Delta G^+-\Delta G' > \Delta G-\Delta G^+$) and vice versa for fast-binding ligands ($\Delta G-\Delta G^+ > \Delta G^+-\Delta G'$).

The reaction coordinate diagrams shown in Fig. 2 are associated with specific tertiary structures. In haemoglobin but not myoglobin the tertiary structure, even within a given quaternary state, responds to ligand binding[23]. For example, upon dissociation the liganded R structure ($\nu_{Fe-His} = 230cm^{-1}$) relaxes ($\tau < 21\mu sec$) to the deoxy R structure ($\nu_{Fe-His} \approx 222cm^{-1}$). This relaxation is slow compared to the processes defined by the reaction coordinate diagrams. It is therefore possible to obtain a value of ν_{Fe-His} for the intermediate occurring within a short window $\Delta\tau$ after some well-defined delay τ subsequent dissociation. This frequency defines the reaction coordinate diagram describing the ligand binding dynamics over this specific short interval $\Delta\tau$. Since this frequency is decreasing, subsequent intervals will exhibit dynamics characterized by progressively higher barriers both for ligand rebinding and subsequent redissociation[23]. Thus we have a hierarchy of dynamical processes built upon these reaction coordinate diagrams. The hierarchy starts with thermal dissociation and geminate rebinding within a "fixed" tertiary structure, and ends with the large-scale changes in reactivity occurring as the quarternary structure changes ($\tau > 20\mu sec$).

REFERENCES

1. M. F. Perutz, Proc. R. Soc. London Ser. B208, 135 (1980).

2. R. G. Shulman, J. J. Hopfield and S. Ogawa, Q. Rev. Biophys. 8, 325 (1975).

3. J. M. Baldwin and C. Chothia, J. Mol. Biol. 129, 175 (1979).

4. N. Alberding, S. S. Chan, L. Eisenstein, H. Frauenfelder, D. Good, I. C. Gunsalus, T. M. Nordlund, M. F. Perutz, A. H. Reynolds, and L. B. Sorensen, Sorensen, Biochemistry 17, 43 (1978); D. D. Dlott, H. Frauenfelder, P Langer, H. Roder, and E. E. Dilorio, Proc. Natl. Acad Sci. US 80, 623 (1983).

5. D. A. Chernoff, R. M. Hochstrasser and A. W. Steele, Proc. Nat. Acad. Sci. USA 77, 5606 (1980); P. A. Cornelius, R. M. Hochstrasser, A. W. Steel J. Mol. Biol. 163, 119 (1983).

6. J. M. Friedman, T. W. Scott, G. J. Fisanick, S. R. Simon, E. W. Findsen, M. R. Ondrias and V. R. Macdonald, Science, in press.

7. a. D.A. Duddell, R. J. Morris, J. T. Richards, J. C. S. Chem. Comm., 75-76 (1979);b. D. A. Duddell, R. J. Morris, N. J. Muttucuman and J. T. Richards, Photochem. Photobiol., 31, 479-484 (1980);c. B. Alpert, S. El Moshni, L. Lindquist and F. Tfibel, Chem. Phys. Lett. 64, 11-16 (1979);d. J. M. Friedman and K. B. Lyons, Nature 284, 570-476 (1980);e. J. Hofrichter, J. H. Sommer, E. R. Henry and W. A. Eaton, Proc. Nat. Acad. Sci. USA 80, 2235 (1983).

8. R. J. Morris and Q. H. Gibson, J. Biol. Chem. 259, 265 (1984); E. R. Henry, J. Hofrichter, J. H. Sommer and W. A. Eaton in Proceedings of International Conference on Photochemistry and Photobiology, (1983) A. Zewail, Ed (Harwood Academic Publishers, New York 1983) pp. 791-809.

9. K. Nagai, T. Kitagawa and H. Morimoto, J. Mol. Biol. 136, 2271-2189 (1980); M. Tsubaki and N.-T. Yu, Biochemistry 21, 1140-1145 (1982); D. L. Rousseau, S. L. Tan, M. R. Ondrias and S. Ogawa, Biochemistry 23, 2857-2865 (1984); D. L. Rousseau and M. R. Ondrias, Biophys. J 47, 726-34 (1985).

10. M. Chance, L. Powers, B. Chance, L. Parkhurst, C. Kumar and Y. Chou Biophys. J. 47, 84a (1985).

11. a. J. M. Friedman, R. A. Stepnoski and R. W. Noble, FEBS Letters 146, 278 (1982);b. J. M. Friedman, "Time Resolved Vibrational Spectroscopy," (G. Atkinson, ed.) Academic Press, 307 (1983);d. T. W. Scott, J. M. Friedman, M. Ikeda-Saito and T. Yonetani, FEBS Letters 158, 68;e. J. M. Friedman, T. W. Scott, R. A. Stepnoski, M. Ikeda-Saito and T. Yonetani, J. Biol. Chem. 258, 10564 (1983); e. J. M. Friedman, Science 228, 1273 (1985).

12. E. W. Findsen, J. M. Friedman, M. R. Onduas, ed S. R. Simon, Science in press.

13. K. Nagai, T. Kitagawa and H. Morimoto J. Mol. Biol. 136, 271 (1980), M. R. Ondrias, D. L. Rousseau, J. A. Shelnutt, and S. R. Simon Biochemistry 21, 3428 (1982).

14. J. L. Martin, A. Migus, C. Poyart, Y. Lecarpentier, R. Astier and A. Antonetti, Proc. Nat. Acad. Sci. USA 80, 173 (1983).

15. E. R. Henry and W. A. Eaton, Biophysical J. 47 208a (1985); E. R. Henry, M. Levitt, An W. A. Eaton, Proc. Nat. Acad. Sci. USA 82, 2034 (1985).

16. B. R. Gelin and M. Karplus, Proc. Nat. Acad. Sci. USA 74, 801-805 (1977); B. R. Gelin, A. W-M. Lee and M. Karplus, J. Mol. Biol. 171, 489 (1983).

17. A. H. Chu and G. K. Ackers, J. Biol. Chem. 256, 1199 (1981).

18. B. D. Olafson and W. A. Goddard, Proc. Natl. Acad. Sci. USA 74, 1315 (1979); W. A. Goddard and B. D. Olafson, in "Biochemical and Chemical Aspects of Oxygen," (Caughy, W., ed.) Academic Press, New York, 87-123 (1979); A. Warshel, Proc Natl. Acad. Sci. USA 74, 1789 (1977).

19. W. R. Scheidt, A. C. Briniger, E. B. Ferro and J. F. Kirner, J. Amer. Chem. Soc. 99, 7315 (1972).

20. A. Szabo, M. F. Perutz Biochemistry 15, 4427 (1976); M. F. Perutz, J. D. Kilmartin, K. Nagai, A. Szabo and S. R. Simon, Biochemistry 15 378 (1976); K. Nagai, G. Wilson, D. Dolphin, T. Kitagawa, Biochemistry 19, 4755 (1980).

21. J. M. Friedman and M. R. Ondrias, unpublished results.

22. J. M. Friedman, S. R. Simon and T. W. Scott, Copeia in press.

23. T. W. Scott and J. M. Friedman, J. Amer. Chem. Soc. 106, 5677 (1984); E. R. Henry, J. Hofrichter, J. H. Sommer and W. A. Eaton In "Hemoglobin"; Schnek, A. G., Paul, C. Eds. Brussels University Press, Brussels, Belgium pp. 193-203 (1984).

24. A. Szabo, Proc. Nat. Acad. Sci USA 75, 2108 (1978).

Resonance Raman Spectra of Reaction Intermediates of Heme Enzymes

T. Kitagawa, S. Hashimoto, and T. Ogura

Institute for Molecular Science, Okazaki National Research Institutes, Myodaiji, Okazaki, 444, Japan

1. INTRODUCTION

Iron porphyrins of heme enzymes including peroxidases, oxidases, monooxidases, and dioxidases are endowed with a specific reactivity toward oxygen, but when the porphyrins are taken out from the protein, the reaction becomes nonspecific. Therefore, the specificity of catalytic activities of the heme iron in the enzyme must be determined by the surrounding amino acid residues of the heme vicinity. To characterize the iron-coordination environments of various heme enzymes, we have investigated their resonance Raman spectra (RRS) [1], and currently reached the stage for discussing the heme structure of reaction intermediates on the basis of transient RRS. Here we report the RRS of Compound II of horseradish peroxidase (HRP) and an oxygenated intermediate of cytochrome c oxidase (CcO).

2. COMPOUND II OF HRP

HRP is a protoporphyrin IX-containing enzyme which catalyzes the oxidation of various organic compounds with hydrogen peroxide as a specific oxidant. The catalytic cycle involves the oxidation of the native ferric enzyme by H_2O_2 and successive two equivalent reduction by substrates via two intermediates called Compound I and Compound II. Compounds I and II have the oxidation states higher than the Fe^{III} state by two and one oxidative equivalents, respectively. Mössbauer study on Compound II revealed the presence of the Fe^{IV} heme. There have been arguments about the sixth ligand of the heme iron; the kinetic study suggested the Fe^{IV}-OH structure for explaining the pH and mass (H/D) dependences of the rate-constant which required involvement of a deprotonating group with $pK_a = 8.6$ in the acid/base catalysis of Compound II, while ENDOR and NMR studies suggested the Fe^{IV}=O structure. Our RRS of Compound II bring conclusive evidence for the Fe^{IV}=O structure and also for a new feature of the oxygen exchange between the Fe^{IV}=O heme and bulk water, which is catalyzed by the proton of the $pK_a = 8.6$ residue.

Raman scattering was excited at 406.7 nm with a Kr^+ ion laser (Spectra Physics, model 165) and detected with an OMA II system (PAR 1420/1215) attached to Spex 1404 double monochromator. The frequency calibration was performed with indene as a standard. Since Compound II was found to be photolabile at this excitation wavelength, all RRS were measured with a spinning cell (1800 rpm, dia. 2 cm) and with a minimal laser power (<10 mW). The measurement started within 30 sec after the formation of Compound II and finished in 3 min. Compound II was prepared by mixing equimolar amounts of hydrogen peroxide, ferric enzyme (Isozyme C, Toyobo, I-C) and ferrocyanide. $H_2{}^{18}O_2$ was given by Dr. Y. Tatsuno of Osaka University.

The RRS of Compound II at pH 11.2 are shown in Fig. 1, where the RRS of Compound II derived from $H_2{}^{18}O_2$ for normal (^{56}Fe) HRP (bottom)

and $H_2{}^{16}O_2$ for the ^{54}Fe-incorporated HRP (top) are compared with that of ordinary Compound II (middle). The Raman line of the middle spectrum at 787 cm^{-1} is shifted downward to 753 cm^{-1} with $H_2{}^{18}O_2$ (bottom) but upward to 790 cm^{-1} with the ^{54}Fe heme, while other Raman lines remain unshifted. These frequency shifts agree closely with those expected for the Fe=O two-body harmonic oscillator (+3.3 and -35 cm^{-1} shifts for ^{54}Fe and ^{18}O substitutions, respectively). Consequently, the Raman line of Compound II at 787 cm^{-1} is assigned to the Fe^{IV}=O stretching mode ($\nu_{Fe=O}$), in agreement with the recent report by Terner et al [3]. This line exhibited no frequency shift in D_2O, indicating the absence of an exchangeable proton attached to the heme-bound oxygen.

Figure 2 shows the RRS of native HRP and of Compound II at pH 7 derived from four combinations of $H_2{}^{18}O_2$ or $H_2{}^{16}O_2$, and $H_2{}^{18}O$ or $H_2{}^{16}O$. When Compound II was formed with $H_2{}^{16}O_2$ in $H_2{}^{16}O$, a new line appeared at 774 cm^{-1} (b), and this line did not exhibit a frequency shift upon replacement of $H_2{}^{16}O_2$ with $H_2{}^{18}O_2$ (c) contrary to the results shown in Fig. 1(bottom). However, when $H_2{}^{16}O$ was replaced with $H_2{}^{18}O$ without using $H_2{}^{18}O_2$, the intensity of the 774 cm^{-1} line decreased significantly and instead a new line appeared at 740 cm^{-1} (d). When $H_2{}^{18}O_2$ was used to derive Compound II in $H_2{}^{18}O$, the 774 cm^{-1} line remained. Therefore, the ^{18}O isotope-sensitive line at 774 cm^{-1}

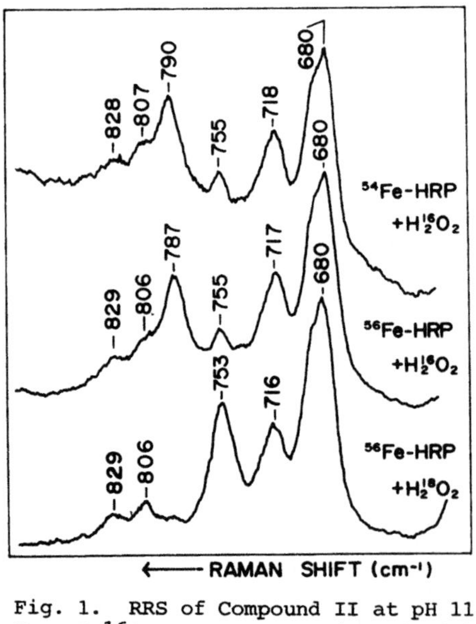

Fig. 1. RRS of Compound II at pH 11.2. Top: $H_2{}^{16}O_2$ + HRP reconstituted with ^{54}Fe-incorporated heme. Middle: $H_2{}^{16}O_2$ + native HRP(^{56}Fe). Bottom: $H_2{}^{18}O_2$ + native HRP(^{56}Fe). laser, 406.7 nm, 10 mW; accumulation time, 3 min.

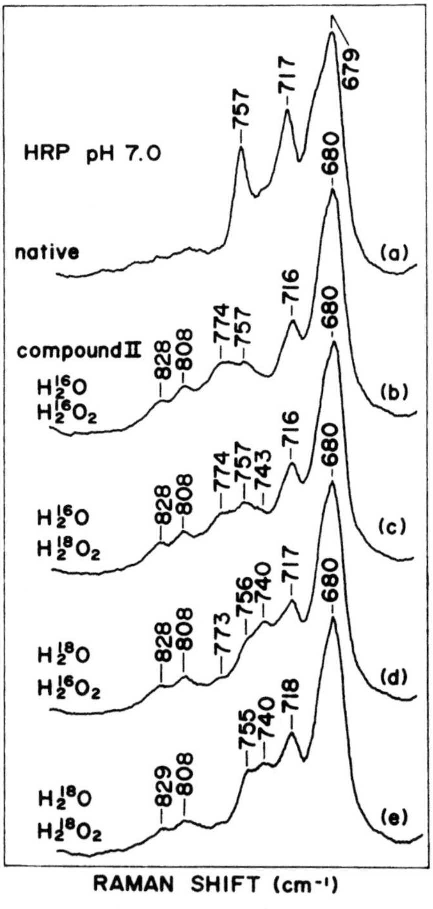

Fig. 2. RRS of native HRP and Compound II at pH 7.0. (a) native, (b) Compound II derived from $H_2{}^{16}O_2$ in $H_2{}^{16}O$. (c) from $H_2{}^{18}O_2$ in $H_2{}^{16}O$, (d) from $H_2{}^{16}O_2$ in $H_2{}^{18}O$, (e) from $H_2{}^{18}O_2$ in $H_2{}^{18}O$. laser, 406.7 nm, 10 mW, accumulation time, 3 min.

is assigned to the $Fe^{IV}=O$ stretching mode of Compound II at pH 7. The fact that this line exhibits the isotopic frequency-shift for the water isotope suggests that the heme-bound oxygen is exchanged with an oxygen atom of bulk water. Such an oxygen exchange has never been imagined for the HRP catalysis until this study. The 774 cm^{-1} line was shifted to 776 cm^{-1} in D_2O at pD 7.0 and appreciably intensified. This is contradictory to the Fe-OH structure, which should give rise to a shift to lower frequency upon deuterium substitution. Consequently, we conclude that there is no exchangeable proton attached to the heme-bound oxygen even at neutral pH.

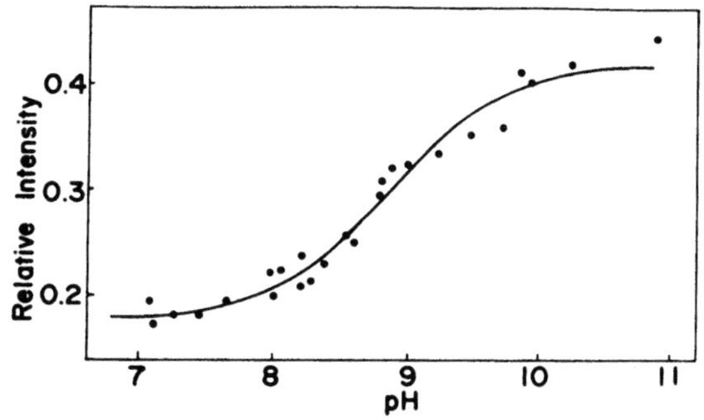

Fig. 3. The pH dependent intensity change of the $Fe^{IV}=O$ stretching RR line of Compound II in H_2O. The maximum intensities at 787 cm^{-1} relative to those at 680 cm^{-1} are plotted against pH. The solid line is a theoretical curve expected for dissociation of one proton with pK_a = 8.8.

The $\nu_{Fe=O}$ mode is somewhat shifted from that at neutral pH. Since the intensity of the Raman line at 680 cm^{-1} remains unaltered upon the pH change, the relative intensity of the 786 cm^{-1} line to the 680 cm^{-1} line was plotted against pH in Fig. 3. This plot can be reproduced well by a theoretical curve (solid line) expected for ionization of a residue with pK_a = 8.8, which is in close agreement with the reported pK_a value (= 8.6) of the heme-linked ionization of Compound II [4].

The $\nu_{Fe=O}$ frequency is considerably higher than the Fe^{III}-OH stretching mode at 495 cm^{-1} [5] and $Fe^{II}-O_2$ stretching mode at 568 cm^{-1} [6]. Simple diatomic approximation gives rise to the stretching force constants; K(Fe-OH)=1.8, K($Fe-O_2$)=3.8, and K(Fe=O)=4.5 mdyn/A. Therefore, the $Fe^{IV}=O$ bond of HRP Compound II has somewhat double bond character but is weaker than the $Fe^{IV}=O$ bond of a model compound, which gave the $\nu_{Fe=O}$ mode at 852 cm^{-1} at 15 K [7].

The lower frequency of the $\nu_{Fe=O}$ mode at neutral pH than at alkaline pH may suggest the presence of a weak hydrogen bond between the protonated pK_a=8.8 residue and the heme-bound oxygen. This hydrogen bonded proton seems to play an essential role in the oxygen exchange reaction, because the reaction does not occur at alkaline pH. In the actual catalysis of HRP, this exchange reaction continues until an electron is donated from a substrate. Thus the structural implication of the pH and H/D dependences of the rate-constant can be reinterpreted consistently with the $Fe^{IV}=O$ structure.

3. OXYGANATED INTERMEDIATES OF CcO

CcO catalyzes reduction of molecular oxygen to water in an energy-transducing membrane. Studies on the catalysis of CcO are currently focused on identification of the oxygenated intermediates. Recently Babcock et al. [8] applied the time-resolved Raman technique to a transient species of CcO, reporting evidence for the existence of a photolabile $Fe-O_2$ type intermediate. Since the high peak power (~100 kW) of the laser pulse used might induce undesirable chemical reactions, such as photodissociation of the ligand, we adopted different approach to the same subject, that is, a combination of a CW laser and a mixed flow apparatus. This allows us to detect an intermediate with its life time of 10-100 μs when the excitation wavelength is close to the absorption maximum of the intermediate.

The mixed flow apparatus consists of two glass syringes (10 mL each), a jet flow mixer, and a rectangular cell (height = 3 mm, thickness = 0.5 mm). The volume from the mixer to the cell is 160 μL. The flow rate of the fluid outgoing from the mixer can be changed between 2 and 20 mL/min, and accordingly the period during which the flowing molecules stay in the laser beam (0.1 mmφ) is changed between 4.5 and 0.45 ms. Since the carbonmonoxy enzyme and oxygen-saturated buffer are mixed and photodissociation of the carbonmonoxy ligand initiates the enzymatic reaction in this experiment, the spectrum should arise from intermediates which are produced while the photodissociated molecules stay in the laser beam.

Raman scattering was excited with a dye laser (Coherent CR-599, stilbene 420) pumped by an Ar^+ ion laser (Coherent Innova20). The scattered light was collected with a lens (f = 80 mm, dia = 40 mm) into a single monochromator (Jasco CT-50C) and detected with an OMA II system (PAR 1215/1420). Frequencies of Raman lines were determined with indene for the photoreduced enzyme in the stationary cell first,and this was used as a secondary standard for the flow cell. CcO was isolated from beef heart and finally dissolved in a 50 mM phosphate buffer, pH 7.4.

The RRS of the resting and fully reduced enzymes excited at 416 nm were essentially the same as the spectra excited at 406.7 nm. The ν_4 line of the aerobically photoreduced enzyme, which has a mixed valence state designated as $\underline{a}^{2+}\underline{a}_3^{3+}$ [9], appeared as a symmetric band at 1371 cm^{-1}, at the frequency of the ferric heme. Therefore, the spectral contribution from the $\underline{a}^{2+}$ heme is insignificant at this excitation wavelength due to the resonance effect, while it becomes increasingly significant upon excitation at longer wavelengths.

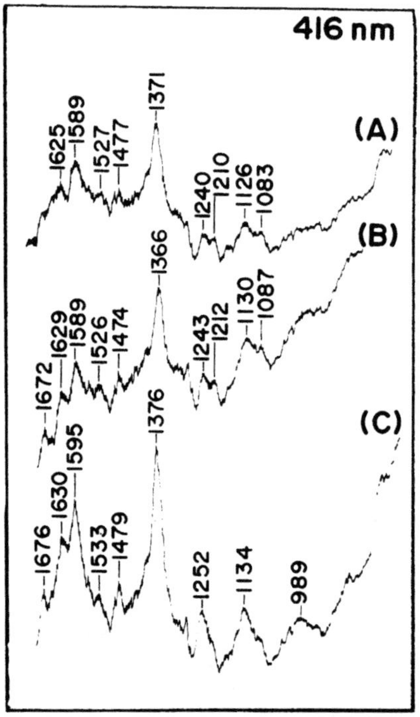

Fig. 4. RRS of CO-bound CcO (A) and partially photodissociated CO-bound form in the absence (B) and presence (C) of oxygen. The flow rate outgoing from the mixer is 4, 10, and 20 mL/min for (A), (B), and (C), respectively. sample concentrations, 144, 144, and 72 μM for (A), (B), and (C), respectively. laser power, 5, 60, and 60 mW for (A), (B) and (C), respectively. accumulation time, 250, 56, and 39 sec for (A), (B), and (C), respectively.

Figure 4 shows the RRS of CO-bound (A), partially photodissociated species in the absence (B) and presence (C) of oxygen. When CO-bound form was measured in the same apparatus but without flowing the sample, the RRS was identical with that of the fully reduced enzyme, indicating occurrence of photodissociation. Upon raising the laser power from 5 to 60 mW under the flowing condition, the Raman line of spectrum (A) at 1371 cm^{-1} was shifted to lower frequency (1366 cm^{-1}) as shown in spectrum (B), which implied an increase of the contribution from the photodissociated species to the unresolved peak. In contrast, upon photodissociation of CO in the presence of O_2, the ν_4 line was shifted to 1376 cm^{-1} as shown by spectrum (C), which is distinct from the CO-bound enzyme. The spectral features at 1676, 1630, 1595, 1252 1134 and 989 cm^{-1} are also distinguishable from those of the CO-bound form.

It is noticed that the S/N ratio of spectrum (C) is much better than those of spectra (A) and (B) despite the shorter accumulation time and lower concentration of the enzyme. This suggests the increased resonance effect and thus that spectrum (C) arises from an intermediate which has the Soret maximum closer to 416 nm than the CO-bound form has. The Soret maximum of the oxygenated intermediate of CcO is likely to be located at slightly shorter wavelength than that of the CO-bound form [10]. When the same intermediate was excited at 425 nm, the ν_4 line was identified at 1355 cm^{-1} similar to the enzyme with the mixed valence state, indicating that cytochrome a is retained in the ferrous state. Since the half-time (~1 ms) of the reaction of the reduced enzyme with oxygen [11] is much longer than the retention time of the photodissociated molecules in the laser beam (0.45 ms), spectrum (C) should arise from an intermediate formed in a single turnover of the enzyme. The intense RR line of the "pulsed" form at 1574 cm^{-1} [12] is hardly recognized in spectrum (C). Furthermore, oxygenated Hb gives the ν_4 line at slightly higher frequency (1380 cm^{-1}) than COHb (1372 cm^{-1}) [13]. These considerations lead us to propose that the most likely candidate responsible for the present intermediate is a transient oxygenated species having the $Fe^{II}-O_2$ or $Fe^{IV}=O$ heme in cytochrome a_3 and the Fe^{II} heme in cytochrome a.

REFERENCES

1. T. Kitagawa & T. Teraoka, "The Biological Chemistry of Iron" (H. B. Dunford, D. Dolphin, K. Raymond, & L. Sieker, eds.) pp. 375-389, D. Reidel, Holland (1982).
2. S. Hashimoto, T. Tatsuno, & T. Kitagawa, Proc. Japan Acad. 60, Ser. B, 348 (1984).
3. J. Terner, A.J. Sitter, & C.M. Reczek, Biochim. Biophys. Acta, 828, 73 (1985).
4. I. Yamazaki, M. Tamura, & R. Nakajima, Mol. Cell. Biochem. 40, 143 (1981).
5. S.A. Asher & T.M. Schuster, Biochemistry 18, 5377 (1979).
6. H. Brunner, Naturwissenschaften 61, 129 (1974).
7. K. Bajdor & K. Nakamoto, J. Am. Chem. Soc. 106, 3045 (1984).
8. G.T. Babcock, J.M. Jean, L.N. Johnston, G. Palmer, & W.H. Woodruff, J. Am. Chem. Soc. 106, 8305 (1984).
9. T. Ogura, S. Yoshikawa, & T. Kitagawa, Biochemistry, in press.
10. G.T. Babcock & C.K. Chang, FEBS Lett. 97, 358 (1979).
11. B.C. Hill & C. Greenwood, Biochem. J. 218, 913 (1984).
12. R.A. Copeland, A. Naqui, B. Chance, & T.G. Spiro, FEBS Lett. 182, 375 (1985).
13. L. Rimai, I. Salmeen, & D.H. Peterings, Biochemistry 14, 378 (1975).

Heme-Linked Ionization of Horseradish Peroxidase Compound II Monitored by Changes in the Fe(IV)=0 Resonance Raman Frequency

J. Terner, A.J. Sitter, and C.M. Reczek

Department of Chemistry, Virginia Commonwealth University, Richmond, VA 23284, USA

Time-resolved resonance Raman spectroscopy has been demonstrated to be a powerful method for elucidation of chemical mechanisms of a number of photo-initiated reactions of biological importance [1-3]. Recently we have turned our attention to the study of activated enzymatic intermediates using mixing techniques.

Peroxidases are members of a class of heme proteins which undergo oxidations of the heme to states above iron(III) in intermediate enzymatic states [4]. Upon reaction with peroxides, the brown resting enzyme, which contains an Fe(III) heme, is converted to a green intermediate known as compound I, which is two oxidation equivalents above the resting enzyme. A one-electron reduction of compound I produces a red intermediate known as compound II. Compound I is believed to contain an Fe(IV) porphyrin π-radical cation [5]. Compound II is an Fe(IV) heme [6].

The configuration of the peroxide-derived oxygen in compound II has been described by several different proposals including Fe(IV)=O, Fe(IV)-OH, or Fe(IV)-OOH, with the predominant formalism being Fe(IV)=O by analogy to the oxo-groups which commonly stabilize the higher oxidation states of the transition metals. Identification by this laboratory, of high iron-oxygen stretching frequencies based on isotopically-induced frequency shifts, for horseradish peroxidase compound II [7], and the peroxidase model, ferryl myoglobin [8], are indicative of an iron-oxygen double bond, Fe(IV)=O.

Though heme proteins such as horseradish peroxidase and myoglobin have identical protoporphyrin IX heme groups at the active site, the physiological functions of these proteins are quite different. Much effort has been focused towards elucidating the amino acid structure of the heme pocket, in order to account for the differing reactivities of these and other heme proteins. We have now observed [9] that Fe(IV)=O stretching frequencies for horseradish peroxidase compound II will switch between two values depending on pH. Similar pH-dependent shifts of the Fe(IV)=O frequency of ferryl myoglobin were not detected in the physiological pH range. These findings have important implications regarding the structure and dynamics of the distinctive heme pockets of myoglobin and horseradish peroxidase.

1. Results

The Fe(IV)=O vibrational band of horseradish peroxidase compound II can exist as either of two frequencies depending on pH. At acidic and neutral pH, the Fe(IV)=O stretching frequency assumes the lower of the two frequencies, 775 cm^{-1}, confirmed by an ^{18}O-induced shift to 743 cm^{-1} as we have previously reported [7]. At alkaline pH the Fe(IV)=O stretching frequency assumes the higher of the two values, 788 cm^{-1}, which is confirmed by an ^{18}O-induced frequency shift to 756 cm^{-1}. The pK of this transition is approximately 8.5 for isoenzymes B and C in agreement with the previously reported pK of the heme-linked ionization of compound II of isoenzyme C [11,12]. For isoenzymes A-1 and A-2, the pK of the transition is approximately 7, in agreement with the reported pK of 6.9 for compound II of isoenzyme A-2 [12].

When the measurements were performed in D_2O, an additional effect was observed. At pD values above the pK no significant deuterium sensitivity (that we can detect) was observable for either set of isoenzymes. At pD values below the pK the Fe(IV)=O vibration was seen to shift from 775 to 779 cm^{-1} for isoenzymes B and C. At pH values below the pK the Fe(IV)=O oxygen atom appears to be hydrogen bonded to an exchangeable proton of an amino acid residue, as it is sensitive to substitution of the amino acid hydrogen by a deuteron. At pH values above the pK, the amino acid is unprotonated, and the Fe(IV)=O stretching frequency is apparently insensitive to deuterium substitution.

Compound X is a horseradish peroxidase intermediate formed upon reaction with chlorite ion. It is believed to be similar but not identical to compound II, containing an Fe(IV)-OCl heme [13]. We have reported a polarized band at 787 cm^{-1} in compound X made from isoenzymes B and C. We suggested that this band was a possible compound X Fe(IV)=O stretching vibration [7]. The 787 cm^{-1} frequency is seen for the Fe(IV)=O group of compound II of isoenzymes B and C at alkaline pH [9]. Our recent isotopic labelling experiments, using ^{18}O-labelled chlorite ion, confirm that compound X contains an Fe(IV)=O rather than an Fe(IV)-OCl [14]. We find that resonance Raman spectra of compound X are remarkably similar to the modified compound II that is formed at alkaline pH. Since compound X is commonly produced by reacting isoenzymes B and C with sodium chlorite at alkaline pH, it is reasonable that the essentially identical oxyferryl structure would be formed for compound X and the alkaline form of compound II.

2. Discussion

The existence of an ionizable group in the proximity of the heme group of horseradish peroxidase has been known for many years. Several possibilities for this group have been proposed. The unique activity of peroxidase compared to other heme proteins has been suggested to be due in part to the proximal histidyl imidazole group,which has been believed to be deprotonated [15] in resting horseradish peroxidase. This view has been supported by comparison of the optical spectrum of reduced horseradish peroxidase with bound CO, to those of model compounds with imidazolate anions [16], as well as the high Fe-imidazole resonance Raman stretching frequency of reduced horseradish peroxidase relative to those of deoxyhaemoglobin and deoxymyoglobin [17]. NMR evidence however, argues that the proximal imidazole of horseradish peroxidase is not deprotonated [18]. It has been stated [19] that the sensitivity of the resonance Raman iron-imidazole frequency to changes in pH does not, in fact, imply that the heme-linked ionization involves a deprotonation of the histidyl imidazole, since an upward frequency-shift is seen upon ionization of a model compound as opposed to the downward frequency-shift observed in the reduced enzyme at alkaline pH. Alternatively, it has been postulated [20] that resonance Raman spectra of horseradish peroxidase may be indicative of a deprotonated axial histidyl imidazole.

Several proposals have been made regarding heme-linked ionizations on the distal side of the heme, where the Fe(IV)=O group is located. The importance of a deionized carboxyl group of a distal aspartate has been demonstrated towards the formation of compound I [21]. While the aspartate evidently plays an important role, the pK of the aspartate group is likely in the acidic, rather than neutral, pH range. A distal histidine [12] is probably of more relevance to the present discussion. The distal ionizing groups have been suggested to be histidines in both isoenzyme C and the A isoenzymes [12]. Though the pK values of compound II for isoenzymes A-1 and C are 6.9 and 8.5 [22], the variation in pK of histidine is well known to be dependent on differences in neighboring charged groups [23].

The ionizations of the distal histidines appear to be responsible for the observed frequency-shifts of the Fe(IV)=O groups of both isoenzymes. The low energy value of the Fe(IV)=O frequency, at 779 cm^{-1} for isoenzymes A-1 and A-2, and 775 cm^{-1} for isoenzymes B and C, is likely to be due to hydrogen bonding of the Fe(IV)=O group to a protonated distal group such as histidine. Above the pK of the ionizing group, the Fe(IV)=O frequency is high, at 789 cm^{-1}, presumably due to the deionization of the distal group and lack of hydrogen bonding. The transition of the resonance Raman Fe(IV)=O band from low-frequency to high is a titration from one form to the other. One frequency appears while the other disappears, with both bands equally apparent when the pH is approximately equal to the pK. This can be interpreted as resulting from a protonation and deprotonation of an amino acid residue. Hydrogen bonding to the protonated amino acid residue appears to be coincident to higher oxidative activity of the Fe(IV) heme at pH values below the pK of the amino acid residue.

Kinetic rate-constants for the reactions of compound II increase with decreasing pH [22]. There is evidence [24] that the pH variation of the rate-constant is caused by a titration of a catalytically important acid group rather than by varying the charge of the protein. Kinetic rate-constants for the reactions of compound II have been shown to be influenced by D_2O [25-27]. The resonance Raman data of this report show a sensitivity of the resonance Raman frequencies to deuteration below the pK of the compound II heme-linked ionization but not above, suggesting the involvement of hydrogen bonding of the oxyferryl group to a distal amino acid group.

Myoglobin in the Fe(III) heme state (metmyoglobin) can be made to react with hydrogen peroxide to form a compound known as ferryl myoglobin which contains an Fe(IV) heme. This compound is similar in structure to horseradish peroxidase compound II, however its peroxidative activity is much lower [28]. Resonance Raman data shows that the heme structures of horseradish peroxidase compound II [7,29] and ferryl myoglobin [8] are quite similar, though with a smaller porphyrin core size in ferryl myoglobin. One of the main dissimilarities in the resonance Raman spectra of horseradish peroxidase compound II and ferryl myoglobin are the Fe(IV)=O stretching frequencies. It has been suggested [12] that the interaction of a distal base with the sixth ligand is weak in myoglobin, but very strong in peroxidases. We observe no shifting of the Fe(IV)=O stretching frequency of ferryl myoglobin as pH is varied from pH = 6 to pH = 12, contrary to our observations for horseradish peroxidase compound II. Additionally, the ferryl myoglobin Fe(IV)=O resonance Raman frequency is high (797 cm^{-1}), showing no detectable sensitivity to deuterium substitution, suggesting a lack of hydrogen bonding to the oxo-group by a distal amino acid group. The alkaline value (788 cm^{-1}) of the Fe(IV)=O stretching frequency of horseradish peroxidase compound II, approaches the ferryl myoglobin Fe(IV)=O frequency. Compound II reactivity tends to a minimum at alkaline pH.

3. Summary

The Fe(IV)=O stretching frequency of horseradish peroxidase can assume one of two possible values in the pH region from pH = 5 to 11. Similar pH dependent frequency-shifts were not observed for ferryl myoglobin, above pH = 6. Based on the observation of deuteration effects, the shifts were interpreted to be due to hydrogen bonding of the oxygen of the compound II Fe(IV)=O group to a distal amino acid group which is likely to be a histidine. An alkaline deprotonation of the distal amino acid group apparently disrupts hydrogen bonding, resulting in the raised Fe(IV)=O stretching frequency, and is proposed to account for the lowering of compound II reactivity at alkaline pH. The transition points of the frequency-shifts corresponded to increases in the kinetic rate-constants measured for horseradish peroxidase compound II that are observed below the pK of the transition. Thus, hydrogen bonding to the oxygen of the oxyferryl groups in peroxidase intermediates appears to play an essential role in peroxidase activity. The operative mechanism for peroxidases may possess features that are similar to the hydrogen abstraction mechanism that has been proposed for cytochrome P-450 [30].

Acknowledgements

Financial support is acknowledged from the Research Corporation, the donors of the Petroleum Research Fund as administered by the American Chemical Society, and the Jeffress Memorial Trust.

References

1. J. Terner, A. Campion, and M.A. El-Sayed, (1977) Proc. Nat. Acad. Sci. USA 74, 5212-5216.
2. T.G. Spiro, and J. Terner, in Time-Resolved Vibrational Spectroscopy, Atkinson, G.H., ed. (Academic Press, New York, 1983) pp. 297-306.
3. J. Terner, T.G. Spiro, D.F. Voss, C. Paddock and R.B. Miles, in Picosecond Phenomena III, Springer Ser. in Chem. Phys. Vol. 23, K.B. Eisenthal, R.M. Hochstrasser, W. Kaiser, and A. Laubereau, eds. (Springer-Verlag, Berlin 1982) pp. 327-330.
4. B. Chance, (1949) Arch. Biochem. Biophys. 22, 224-252.
5. D. Dolphin, A. Forman, D.C. Borg, and J. Fajer, (1971) Proc Nat. Acad. Sci. USA 68, 614-618.
6. P. George, (1953) Biochem. J. 54, 267-276.
7. J. Terner, A.J. Sitter, and C.M. Reczek, (1985) Biochim. Biophys. Acta 828, 73-80.
8. A.J. Sitter, C.M. Reczek, and J. Terner, (1985) Biochim. Biophys. Acta 828, 229-235.
9. A.J. Sitter, C.M. Reczek, and J. Terner (1985) J. Biol. Chem. in press
10. J.E. Critchlow, and H.B. Dunford, (1972) J. Biol. Chem. 247, 3703-3713.
11. H.B. Dunford, and M.L. Cotton, (1975) J. Biol. Chem. 250, 2920-2932.
12. I. Yamazaki, T. Araiso, Y. Hayashi, H. Yamada, and R. Makino, (1978) Adv. Biophys. 11, 249-281.
13. S. Shahangian, and L.P. Hager, (1982) J. Biol. Chem. 257, 11529-11533.
14. A.J. Sitter, C.M. Reczek, and J. Terner (1985) manuscript in preparation
15. M. Morrison, and G.R. Schonbaum, (1976) Ann. Rev. Biochem. 45, 86-127.
16. T. Mincey, and T.G. Traylor, (1979) J. Amer. Chem. Soc. 101, 3396-3398.
17. J. Teroaka, and T. Kitagawa, (1980) Biochem. Biophys. Res. Commun. 93, 694-700.
18. G.N. La Mar, V.P. Chacko, and J.S. de Ropp, in The Biological Chemistry of Iron, Dunford, H.B. et al. eds., (D. Reidel Publ. Co., 1982) pp. 357-373.
19. J. Teroaka, and T. Kitagawa, (1981) J. Biol. Chem. 256, 3969-3977.
20. A. Desbois, G. Mazza, F. Stetzkowski, and M. Lutz, (1984) Biochim. Biophys. Acta 785, 161-176.
21. H.B. Dunford, and T. Araiso, (1979) Biochem. Biophys. Res. Commun. 89, 764-768.
22. Y. Hayashi, and I. Yamazaki, (1978) Arch. Biochem. Biophys. 190, 446-453.
23. O. Jardetzky, and G.C.K. Roberts, NMR in Molecular Biology, (Academic Press, New York, 1981).
24. H. Steiner, and H.B. Dunford, (1978) Eur. J. Biochem. 82, 543-549.
25. G.L. Kedderis, and P.F. Hollenberg, (1984) J. Biol. Chem. 259, 3663-3668.
26. B.B. Hasinoff, and H.B. Dunford, (1970) Biochemistry 9, 4930-4939.
27. C.D. Hubbard, H.B. Dunford, and W.D. Hewson, (1975) Can. J. Biochem. 53, 1563-1569.
28. D. Keilin, and E.F. Hartree, (1951) Biochem J. 49, 88-104.
29. J. Terner, and D.E. Reed, (1984) Biochim. Biophys. Acta. 789, 80-86.
30. J.T. Groves and T.E. Nemo (1983) J. Amer. Chem. Soc. 105, 6243-6248.

Resonance Raman Studies of Transient States in Bacterial Reaction Centers

B. Robert, W. Szponarski, and M. Lutz

Service de Biophysique, Département de Biologie, CEN Saclay, F-91191 Gif-sur-Yvette Cedex, France

INTRODUCTION

In photosynthetic bacteria, transduction of the energy of incoming photons into chemical potential energy occurs in specialized protein-pigment complexes of the photosynthetic membrane named reaction centers (RC). These complexes have been isolated and well characterized biochemically these last years : in purple bacteria (Rhodospirillales) they consist in a ca 90000 dalton proteic core constituted of three polypeptides containing four bacteriochlorophylls (BChl), two bacteriopheophytins (Bpheo), and one carotenoid. Recent success in crystallizing a RC from the BChl b-containing bacteria Rhodopseudomonas viridis permitted to provide an electron density map of this complex with a 3 A resolution (1). The primary charge-separation occurs at the level of two strongly electronically interacting BChls, and is followed by an electron-transfer involving several transient states. If most of these transient states have now been well characterized by fast absorption spectroscopy, there still is a lack of structural information about their structure and its evolution with time. In particular, very little is known about the ground-state, the pigments-pigments and pigment-protein interactions and about the evolution of these interactions during the charge-separation and the primary steps of the electron-transfer.

Resonance Raman spectroscopy (RR) has been shown to bring information on the conformation and intermolecular binding states of the pigments of the RC (2). For the bacteriochlorophylls and -pheophytins the highest information content of the RR spectra is obtained when resonance is with the Soret electronic transition (2,3). In particular, RR spectra of these molecules excited in these conditions inform on the intermolecular binding states of their conjugated acetyl and ketone carbonyls, and of the $Mg\text{-}N_4$(pyrrole)LL' group when present. In these conditions, however, a limited selectivity is obtained amid the contributions of the six bacteriochlorin pigments of the RC. Similar information can be obtained about transient states of RC by using time-resolved resonance Raman spectroscopy (TR3) and conventional resonance Raman spectroscopy, playing on the actinic properties of the analysis beam. In addition, RR spectra may be quite useful in merely identifying certain of these states.

TRIPLET TRANSIENTS IN BACTERIAL REACTION CENTERS

When a reaction center is excited by an incoming photon, or by a singlet exciton provided by the antenna system, the primary electron donor P is raised to its first excited singlet state P* and undergoes a charge separation. Then in 100 picoseconds, the electron moves through a Bpheo to its primary stable acceptor, a quinone. The question whether a BChl is involved in the electron-transfer is still a matter of debate. When reaction centers are treated e.g. with dithionite this quinone is reduced before illumination and the electron-transfer is blocked at the level of the acceptor Bpheo. The lifetime of the P^+ $Bheo^-$ state is approximately 10 ns.

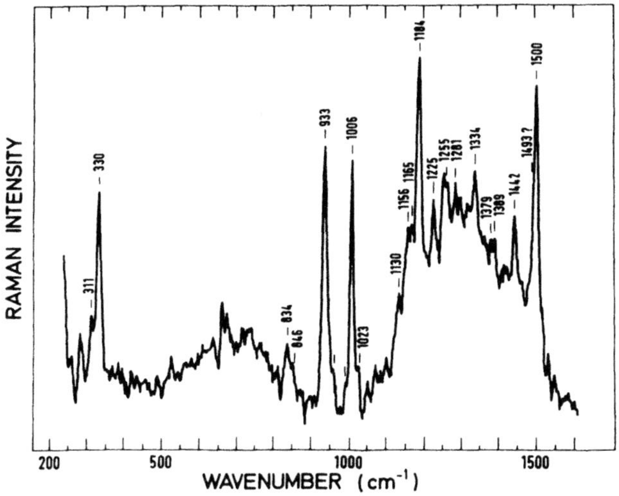

Figure 1 : Resonance Raman spectrum of the triplet state of the spheroidene molecule present in reaction centers of Rps. sphaeroides, strain 2.4.1., obtained at 60 K and with 545 nm excitation, using a single cw beam, pump-probe technique. The spectrum shown is a difference between RR spectra of chemically reduced and oxidized centers, normalized on the contributions from ground-state bacteriopheophytins.

Desexcitation of the RC occurs through a triplet state involving both P^+ and Bpheo$^-$. This state is converted into another triplet state which, at room temperature, involves an equilibrium between a $^3P^+BChl^-$ form and a triplet state of the primary donor alone. A rapid conversion between P^R and the carotenoid occurs above 60 K, producing a carotenoid triplet state (^{T}Car) with a high yield, which lifetime is 5 microseconds. At lower temperatures this conversion does not occur, and the lifetime of P^R is about 100 microseconds (for a review concerning these states see e.g. ref 4).

TRIPLET STATE OF THE CAROTENOID BOUND TO RC

Resonance Raman data showed that for different strains of purple bacteria containing different carotenoids, the latter was present in a particular cis configuration when bound to the RC (5,6). Experiments using TR^3 were conducted by LUTZ et al (7) on triplet states of RC-bound carotenoids. The frequency-shifts measured on the ν_1 band between triplet spectra obtained from carotenoids bound to RC and triplet spectra obtained from carotenoids in vitro indicated that the native conformation was most probably retained by the molecule during excitation and that no cis-trans isomerization occurred during the $S_0 \rightarrow T_0$ transition. Interest in this kind of study developed, because it was possible to obtain RR spectra of triplet states of carotenoids stabilized in a precise twisted cis conformation. Indeed, in vitro work has shown that self-isomerisation during TR^3 experiments makes difficult observations on triplets of carotenoids in precisely determined conformations (8).

More recently, it was shown that exciting reduced RCs with a cw laser line at 545 nm permitted to obtain RR spectra of the RC bound-carotenoids in the triplet state by pump-probe excitation (9). This method permitted to increase the signal-to-noise ratios in the RR spectra, and, in particular, to observe an intense band at 330 cm^{-1} which could be

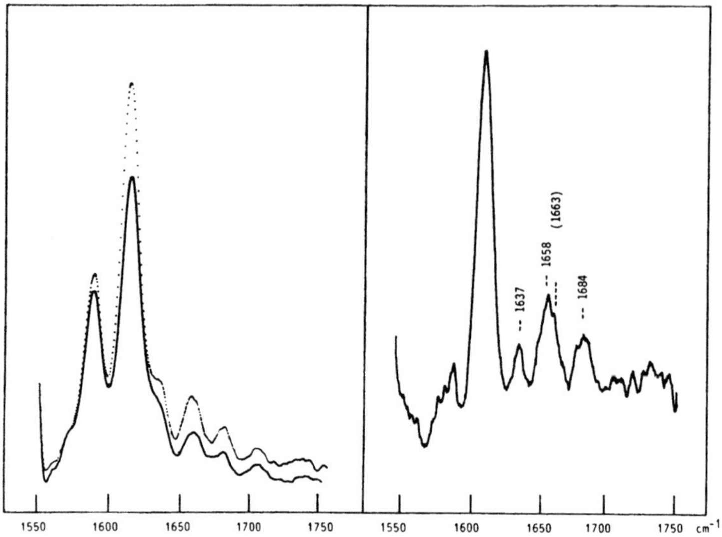

Figure 2 : (left) : Resonance Raman spectra (1500-1700 cm^{-1} region) of reaction centers from Rps. sphaeroides 2.4.1., treated with excess dithionite. Excitation 363.8 nm. Temperature 30 K (solid line) and 75 K (dotted line). These spectra were normalized on the 1590 cm^{-1} band intensity. (Right) : difference RR spectrum, attributed to the primary donor alone.

conformation-sensitive (9). Studies currently are under way on carotenoids bound to antenna complexes of photosynthetic bacteria, which presumably are stabilized in all-trans conformations.

PRIMARY DONOR

As the triplet state of carotenoids above 60 K, the triplet state of the primary donor is accumulated in RCs when steady-state RR experiments are conducted below 60 K, the lifetime of this state being 100 µs. However, up to now, any attempt to obtain RR spectra of P in its triplet state failed because of spurious fluorescence, typically arising when reducing the RCs. It is however possible to play with the steady states of chemically reduced RC during Raman experiments in two ways : first, it is possible to induce or forbid the triplet-triplet conversion between P^R and the carotenoid by changing the sample temperature : at 80 K, the rate of desexcitation of the RC is much higher, according to the 20 fold shorter lifetime of ^{T}Car relative to P^R. In these conditions, it is possible to obtain RR spectra of RC at 363.8 nm with the two BChl constituting the primary donor in their neutral, ground-state, participating to the resonance. If temperature is lowered below 50 K the yield of formation of TCar rapidly decreases, and more primary donor molecules remain in the P^R state, which is weakly or non-resonating under 363.8 nm excitation. It is also possible to decrease the amount of P^R present in the sample at a fixed temperature, by modifying the conditions of excitation, e.g. by lowering the power of the pump-probe beam, or by defocusing it at constant radiation power.

RR spectra of RCs (1550-1750 cm^{-1} region) obtained at 30 and 80 K are presented in figure 1, as well as the corresponding difference spectrum in the same region, taking the 1590 cm^{-1} band arising from Bpheo alone as an internal standard. At 80 K the 1615 cm^{-1} band arising from the stretching of the methine bridges of both the BChl and Bpheo molecules is enhanced, indicating that more BChl molecules are contributing at 80 K than at 20 K. The difference spectrum is clearly constituted by positive contributions of BChl molecules, and no negative contributions which might have arisen from P^R. Taking into account the relatively low signal-to-noise ratios of these difference spectra, the minimal number of carbonyl bands observed in the 1630-1700 cm^{-1} region is three, indicating that at least two BChl molecules contribute. A first conclusion is that there is no need of any additional molecule to account for this difference spectrum. This confirms that, at 20 K, the P^R state involves the primary donor alone, as previously indicated by differential electronic absorption spectroscopy (10). The difference spectrum of figure 1 thus arises from the two BChls constituting P in their neutral states.

The high-frequency value (1612 cm^{-1}) and the small halfwidth (13 cm^{-1}) of the methine bridge stretching mode indicate that, in neutral P 870, the Mg atom of both molecules constituting the primary donor should be five-coordinated. Indeed, in-vitro work has established that the νC_aC_m frequency occurs around 1605 cm^{-1} for BChls with a six-coordinated Mg, and at 1615 cm^{-1} when this atom assumes five coordination (11,12).

In the carbonyl stetching region, it appears reasonable to attribute the bands at 1637 cm^{-1} and at 1659 cm^{-1} to stetching of acetyl carbonyl groups (3). The 9-keto carbonyls should either both vibrate at 1683 cm^{-1}, or, taking into account the 1663 cm^{-1} shoulder, one of them might vibrate at this latter frequency, the other one then contributing alone to the 1683 cm^{-1} band.

Some inconsistency appears between these results and X-ray data on the primary donor, inasmuch as RCs from the two species considered (*Rps. viridis* in X-ray studies and *Rps. sphaeroides* in RR studies) should have closely similar structures. Indeed, X-ray data indicate that the Mg of both BChls constituting the primary donor should be six-coordinated, simultaneously being interacting with the protein and being in close contact with the acetyl C=O group of the adjacent molecule (1). In our interpretation, BChl molecules constituting the primary donor should assume markedly different intermolecular bonding, and particularly through their acetyl carbonyl groups : one of these groups indeed should be strongly bound, assuming a frequency close to that observed in $BChl_n$ oligomers in vitro (3), and the second one should be only very weakly interacting if at all, its frequency corresponding to that observed for a free vibrator. On the other hand, the frequency of the νC_aC_m band is only indirectly sensitive to the coordination of the central Mg : a 1612 cm^{-1} frequency conceivably might arise from a markedly unsymmetrical liganding of the Mg, allowing an out-of-plane of this atom. However, a symmetrical structure for P, involving equivalent C=O bonds, appears inconsistent with RR data. This situation conceivably may result from a non-unicity of the ground-state structure of P in a macroscopic sample, an average picture being given by X-ray data, or from a slight asymmetry in the structure, which could imply large differences in the interactions assumed by the pigments.

REFERENCES

1 DEISENHOFER, J., EPP, O., MIKI, K., HUBER, R. and MICHEL, H. (1984) J. Mol. Biol. 180, 385

2 LUTZ, M., KLEO, J. and REISS-HUSSON, F. (1976) Biochem. Biophys. Res. Comm. 69, 711

3 LUTZ, M. in : Advances in Infrared and Raman spectroscopy (1984) (R.J.H. Clark and R.E. Hester eds) Wiley Heyden, Chichester, vol. 11, chap. 5

4 PARSON, W.W. (1982) Ann. Rev. Biophys. Bioeng. 11, 57
5 LUTZ, M., AGALIDIS, I., HERVO, G., GOGDELL, R.J. and REISS-HUSSON, F. (1978), Biochim. Biophys. Acta 503, 287
6 LUTZ, M., SZPONARSKI, W., NEUMANN, J.M., BERGER, G. and ROBERT, B. (1985) unpublished
7 LUTZ, M., CHINSKY, L. and TURPIN, P.Y. (1982) Photochem. Photobiol. 36, 503
8 WILBRANDT, R. and JENSEN, N. in : time resolved vibrational spectroscopy (1983) (G.H. Atkinson ed.) Academic Press, New York p. 273
9 SZPONARSKI, W., ROBERT, B., SZALONTAI, B. and LUTZ, M. (1984) 7th Int. Symp. on Carotenoids, München, abstract 37
10 SHUVALOV, V.A. and PARSON, W.W. (1981) Proc. Natl. Acad. Sci. USA, 78, 957
11 COTTON, T.M. and VAN DUYNE, R.P. (1981) J. Am. Chem. Soc. 103, 6020
12 ROBERT, B. and LUTZ, M. (1985) Biochim. Biophys. Acta. 807, 10

Energy-Transfer Between Dimeric Chlorophyll *a* and All-Trans β-Carotene *in vitro* as Resolved by Fluorescence Photoquenching

*N.E. Binnie, L.V. Haley, and J.A. Koningstein**

The Ottawa-Carleton Chemistry Institute, Department of Chemistry, Carleton University, Ottawa, Ontario, K1S 5B6, Canada

1. INTRODUCTION

Thirty years ago Livingston [1] reported that the fluorescence yield from a solution of Chlorophyll a (Chl a) in dry hydrocarbon solvents was less than 10% that for a solution where monomeric fluorescence was activated by addition of water or other nucleophilic species. At that time, the aggregation behavior of Chl a in solution had not been established, and Livingston hypothesized the presence of a non-fluorescent dimer in favour of a non-activated monomeric species of Chl a to explain the weak fluorecence of the dry solutions. More recent work pointed to the fact that such solutions contain monomeric, dimeric and higher aggregates of Chl a, and that the dimer is dominant in dry solutions which are 10^{-4}M [2,3]. The fluorescence of the dimer is quite easily observed,if induced by lasers which are tuned to specific wavelength regions [4-6]. Some reports in the literature indicate that dimeric forms of Chl a could play a role in photophysical processes,and it appears to the present investigators that the dimers which exist in dry hexane solutions can serve as good analogs to these biological systems in order to study their spectroscopic properties before more complicated systems are investigated. It is,however, curious that although we can study properties such as energy relaxation etc., the actual structure of the dimers (whose spectroscopic properties are presented here) is not known to us. It is thought that the more complete the spectroscopic studies, the easier it will be to choose between already proposed structural models [7-9].

One of the more important questions related to biological systems is the process by which energy is captured by pigment molecules and is transferred to the reaction centre. Studies of in vivo systems have shown a variety of results which indicate detectable [9-13] energy-transfer. Some experiments carried out on model systems have produced controversial results [14-16]. At this time, an investigation of a model system consisting of monomeric and dimeric Chl a and all trans β-carotene is reported.

The interpretation and assignment of spectroscopic data from the monomer and dimers (Figs. 1-3), based on earlier published and unpublished work [4-6,17, 18],is

*Killam Fellow

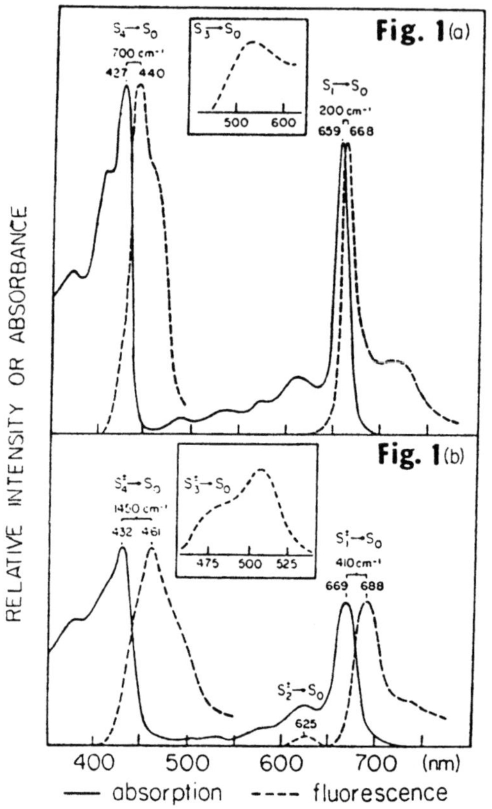

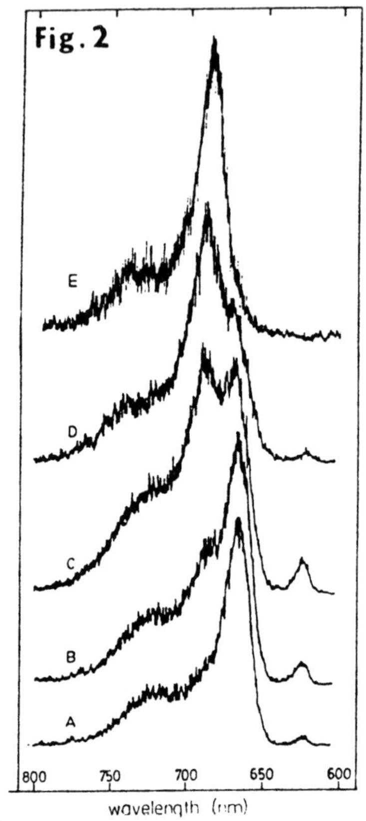

Figure 1: The absorption (——) and fluorescence (----) spectra of (a) monomeric and (b) dimeric Chl a. The inserts show the blue-green emissions, intensity scale enlarged from that of the main diagram.

Figure 2: The wavelength selective fluorescence spectrum of monomeric (668 nm) and dimeric (688 and 625 nm) for Chl a in dry hexane. λ_L(nm): (A) 430.5, (B) 438.5, (C) 444.5, (D) 450.5, and (E) 457.9 nm.

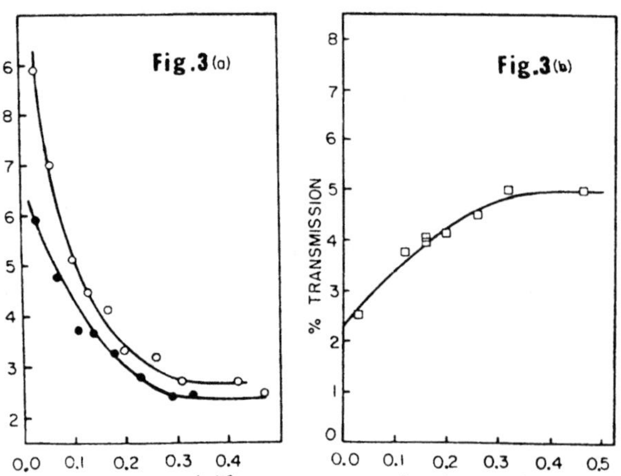

Figure 3

(a) Fluorescence quenching with excitation at 629 nm for monomeric (●) and dimeric (O) Chl a detected at 668 nm and 688 nm respectively. (b) Monochromatic transmissions experiment for a solution of Chl a in dry hexane. (0.5 mW corresponds to 10^{24} photons s^{-1} cm^{-2} at 629 nm).

Non-linear Raman studies of the wavelength-dependent intensity of normal modes of the groundstate of dimeric Chl a revealed [1] that electronic origins are located at 434.1 nm (23036 cm^{-1}) and 435.3 nm (22972 cm^{-1}), assigned to the S_4^+ and S_4^- respectively. Experimental data point to the presence of an effective non-radiative channel for energy relaxation between $S_4^{\pm}$ and $S_2^{\pm}$ because of the $S_1^{\pm} \dashrightarrow S_0$ fluorescence is very weak if excited with 420 nm $< \lambda_L <$ 450 nm whereas the $S_2^{\pm} \dashrightarrow S_0$ fluorescence is not. The wavelength selective fluorescence spectrum between 600 nm and 800 nm given below. Laser excitation into the shoulder of the Soret band of the monomer and dimer produces the following fluorescences: in the blue at 440 nm (460 nm), blue-green at 455 nm (477 nm, 507 nm), red at (625 nm), and deep red at 667 nm (688 nm), where the first wavelength is assigned to monomeric emission and the wavelength between brackets is assigned to dimeric emission. These assignments (Fig. 4) are based on allowed transitions between the electronic surfaces of respectively S_4---S_0 ($S_4^{\pm}$---S_0); S_3---S_0 ($S_3^{\pm}$---S_0);($S_2^{\pm}$---S_0) and S_1---S_0 ($S_1^{\pm}$---S_0). (Fig. 2) illustrates this. Goutermans theoretical calculations [19] gave the result that another singlet (S_3 in the present notation), is situated at 20000 cm^{-1} above the groundstate. The $S_3 \dashrightarrow S_0$ fluorescence for pyridinated monomeric Chl a was reported earlier from this laboratory,and is shown in Fig. 1(a). Similar blue-green fluorescence from the dimer can also be recorded with excitation over the range 450 nm -- 470 nm. As well, the $S_1^{\pm} \dashrightarrow S_0$ emission intensity at 688 nm undergoes a maximum if induced with pulsed laser radiation tuned to this interval (Fig. 2). The contour of the $S_3^{\pm} \dashrightarrow S_0$ emission band shows maxima at ~477 nm and ~507 nm (Fig. 1(b)). These data indicate the presence of electronic origins at ~477 nm (20950 cm^{-1}) and

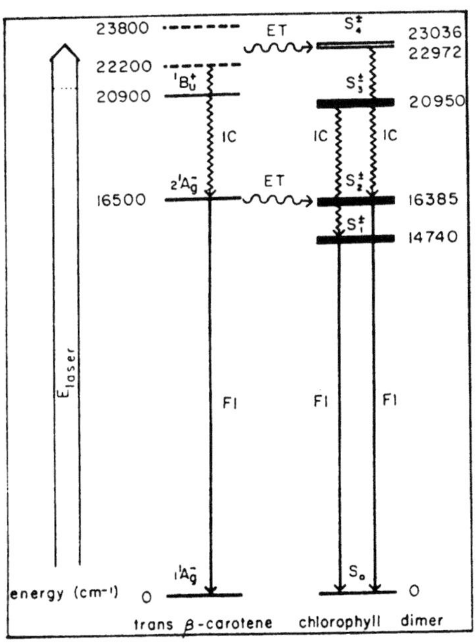

Figure 4

The Energy-level diagram of dimeric Chl a and all trans β-carotene in dry hexane. IC(internal conversion) and Fl(fluorescence) represent intra-molecular energy relaxation paths while ET(energy-transfer) represents inter-molecular energy relaxation. In the carotene diagram, dotted lines represent maxima in the absorption spectra due to vibronic side bands.

point to the existence of an effective channel for non-radiative energy relaxation between $S_3^{\pm}$ and $S_1^{\pm}$ states (Fig. 4).

2. MONOCHROMATIC TRANSMISSION STUDIES

The radiative lifetimes of the lowest-lying dimer $S_1^{\pm}$ and $S_2^{\pm}$ excited states and of the monomer S_1 are of nanoscecond duration. If a nanosecond pulsed laser is used to directly populate these states, with radiation which is in resonance with $S_1^{\pm} \leftarrow -- S_0$, $S_2^{\pm} \leftarrow -- S_0$, or $S_1 \leftarrow -- S_0$ absorptions, or indirectly via the non-radiative channels $S_4^{\pm} \rightsquigarrow S_2^{\pm}$, or $S_3^{\pm} \rightsquigarrow S_1^{\pm}$ or $S_4 \rightsquigarrow S_2$, $S_3 \rightsquigarrow S_1$, by pumping into $S_4^{\pm}$, $S_3^{\pm}$ or S_4, S_3 respectively, then the time evolution of the $S_1^{\pm}$, $S_2^{\pm}$ and S_1 population follows the laser pulse intensity. In general, not all photons of the laser beam are involved in the creation of excited states. The remaining photons can also be absorbed by the excited molecules. This consecutive absorption of laser photons can be monitored via monochromatic transmission (MT) experiments [17, 20]. In earlier publications, experiments have been used to show that excited states of monomeric Chl a and β-carotene can be created by absorption of nanosecond light pulses [21, 22].

3. FLUORESCENCE INTENSITY MEASUREMENTS

It is evident from the previous sections that the use of MT experiments for studies in energy-transfer is rather limited in distinguishing the effects due to specific species. The method of using the fluorescence intensity from particular species was used for the following reasons: (i) the main fluorescences of monomeric, dimeric Chl a and all trans β-carotene in hexane occur at different wavelengths allowing us to study these species independently, and (ii) the decay of the fluorescence intensity of the dimer and monomer are of nanosecond duration, while the decay for β-carotene is much shorter (<1 ps) thus allowing the study of these fluorescences in time. Also, fluorescence intensities are dependent on excitation wavelength, pulse duration, laser power, and radiative lifetime with respect to the resolution capabilities of the detectors.

Fluorescence quenching with red laser light for the monomer and dimer in the pure solution is shown in Fig. 3. The excitation wavelength 629 nm is in resonance with dimer and monomer vibronic absorptions for $S_1^{\pm} \leftarrow -- S_0$ and $S_1 --- S_0$. Also shown are the laser intensity-dependent values of transmission for the solution, composed of contributions from absorption due to both species at 629 nm. The behaviour of the transmission of the solutions and the fluorescence quenching provide analagous information, but the transmission data can be somewhat ambiguous, because it does not distinguish clearly between effects due to the species present. The fluorescence data clearly indicate when the saturation of an excited state occurs (no change in fluorescence intensity with laser power) for a specific species, since only the fluorescence of that species is monitored.

4. ENERGY LEVEL DIAGRAM OF β-CAROTENE

Part of the absorption spectrum and the Stokes-shifted fluorescence spectrum of all-trans β-carotene are shown in Fig. 5. If compared to similar experimental data recorded for all-trans β-carotene in CS_2, we find [21] that the spectra are blue shifted but the overall features remain the same. From the excitation spectra of the Raman intensity of normal modes for the CS_2 solution, it was found [21] that the origin for electronic transitions between the $^1B_u^+$ and the $1^1A_g^-$ ground states lies close to the peak of the absorption at 475 nm (21652 cm^{-1}) in hexane. The other absorption bands towards shorter wavelengths are vibronic side bands. For β-carotene in KBr a lower-lying $2^1A_g^-$ state is at 16500 cm^{-1} above the groundstate [23]. Theoretical and experimental evidence already has been presented with regard to the relative positions of $^1B_u^+$ and $2^1A_g^-$ for polyene-type molecules [24]. The former is more sensitive to the nature of the environment than the latter. An approximate energy-level diagram for all-trans β-carotene in hexane based on these arguments is shown in Fig. 4.

5. ENERGY-TRANSFER

In the previous section we have discussed the energy-level diagrams of the components of our model systems. Evidence is presented, as follows, for selective

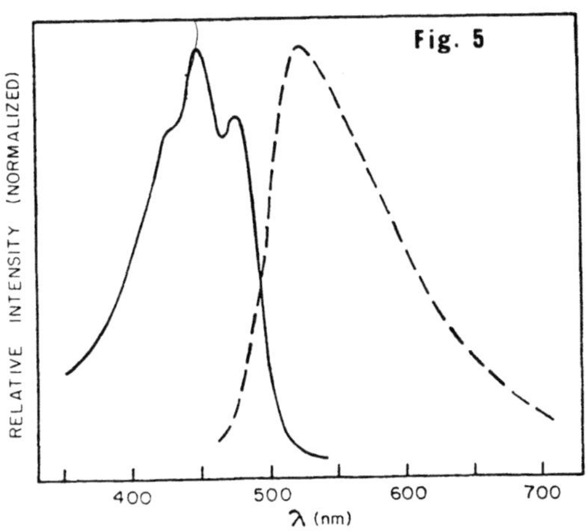

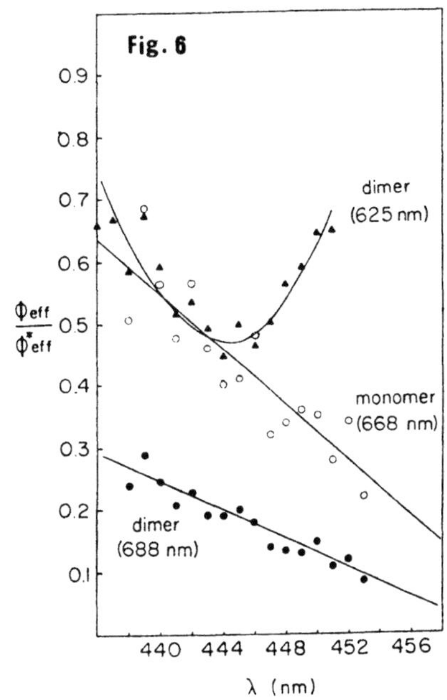

Figure 5: The normalized absorption (——) and fluorescence (---) spectrum of all trans β-carotene in dry hexane.

Figure 6: Effective quantum efficiency ratios for a mixed solution of Chl a and all trans β-carotene (ϕ_{eff}) and a pure Chl a solution (ϕ^*_{eff})for fluorescence emissions at 668 nm (monomer), 688 nm (dimer Q_y) and 625 nm (dimer Q_x).

energy-transfer from all-trans β-carotene to Chl a using fluorescence intensity measurements normalized for laser power.

In Fig. 6 values are shown for the ratio of effective quantum efficiencies, $\phi_{eff}/\phi^{*}_{eff}$ for the 668 nm, and 625 nm fluorescences as a function of the pulsed laser wavelength. In the absence of energy-transfer between β-carotene and monomeric and/or dimeric Chl a, the ratio $\phi_{eff}/\phi^{*}_{eff}$ is expected to change gradually with excitation wavelength over the spectral interval where the carotene molecule absorbs. From the transient S_1 absorption spectrumn of monomeric Chl a we find [24] that the cross-section for absorption of laser radiation with λ_L=434 nm (Fig. 5(a) is larger than the cross-sections for S_m---S_1 excited state transitions at 460 nm. Based on cross-sections alone, the value of the ratio $\phi_{eff}/\phi^{*}_{eff}$ at λ_L=434 nm should be smaller than at longer wavelengths. The data for the monomeric Chl a 668 nm band shows this to be true. The data for the dimeric 688 nm band is similar to the monomeric 668 nm band, however the transient absorption spectrum of the dimer is not known and we cannot draw any conclusions on the predicted behavior of the ratios of the quantum efficiencies. At short wavelengths the ratio for the 625 nm band follows the behaviour of the monomeric band, whereas towards longer wavelengths the ratio increases into the absorption band of the β-carotene (which reaches a maximum intensity at 450 nm). We interpret the data for the 625 nm band in terms of selective energy-transfer from all trans β-carotene to the Chl a dimer in hexane and conclude that such a transfer is not detectable between β-carotene and monomeric Chl a when the mixture is exposed to laser radiation between 430 nm and 460 nm.

As explained previously, the carotene absorption band due to vibronic transitions between the $^1B_u^+$ and $1^1A_g^-$ electronic (0---0) surfaces is estimated at 475 nm (21652 cm^{-1}). If excited with pulsed laser radiation in the blue spectral region, fluorescence of all-trans β-carotene is recorded with a maximum intensity at 526 nm (Fig. 5). In mixed Chl a and β-carotene solutions, this fluorescence could in principle be re-absorbed by dimeric Chl a at transition wavelengths resonant with $S_1^{\pm}$<--S_0 or $S_n^{\pm}$<--$S_1^{\pm}$ and $S_n^{\pm}$<--$S_2^{\pm}$. However, the Chl a groundstate absorption spectrum is minimal at 526 nm. Even if the pulsed laser source illuminating a pure Chl a sample is tuned to that wavelength, the $S_2^{\pm}$-->S_0 fluorescence is too weak to be detected. We conclude that re-absorption of emission from the $^1B_u^+$ state of all-trans β-carotene by the Chl a dimer in the mixed solution is not responsible for the increase in the intensity of the 625 nm fluorescence of the dimer.

A second method for energy-transfer might involve energy-relaxation within the vibronic band system of $^1B_u^+$ and resonant coupling of the $^1B_u^+$ state at 21052 cm^{-1} and the $S_3^{\pm}$ states of the dimer at 20450 cm^{-1}. This would result in an enhancement of the dimer $S_1^{\pm}$-->S_0 fluorescence,because of radiationless energy-relaxation from $S_3^{\pm}$ to $S_1^{\pm}$. This in obvious disagreement with the experimental results.

However, coupling between vibronic states of $^1B_u^+$ and the $S_4^{\pm}$ states of the dimer would result in enhancement of the intensity of the 625 nm band,while leaving the 688 nm fluorescence unchanged because of the highly effective energy-relaxation channels between $S_4^{\pm}$ and $S_2^{\pm}$. From the energy-level diagram, it can be seen that the $S_4^{\pm}$ states at 434.1 nm and 435.3 nm are resonant with $^1B_u^+$ vibrational levels at 1983 cm^{-1} and 1920 cm^{-1}. However, the experimental results do not support this mechanism for energy-transfer. Given the large energy-gap between the $S_4^{\pm}$ and $^1B_u^+$, the probability is small for resonant energy-transfer between vibronic levels of the $^1B_u^+$ and the $S_4^{\pm}$ states. Assuming that the absorbed laser radiation results in a Boltzmann distribution of population in the vibronic levels, tuning the excitation wavelength from 430 nm to 450 nm results in a decrease in the population by about 20%, which is offset by an increase in the absorption of β-carotene by the same amount. Presently,our experimental data are not complete enough to comment on the details of efficiency of this process.

Recently [23] we established the position of the $2^1A_g^-$ state of trans β-carotene in KBr at 16150 cm^{-1} above the $1^1A_g^-$ groundstate. The $2^1A_g^- \dashrightarrow 1^1A_g^-$ fluorescence is induced with visible laser light exciting the $^1B_u^+$ which suggests radiationless energy-transfer exists between $^1B_u^+$ and $2^1A_g^-$ for trans β-carotene in KBr at 77 K. As the estimated position of the $S_2^{\pm}$ state of the Chl a dimer in hexane is at 16365 cm^{-1}, the probability for resonant energy-transfer from $2^1A_g^-$ to $S_2^{\pm}$ is optimized [25,26].

REFERENCES

/1/ R. Livingston, W.F. Watson and J. McArdle, J. Am. Chem. Soc., 71 (1949) 1542.

/2/ M.J. Yuen, L.L. Shipman, J.J. Katz and J.C. Hindman, Photochem. Photobiol., 36 (1982) 211.

/3/ W.W.A. Keller, B.A. Alberts, E. Straus and K. Maier-Schwartz, J. Lumin., 31 (1984) 892.

/4/ A.C. de Wilton, L.V. Haley and J.A. Koningstein, J. Phys. Chem., 88 (1984) 1077.

/5/ A.C. de Wilton and J.A. Koningstein, J. Am. Chem. Soc., 106 (1984) 5088.

/6/ A.C. de Wilton, L.V. Haley and J.A. Koningstein, J. Phys. Chem., 87 (1983) 185.

/7/ F.K. Fong and V.J. Kvester, J. Am. Chem. Soc., 97 (1975) 6888.

/8/ J.J. Katz, J.R. Norris, L.L. Shipman and M.C. Thurnaner, Ann. Rev. Biophys. Bioenerg., 7 (1978) 343.

/9/ S.G. Boxer and G.L. Closs, J. Am. Chem. Soc., 98 (1976) 5406.

/10/ H.A. Frank, J. Machnicki, and R. Friesner, Photochem. Photobiol., 38 (1983) 451.

/11/ D. Siefermann-Harms, and H. Ninnemann, Photochem. Photobiol., 35 (1982) 719.

/12/ R. van Grondelle, H.J.M. Kramer, and C.P. Rijgersberg, Biochim. et Biophys. Acta, 682 (1982) 208.

/13/ J.C. Goedheer, Biochim. et Biophys. Acta, 172 (1969) 252.

/14/ R.J. Thrash, H. L-B Fang and G.E. Leroi, Photochem. Photobiol., 29 (1979) 1049.

/15/ (a) N.R. Murty and E. Rabinowitch, J. Chem. Phys., 41 (1964) 602.

(b) G.S. Singhal, J. Hevesi, and E. Rabinowitch, J. Chem. Phys., 49 (1968) 5206.

/16/ G.S. Bedard, R.S. Davidson and K.R. Trethewey, Nature 267 (1977) 373.

/17/ M. Asano and J.A. Koningstein, Chem. Phys., 57 (1981) 1.

/18/ A.C. de Wilton and J.A. Koningstein, Chem. Phys. Lett., 114 (1985) 161.

/19/ M. Gouterman, G.H. Wagniere and L.C. Snyder, J. Mol. Spect., 11 (1965) 108.

/20/ B. Halperin and J.A. Koningstein, Can. J. Chem., 59 (1981) 2792.

/21/ L.V. Haley and J.A. Koningstein, J. Phys. Chem., 87 (1983) 621.

/22/ I.W. Wylie and J.A. Koningstein, J. Phys. Chem., 88 (1984) 2950.

/23/ L.V. Haley and J.A. Koningstein, J. Phys. Chem., in press.

/24/ R.R. Chadwich, D.P. Gerrity, and B.S. Hudson, Chem. Phys. Lett., 115 (1985) 24.

/25/ N.E. Binnie, L.V. Haley and J.A. Koningstein, Chem. Phys. Lett., in press.

/26/ N.E. Binnie, L.V. Haley and J.A. Koningstein, J. Lumin., submitted for publication.

Using Time-Resolved Resonance Raman Spectroscopy to Map the Reaction Pathway for Enzyme-Substrate Transients

P.R. Carey

Division of Biological Sciences, National Research Council of Canada,
Ottawa, Canada K1A 0R6, Canada

Because of its ability to relate structural and dynamical details for short-lived species, time-resolved resonance Raman spectroscopy can help solve one of the major problems in modern biochemistry - the way in which enzymes work. One strategy in the use of the RR approach is to create a unique chromophore at the time and place of catalysis - as the enzyme and substrate come together in the active site. This allows us to focus on the critical region of a large molecular complex at the critical time. A prototype reaction involving chromophore creation is the hydrolysis of thionoesters by the cysteine proteinase papain:

$RC(=O)NHCH_2C(=S)OCH_3$ + HS-papain

$K_S \rightleftharpoons$

$[RC(=O)NHCH_2C(=S)OCH_3 \cdot$HS-papain] Michaelis Complex

$k_2 \downarrow$

$RC(=O)NHCH_2C(=S)$-S-papain + CH_3OH Dithioacyl enzyme λ_{max} 315 nm

$k_3 \downarrow H_2O$

$RC(=O)NHCH_2C(=S)$-OH + HS-papain

In this reaction mixture a transient dithioester chromophore is formed from the C=S group of the substrate and the HS- group from a cysteine residue in papain's active site. The dithioester chromophore has a λ_{max} near 315 nm and is the only species in solution with an electronic absorption band to the red of 300 nm. Thus, it is possible to excite specifically the RR spectrum of the dithioester group. The RR spectrum is very sensitive to conformational events in the RC(=O)-NH-CH_2-C(=S)-S-CH_2 bonds and, combined with X-ray crystallographic and kinetic studies, has been extensively used to elucidate the details of catalytic events [1-5, and references therein].

Until recently, the experimental protocol for studying such reactions involved mixing enzyme in a stationary quartz cuvette with a large excess of substrate,so that a quasi-steady state population of dithioacyl enzyme could be formed for the 30-300 seconds required for spectroscopic characterization. Resonance Raman spectra were excited using the cw 324 nm line of Kr^+ and the spectra detected via a Spex triplemate and a TN-6132 Tracor-Northern intensified diode array. With this instrumentation 10-20 sec are required to accumulate a complete RR spectrum at acceptable signal-to-noise. Depending on the nature of the R group in the substrate,the half-life of the dithioacyl enzyme is 1-10 secs. Thus, from a reaction mixture containing steady-state population of intermediates,we are obtaining RR data for a temporally heterogeneous set of dithioacyl enzymes stretching from those just formed on the acylation pathway, to those about to be hydrolysed.

Precise detail for the reaction pathway, essential for a complete picture of mechanism, requires that we obtain sets of RR data for temporally homogeneous dithioacyl enzyme populations starting at the moment of formation. To this end, we have developed a rapid mixing-rapid flow system for RR applications which works on the 20 millisecond, and longer, time-scales. The system is depicted in Fig. 1. Enzyme solution is contained in two syringes and substrate solution in the remaining two syringes. When the syringes are driven by a syringe pump, the solutions are mixed in a four-jet mixing cell and the resulting enzyme-substrate solution is forced out through a square quartz tube, which acts as the Raman cell (Fig. 1). The quartz tube has an internal diameter of 1 mm and, typically, 5-10 mls of enzyme (2×10^{-4} M) are required per RR spectrum, at 50 msec time-resolution.

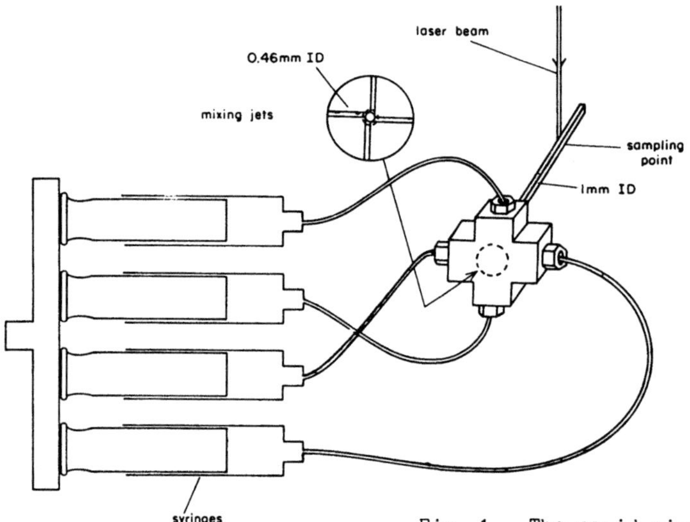

Fig. 1. The rapid mixing-rapid flow system.

The mixing system required careful development in order to deliver a temporally homogeneous population of known age at the RR sampling point. The mixing system was calibrated using the light source and detection system from a stopped-flow apparatus. In this set-up indicator, solutions plus acid or base were used as a test of mixing efficiency. Moreover, progress in a standard reaction (involving potassium ferricyanide and ascorbic acid) with a known rate-constant was monitored at the sampling point, and used to compare the time after mixing with that predicted from simple volumetric considerations.

The time-scale for acylation (rate-constant k_2, see reaction scheme) is one to two orders of magnitude faster than that for deacylation (rate-constant k_3). Thus, it is of interest to use the rapid mixing system to obtain RR data for the dithioacyl enzyme population during the time-scale of acylation. The latter can be measured in a conventional stopped-flow instrument by observing the build-up of absorbance, due to the dithioacyl enzyme, at 315 nm:

Two reactions which have acylation time-scales suitable for study by the rapid mixing system involve papain and the substrates $C_6H_5(CH_2)_2C(=O)NHCH_2C(=S)OCH_3$ or $CH_3OC(=O)PheAlaC(=S)OCH_3$. The half-life for acylation for both substrates is in the 100 millisecond range depending, for an exact value, on temperature and substrate concentration.

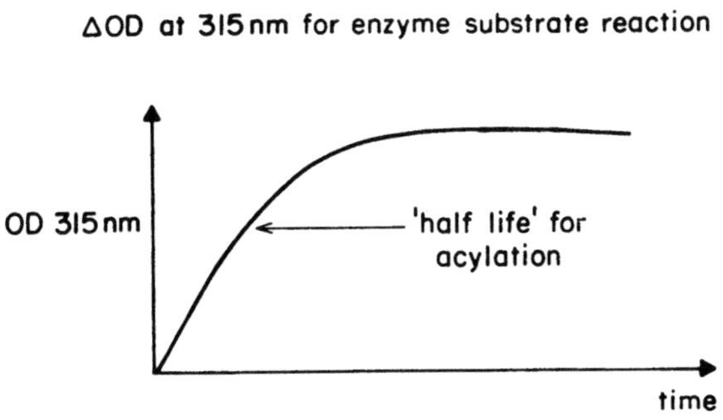

Fig. 2. Acylation observed in stopped-flow instrument.

The RR spectra in Fig. 3 compare a steady-state reaction mixture containing $C_6H_5(CH_2)_2C(=O)NHCH_2C(=S)$-S-papain (acylation half-life ≈100 msecs, deacylation half-life ≈4.5 secs) with a reaction 100 milliseconds after mixing. The major features of the steady-state RR spectrum can be seen at 100 milliseconds,but in the latter spectrum there is increased intensity near 1165 cm^{-1} and a new peak near 615 cm^{-1} due to conformational states,which are too sparsely populated in the steady-state mixture to be detected. Moreover, the steady-state feature near 1090 cm^{-1} is undetected at 100 msecs.

The important generalisations beginning to emerge from data such as these are that it is possible to detect new species during the acylation process. In a few hundred milliseconds these species 'relax' to those detected in steady-state reaction mixtures. The ability to detect and characterise non-relaxed intermediates during acylation provides a valuable new handle on the acylation portion of the reaction pathway.

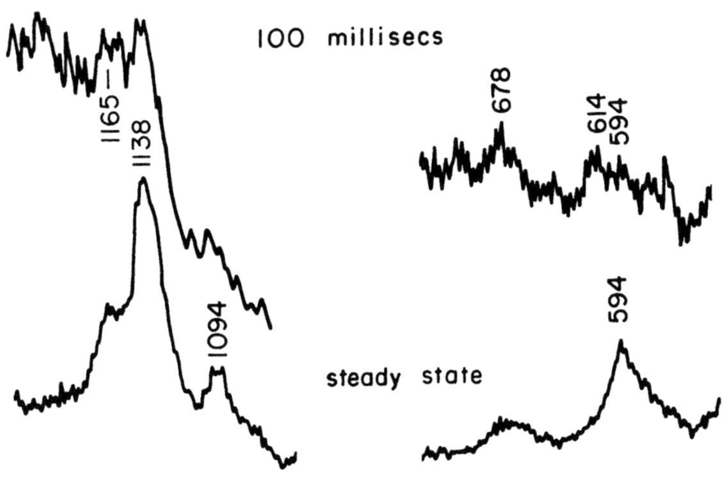

Fig. 3. Comparison of RR spectra at steady-state with population 100 msecs after mixing.

Acknowledgement The rapid mixing-rapid flow system discussed here was developed in collaboration with Mr. L. Sans Cartier and Dr. A. Storer. It is a pleasure to acknowledge their assistance.

References

1. P.R. Carey and A.C. Storer: Pure and Appl. Chem. 57, 225-234 (1985)
2. P.R. Carey and A.C. Storer: Ann. Rev. Biophys. Bioeng. 13, 25-49 (1984)
3. K.I. Varughese, A.C. Storer, and P.R. Carey: J. Am. Chem. Soc. 106, 8252-8257 (1984)
4. P.R. Carey, H. Lee, Y. Ozaki, and A.C. Storer: J. Am. Chem. Soc. 106, 8258-8262 (1984)
5. C.P. Huber, P.R. Carey, S.C. Hsi, H. Lee, and A.C. Storer: J. Am. Chem. Soc. 106, 8263-8268 (1984)

UV, IR, and Raman Spectroscopic Tracings of the B⇌Z Transitions of Poly[d(GC)]·poly[d(GC)]

H. Takashima, S. Tate, M. Nakanishi, Y. Nishimura, and M. Tsuboi

Faculty of Pharmaceutical Sciences, University of Tokyo,
Hongo, Bunkyo-ku, Tokyo 113, Japan

1 Introduction

A salt-induced co-operative conformational change of a synthetic DNA, namely poly [d(GC)]•poly[d(GC)], was found in 1972 by POHL el al. [1] and finely investigated by them [1-3]. This is now known to be a transition from a right-handed B-form helix to a left-handed Z-form helix [4]. We have recently examined, somewhat more in detail, these B-to-Z and Z-to-B conformational changes of poly[d(GC)]•poly[d(GC)] in solutions, by means of ultraviolet, infrared, Raman, and circular-dichroic spectrophotometries, through a measurement of the deuteration rates, and by applying a theory of polymer kinetics of SCHWARZ [5].

2 B⇄Z Equilibrium

The sample of poly[d(GC)]•poly[d(GC)] was purchased from P-L Biochemicals Inc., and according to them its sedimentation constant s_{20} is 9.8. Therefore it is supposed to be as long as 865 base-pairs an average. By changing the NaCl concentration in the solvent, the B⇄Z transition of this long duplex was examined by means of the ultraviolet as well as circular dichroic spectrophotometry. As is shown in the upper part of Fig. 1, the transition was found to take place at [NaCl] =2.25 M very sharply. By comparing this with similar B⇄Z transitions observed of shorter poly[d(GC)]•poly[d(GC)] [1,3], and by applying ZIMM and BRAGG theory [6] here, the σ value has been estimated to be about 10^{-4}. Here $1/\sigma$ is considered to be a parameter indicating the cooperativity of the transition [6].

3 Static Vibrational Spectra

Raman difference spectrum of B-Z was recently reported by BENEVIDES and THOMAS [7]. What was obtained by NISHIMURA et al. [8] is similar but is slightly different from the former. Here, sharp negative peaks are found at 1482, 1316, 1264, 798, 784, and 624 cm^{-1}, as well as sharp positive peaks at 1492, 1336, 1254, 830, 790, 776, and 684 cm^{-1}. The 1336→1316 cm^{-1} and 684→624 cm^{-1} shifts (on B→Z) are interpreted as indicating a conformational change of the guanosine moiety of this DNA, from O4'endo-anti to C3'endo-syn [9]. A somewhat complicated spectral change in the 1200-1300 cm^{-1} range, including the 1264→1254 cm^{-1} shift, on the other hand, is attributable to a conformational change of the cytidine moiety from an inbetween form of C2'endo-Cl'exo (with anti-glycosidic bond) to C2'endo-anti form [8]. In the P-O stretching frequency region, a Raman band at 830 cm^{-1}, characteristic of B-form mainchain, disappears on B→Z transition. Its counterpart in Z form seems to be located at 798 cm^{-1} [8]. It should be pointed out, in addition, that the relative intensity of the PO_2^- symmetric stretching line at 1096 cm^{-1} is appreciably lowered on B→Z transition.

For poly[d(GC)]•poly[d(GC)] in D_2O solution, infrared absorption bands are observed at 1686 and 1656 cm^{-1} and Raman bands [7] at 1683 and 1635 cm^{-1} when it is in B-form. In Z-form, on the other hand, they are at 1669 and 1636 cm^{-1} (IR) and at 1662 and 1634 cm^{-1} (Raman). Thus, each set (B-IR, B-Raman, Z-IR, or Z-Raman) is totally different from another set. This fact was interpreted by considering that carbonyl-carbonyl transition dipole couplings are of appreciable amount among guanine residues and among cytosine residues in the duplex structures now in question.

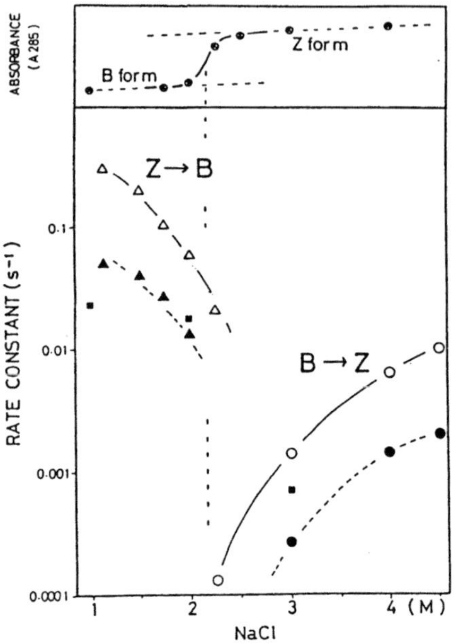

Fig. 1. Upper: Degree of salt induced B→Z transition of poly[d(GC)]•poly[d(GC)] as a function of NaCl concentration deduced from ultraviolet absorbance. Lower: Reaction rates of Z→B and B→Z transitions as a function of final NaCl concentration.

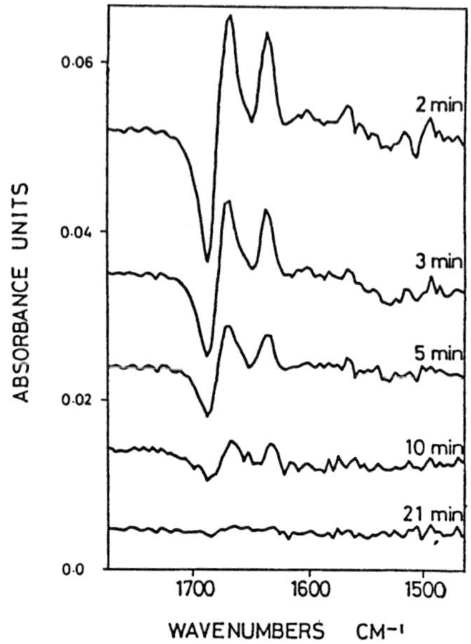

Fig. 2. Infrared difference spectra (30 min-t min) observed in the course of the salt-induced B→Z transition of poly[d(GC)]•poly[d(GC)] on NaCl concentration jump from 1 to 4 M at pH 7.0 and 23°C

4 Time-Resolved Vibrational Spectra

Figure 2 illustrates a series of infrared difference spectra observed in the course of the B→Z conformational change of the duplex in question. A semilogarithmic plot of the intensity of a positive peak (1686 cm^{-1} peak, for example) assignable to the B-form versus time indicates that the B→Z transition is not a simple first-order reaction. For the final NaCl concentration 4M, at pH 7 and at 25°C, two effective rate-constants have been estimated to be 0.006 and 0.0014 sec^{-1}. For this B→Z transition a series of Raman difference spectra was also observed. The rate of the reaction was found to be nearly equal to what was found by the infrared spectroscopy, as long as the duplex is in solution. It was found to be very slow (takes 10 hours or so), however, when it is in a gel. In the course of this B→Z transition, no Raman and infrared bands assignable to an intermediate form have so far been found.

5 B⇌Z Kinetics

By the use of a stopped-flow ultraviolet spectrophotometry, Z→B reaction was found to be much faster than the B→Z reaction. Thus, as is shown in Fig. 1, rate-constant k=0.3→0.05 sec^{-1} at [NaCl]=1.125 M, 25°C. At the midpoint ([NaCl]=2.25 M) of the transition, the conformational changes are extremely slow. Here again, however, the Z→B reaction (on [NaCl]=3→2.25 M) is facter (k=0.02 sec^{-1}) by two orders of magnitude than the B→Z reaction (k=0.0001 sec^{-1}, on [NaCl]=1.5→2.25 M).

6 Deuteration Rates

The results are partly shown by small squares in Fig. 1. The deuteration rate is high for B and low for Z, suggesting that the chance of base-pair opening is very low in Z.

7 Discussions

First of all, it should be pointed out that a long poly[d(GC)]•poly[d(GC)] duplex tends to form a gel, so that its "solution" tends to become heterogeneous. Perhaps one must be cautious of it in interpreting a kinetic data, especially for a concentrated solution. Even by taking this into account, it is quite certain that the rate of reaction at [NaCl]=2.25 M, where [B]/[Z]=1, to reach the equilibrium from the Z-form side is much faster than that from the B-form side. The free energy of the B-unit and Z-unit must be equal to each other here. Therefore, the rate of sliding of the B-Z junction along the duplex must be equal for B growth to that for Z-growth. Therefore, the difference in the overall reaction rates between Z→(B+Z) and B→(B+Z) must be ascribed to the difference in the rate of nucleation. By applying the theory of SCHWARZ [5], the ratio of such a nucleation rate constant γ_B/γ_Z has been estimated to be 1.6×10^2.

This work was supported by a grant (No. 57060004) from Ministry of Education, Science, and Culture of Japan.

References

1. F.M. Pohl and T.M. Jovin: J. Mol. Biol. 67, 375 (1972)
2. F.M. Pohl, A. Rande, and M. Stockburger: Biochim. Biophys. Acta 335, 85 (1973)
3. D.J. Patel, L.L. Canuel, and F.M. Pohl: Proc. Nat. Acad. Sci. USA 76, 2508 (1979)
4. T.J. Thaman, R.C. Lord, A.H.J. Wang, and A. Rich: Nucleic Acids Res. 9, 5443 (1981)
5. G. Schwarz: J.Mol.Biol. 11, 64 (1965)
6. B.H. Zimm and J.K. Bragg: J. Chem. Phys. 31, 526 (1959)
7. J.M. Benevides and G.J. Thomas,Jr.: Nucleic Acids Res. 11, 5747 (1983)
8. Y. Nishimura, C. Torigoe, M. Katahira, and M. Tsuboi: Nucleic Acids Res., Symp. Ser. No. 15, 147 (1984)
9. Y. Nishimura, M. Tsuboi, and T. Sato: Nucleic Acids Res. 12, 6901 (1984)

Kinetic Studies on Bacteriorhodopsin by Resonance Raman Spectroscopy

M. Stockburger, Th. Alshuth, and P. Hildebrandt
Max-Planck-Institut für Biophysikalische Chemie, D-3400 Göttingen, F.R.G.

1 Photochemical Cycle of Bacteriorhodopsin

Bacteriorhodopsin (BR) is a retinal-binding protein in the so-called *purple membrane* (PM) of *Halobacterium halobium*. It acts as a "light-driven" trans-membrane proton pump. The proton gradient established in this way is used by the cell as energy source to drive metabolic processes under unaerobic conditions [1]. The chromophoric site consists of a retinal molecule which is bound to the protein backbone via a Schiff base (C=N) linkage (Fig. 1). Light absorption initiates a cyclic reaction process ("photochemical cycle") in which the parent chromophore BR-570 is converted in a primary photochemical event to the intermediate K-590 which subsequently relaxes in thermal processes via various intermediate states (L-550, M-412) to the original chromophore (Fig. 1) [1].

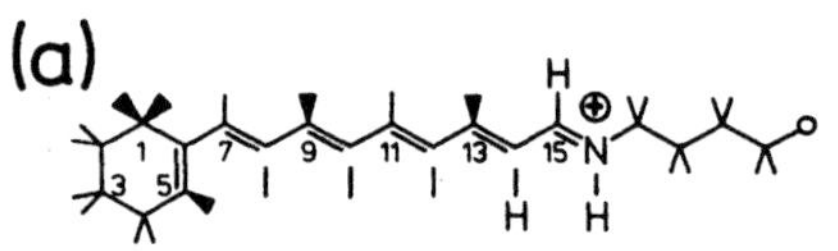

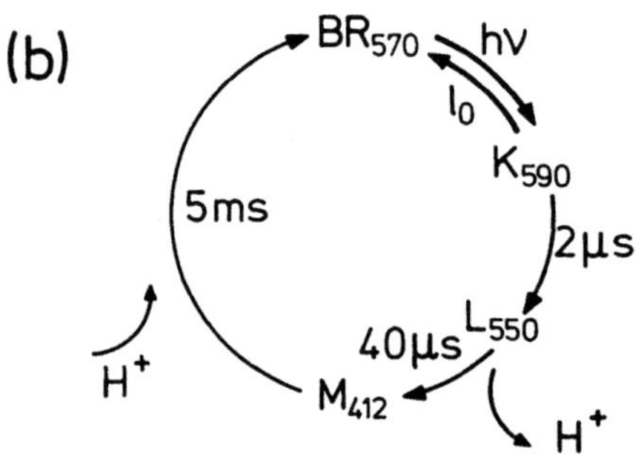

Fig. 1. (a) Retinylidene Schiff base chromophore with retinal in the all-trans configuration. (b) Photochemical cycle of bacteriorhodopsin. The parent chromophore (BR-570) and the intermediates are designated by different capitals and numbers which give the wavelengths of maximum absorption. The thermal decay constants refer to room-temperature and to neutral pH

Resonance Raman (RR) spectroscopy is an ideal tool to study BR's photochemical cycle, since it selectively probes the chromophoric site. In time-resolved experiments Raman spectroscopists succeeded to record spectra of the photolabile parent chromophore and its intermediates [2]. From the spectra it could be definitely concluded that in BR-570 retinal is covalently bound to the protein via a protonated Schiff base (SB) linkage. It was further confirmed that during the photochemical step from BR-570 to K-590 the retinal chain isomerizes from the all-trans to the 13-cis configuration, which is then preserved in the two following intermediates L-550 and M-412.

The RR spectra also clearly revealed that during the transition from L-550 to M-412 a proton is removed from the nitrogen of the SB group. It is tempting to assume that this reaction might trigger the proton pump [1]. In the present study we have therefore concentrated on the two states L-550 and M-412. Not only the RR spectra were recorded but also the kinetic behaviour of the two species was followed on the basis of their characteristic RR bands.

2 Time-Resolved Raman Experiment

When separated from whole cells, one obtains relatively large sheets (≈ 1 μm extension) of purple membrane which contain many thousands of BR molecules in a regular array [3]. The functional properties of BR can be studied conveniently in diluted aqueous suspensions of PM sheets. Therefore, time-resolved RR studies on the photolabile BR chromophore can be performed in the most simple way by using two CW laser beams in combination with a flowing sample of aqueous PM suspension (Fig. 2).

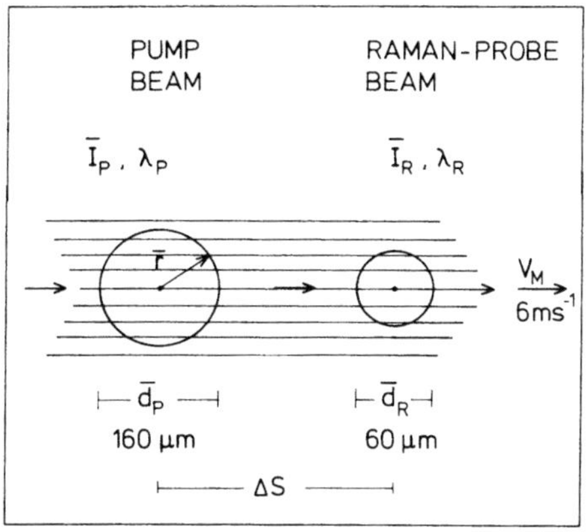

Fig. 2. Design of a pump/probe flow experiment. The focal diameters of the two CW beams ($\bar{d}_P$, $\bar{d}_R$) and the flow velocity (v_M) given in the diagram are those used in the experiments

The primary photo-reaction which starts the cyclic process is governed by the "photoconversion parameter", $l_0\Delta t$, where l_0 is a "rate-constant" and Δt is the residence time of the sample in the pump laser beam [4]. Since l_0 is proportional to the laser power density ($\bar{I}_P$) and Δt is kept constant, the photoconversion parameter can be varied by changing the laser power. In the pump beam $\bar{I}_P$ has to be sufficiently high to form products ($l_0\Delta t > 1$), while in the probe beam $\bar{I}_R$ has to be low enough to avoid largely photoreactions ($l_0\Delta t \ll 1$). In the pump-probe configuration of Fig. 2 power and wavelength as well as the delay time ($\delta = \Delta s/v_M$) can be varied so as to obtain optimum conditions for accumulating and probing the intermediates L-550 and M-412. The RR spectra of the parent BR-570 can be obtained in a probe-only experiment.

A spinning cell (usually 50 cps) of cylindrical shape (40 mm in diameter) was used as a flow system [4, 5]. After a single rotation of the cell (20 ms or longer), the photocycle, whose period under normal conditions (Fig. 1) is 5 ms, is completely finished, so that unphotolyzed sample always enters the laser beam. A time-resolved RR experiment is thus reduced to a quasi-stationary experiment, which allows to accumulate continuously the weak scattered RR radiation and thus to obtain spectra of high quality.

3 RR Spectra of BR-570 and the Intermediates L-550 and M-412

In order to obtain RR spectra of BR-570 and its intermediates, their different absorption spectra have to be considered for selective pumping and probing (Fig. 3). There is no principle problem to obtain the pure RR spectrum of BR-570 in a probe-only experiment. A pure spectrum of M-412 can be also adequately obtained by pumping within the absorption band of BR-570 and probing in the violet (413 nm).

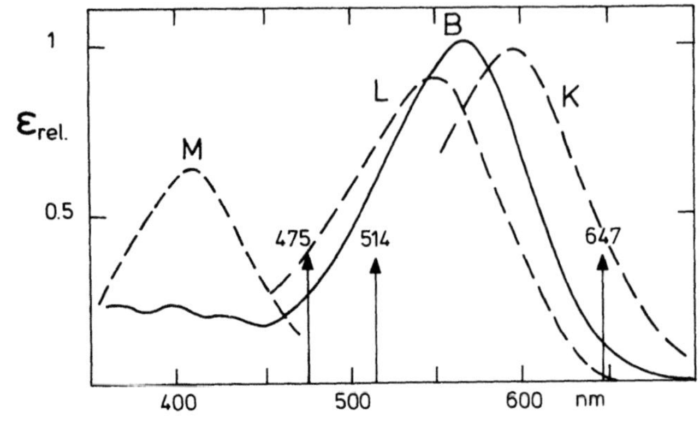

Fig. 3. Absorption spectra of BR chromophores. Spectra of K and L are given in an idealized way based on literature data [6]

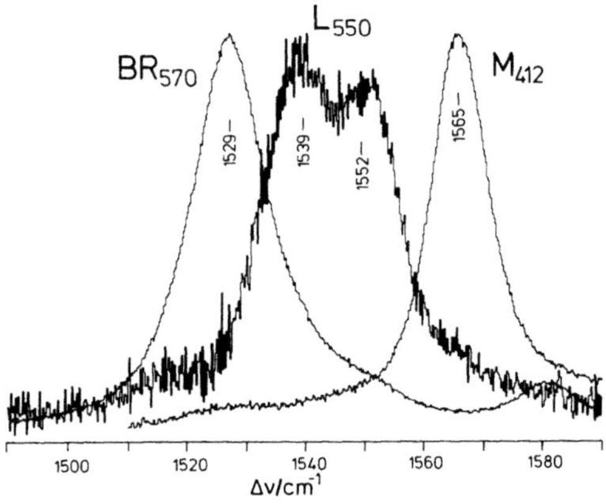

Fig. 4. RR spectra in the C=C stretching region

With respect to L-550, however, two main problems arise. Firstly, not more than 35 percent of BR-570 can be converted into L-550. The reason for this is that photoisomerization in the pump beam also occurs in backward direction from K and L to BR. The second problem arises from the fact that the absorption bands of L-550 and BR-570 strongly overlap, and thus do not allow a spectral separation by selective probing. Pure RR spectra of L, therefore, can only be obtained by subtracting the 65 percent contribution of the parent from the mixed spectrum.

During the photochemical cycle, the retinylidene chromophore isomerizes from all-trans in BR-570 to 13-cis in L-550 and M-412. In addition, deprotonation occurs between L and M and also the electrostatic interaction of the positively charged SB group and the protein environment is modified. All these structural changes are reflected in the RR spectra of the three species. Here we only consider the strongest vibrational bands between 1490 and 1590 cm^{-1} (Fig. 4) which refer to C=C stretching vibrations of the retinal moiety. Shape and width of these bands are characteristic of each chromophore, and thus can be used as a diagnostic tool. This is particularly important for the detection of L-550, which is well characterized by a double peak [7, 8] and therefore can be more accurately separated from BR-570 than is possible on the basis of the strongly overlapping optical absorption bands. We found that the shapes of the C=C stretching bands can be adequately fitted by Lorentzians. With the help of the spectral parameters which were deduced from the fitting procedure, the time-evolution of the three species BR, L and M will be analyzed.

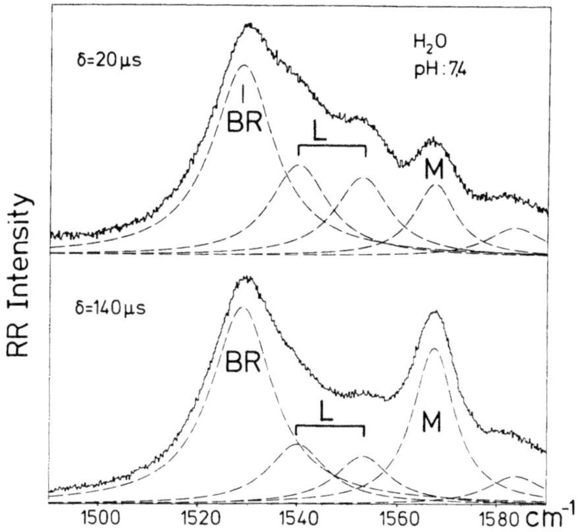

Fig. 5. RR spectra in the C=C stretching region at two different delay times. Pump beam: 647 nm, probe beam: 475 nm

4 Temporal Evolution of BR-570, L-550 and M-412

RR spectra in the C=C stretching region were recorded for various delay times. As pump radiation the 647 nm line of a krypton laser was used, since the red pump radiation does not photolyze the L-550 intermediate. As probe beam, the blue line of an argon laser at 475 nm was chosen. At this wavelength the chromophores BR, L and M have RR cross-sections of comparable magnitude. Representative spectra are shown in Fig. 5 together with the result of the fitting procedure. For delay times $\delta > 10$ µs which are only of interest here, the intermediate K-590 can be neglected and for the evaluation of the RR spectroscopic data only the three species BR, L and M have to be considered. Therefore their relative concentrations from the integrated RR intensities of the diagnostic C=C bands can be obtained, under the condition that they add to unity for each value of δ. The temporal evolution for $0 < \delta < 400$ µs is shown in Fig. 6. As long as the probe beam interferes with the pump beam, BR-570 de-

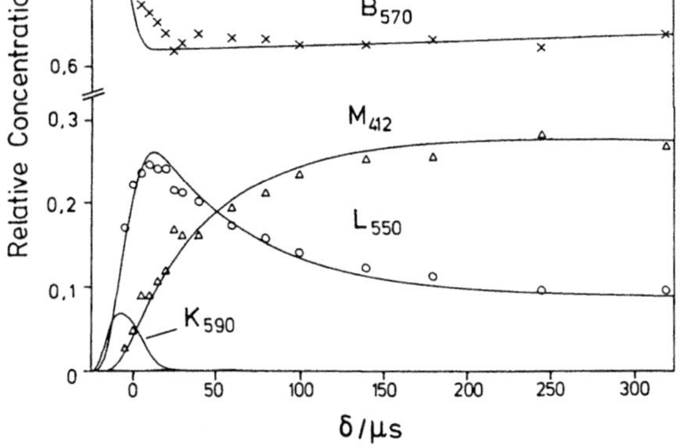

Fig. 6. Time-behaviour of BR-570, L-550 and M-412 in aqueous PM suspensions at pH 7.4 and room temperature

creases and the intermediates increase in concentration. The sharp levelling off in the decay of BR at about 10 µs indicates the separation of the two beams. From there on BR remains constant, while L-550 decays and at the same rate M-412 increases.

New features, however, appear in the time-behaviour of L-550 which had not been reported previously, and which are inconsistent with the reaction scheme in Fig. 1. Taking into account additional data reported by Alshuth and Stockburger [7, 8], the following picture emerges:

(1) The amplitude of L-550 does not decay to zero, as would be expected from a linear reaction chain. Instead, a residual limit is approached between 200 and 500 µs.
(2) In the ms time-domain the residual amplitude of L-550 decays synchroneously with the intermediate M-412 (decay time of 5 ms at pH 7.4).

(3) In the range between pH 7.4 and 4.6 the residual amplitude of L-550 does not depend significantly on the proton concentration in bulk solution.

A straightforward explanation for the peculiar behaviour of L-550 would be that L-550 and M-412 are mutually coupled to each other by forward and backward reactions, which leads to the establishment of an intermediate equilibrium and to an oscillatory motion of the proton from the SB group of the chromophore to an acceptor in the protein environment. Indeed, the solid curves in Fig. 6, which well fit the experimental data, were calculated on the basis of this intermediate equilibrium [7, 8].

5 A Molecular Model

In this section, a molecular model is suggested for the light-induced cyclic reaction of the BR chromophore which takes into account our new data (Fig. 7). The structure of the parent chromophore BR-570 is essentially determined by the electrostatic interaction of the positively charged SB group and a carboxylate counterion. It was argued by Hildebrandt and Stockburger, on the basis of RR spectroscopic studies, that this ion-pair structure is stabilized by surrounding water molecules [9]. After photoisomerization to K-590 the chromophore relaxes in 4 µs to the second metastable state L-550.

At this stage we follow the arguments of Tavan et al. [10]. From quantum chemical calculations, they concluded that in a protonated chromophore like L-550 the barrier for rotation about the 14-15 single bond is much higher than in the unprotonated state. This implies that a torsion about this bond would decrease the pK_a value of the SB group dramatically, and thus could initiate the removal of the SB proton. - For sake of simplicity, a full rotation about the 14-15 bond is indicated in the diagram of Fig. 7. The role of an intermediate proton acceptor is played in our model by the carboxylate counterion. This assumption obtains support from time-resolved

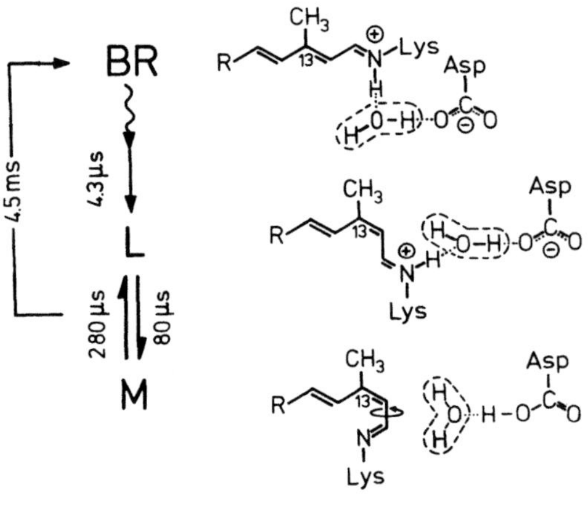

Fig. 7. A molecular model of the cyclic reaction of the BR chromophore. The given rate constants are from the present study and refer to room temperature and pH 7.4

infrared measurements. Thus it was found that synchroneously with M-412, a carboxylate group in the protein becomes protonated [11]. The establishment of an intermediate L $\rightleftharpoons$ M equilibrium then would mean that the proton carries out an oscillatory motion between the SB group and the counterion, which is triggered by a twist about the 14-15 single bond of the retinal moiety.

Until now it was difficult to understand that the thermal isomerization about the 13-14 double bond in the transition from M-412 to BR-570 could occur at the high rate of 200 s^{-1} (Fig. 1), since the rotation barrier in the unprotonated M-412 for this double bond is fairly high [10]. Our model now offers the possibility that the back-reaction occurs from the protonated L-550 state in which the rotational barrier for the 13-14 double bond is expected to be much lower [10].

It should be emphasized that the model of Fig. 7 can explain the experimental results but cannot answer the question if the SB proton is involved in the proton pump mechanism or not. However, it cannot be excluded from our data that the oscillatory proton motion we invoked may trigger the translocation of protons in the vicinity of the chromophore, as was already suggested previously [11].

In conclusion, we wish to emphasize that new and important information on the photochemical cycle of bacteriorhodopsin could be obtained from kinetic vibrational spectroscopy. Basically, we owe this to the fact that the various BR chromophores can be sensitively identified by their characteristic vibrational bands in the C=C stretching region.

1 W. Stoeckenius, R. H. Lozier, and R. Bogomolni: Annu. Rev. Biochem. 51, 587 (1982)
2 S. O. Smith, J. Lugtenburg, and R. A. Mathies: J. Membrane Biol. 85, 95 (1985)
3 R. Henderson: Annu. Rev. Biophys. Bioeng. 6, 87 (1977)
4 M. Stockburger, W. Klusmann, H. Gattermann, G. Massig, and R. Peters: Biochemistry 18, 4886 (1979)
5 T. Alshuth, P. Hildebrandt, and M. Stockburger: Spectroscopy of Biological Molecules, C. Sandory and T. Theophanides (eds.), 329 (1984) (D. Reidel Publishing Company)
6 J. F. Nagle, L. A. Parodi, and R. H. Lozier: Biophys. J. 38, 161 (1982)
7 T. Alshuth, and M. Stockburger: Photochem. Photobiol. (1985), in press
8 T. Alshuth, Ph.D. Thesis, University of Göttingen (1985)
9 P. Hildebrandt, and M. Stockburger: Biochemistry 23, 5539 (1984)
10 P. Tavan, K. Schulten, and D. Oesterhelt: Biophys. J. 47, 415 (1985)
11 F. Siebert, W. Mäntele, and W. Kreutz: FEBS Lett. 141, 82 (1982)

Resonance Raman Study on the Role of Water in Bacteriorhodopsin

P. Hildebrandt, A. Hagemeier, and M. Stockburger

Max-Planck-Institut für Biophysikalische Chemie, Abteilung Spektroskopie, D-3400 Göttingen, F.R.G.

Bacteriorhodopsin (BR), the major component of the purple membrane (PM) of Halobacterium halobium acts as a light-driven proton pump (for a review see ref. [1]). The absorption of BR in the visible is due to its chromophoric center, which contains a retinal molecule bound to the protein via a Schiff's base (SB) linkage. On illumination, BR runs through a cyclic process which controls the proton pumping across the membrane (Fig. 1). In the dark BR exists in the two equilibrated forms BR-570 and BR-548 ("570" and "548" denote the absorption maxima). Under light-adapted conditions BR-570 predominates. It is well known that in BR-570 and BR-548 the retinal moiety is in the all-trans and 13-cis configuration, respectively. The big red shift of BR chromophores with respect to the absorption of related model compounds in solution results from the strong interaction the protein exerts on the bound retinal. It could be demonstrated by spectroscopic and kinetic measurements that water influences the retinal-protein interactions [2, 3].

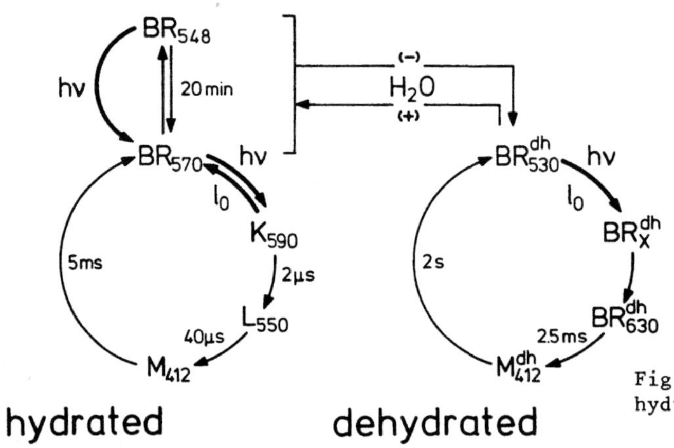

Fig. 1. Photocycles of BR in the hydrated and dehydrated state.

In the present paper the role of water in BR is studied by resonance Raman (RR) spectroscopy which selectively probes the chromophoric site. Most of our experiments were performed with membrane films. From hydrated PM films we obtained RR spectra that were identical with those from aqueous dilute PM suspensions [4]. For time-resolved experiments the PM films were prepared on a rotating disk which was deposited in a vacuum chamber. Time-resolved RR experiments of PM suspensions were performed using a conventional rotating cell. Further description of the experimental device, including the low-temperature equipment for stationary measurements, is given in ref. [4].

Dehydration of PM films by evacuation leads to a blue-shift of the absorption maximum from 570 to 530 nm (BR^{dh}-530). On illumination, the water-free BR runs through a photocycle which is completely different from the hydrated BR. Figure 3 displays the RR spectrum of the parent complex BR^{dh}-530 obtained at low temperature (77 K)

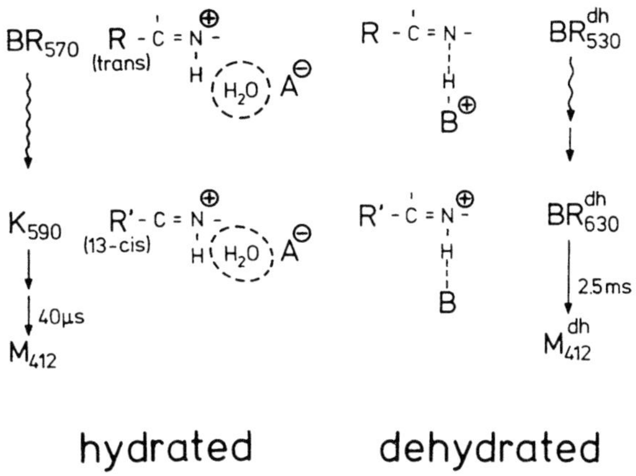

Fig. 2. Primary photochemical events of hydrated and dehydrated BR.

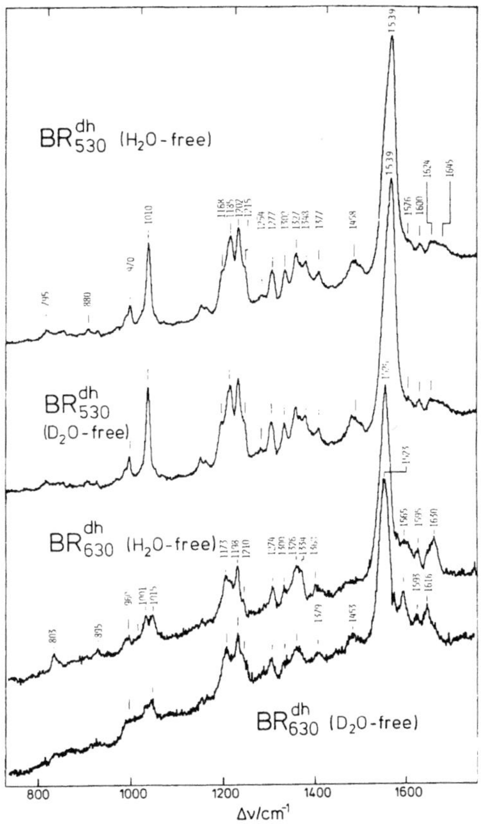

Fig. 3. RR spectra of BR^{dh}-530 (stationary at T = -196 °C; probe beam 514 nm, pump beam 647 nm) and BR^{dh}-630 (time-resolved at room temperature, probe beam 514 nm) of dehydrated PM films prepared from H_2O and D_2O suspensions.

with 514 nm excitation. An additional pump beam (647 nm) was used to convert the primary photoproduct BR^{dh}-X back to the unphotolyzed complex. From the intensity distribution in the fingerprint region,which shows far-reaching similarities with the RR spectrum of BR-548 [4, 5],it is concluded that the chromophore of BR^{dh}-530 mainly exists in the 13-cis configuration. However, in contrast to BR-548 no isotopic shifts are observed in the RR spectra of BR^{dh}-530 prepared from H_2O and D_2O suspensions (Fig. 3). From the identity of both spectra,it is concluded that the SB group in BR^{dh}-530 is deprotonated.

The situation is different in the first thermal intermediate of the BR^{dh}-530 photocycle, BR^{dh}-630, which was characterized by time-resolved RR spectroscopy (Fig. 3). Here,H_2O- and D_2O-dehydrated samples give rise to significant isotopic effects. For example, the C=N stretching vibration shifts from 1630 to 1616 cm^{-1}. These results provide strong evidence for a protonated SB in BR^{dh}-630. No unequivocal conclusion is possible concerning the retinal configuration. Due to the strong bands in the hydrogen out-of-plane bending region and the pattern in the fingerprint region, it is likely that the retinal chain is twisted around the C_{13}=C_{14} bond.

The RR spectrum of the long-living intermediate M^{dh}-412 - obtained in a pump/probe (514/406) double beam experiment - closely resembles the RR spectrum of the corresponding intermediate in the hydrated state, M-412 (13-cis, deprotonated SB) [4].

So far we have demonstrated that water controls the structure and function of BR chromophores. The RR spectroscopic analysis of the C=N stretching vibration in hydrated BR complexes provides evidence that single water molecules are placed in close vicinity to the SB group. Figure 4 shows the bands of BR-570 (protonated SB) from PM suspensions in H_2O, D_2O and in a 1 : 1 mixture of H_2O and D_2O. Besides the

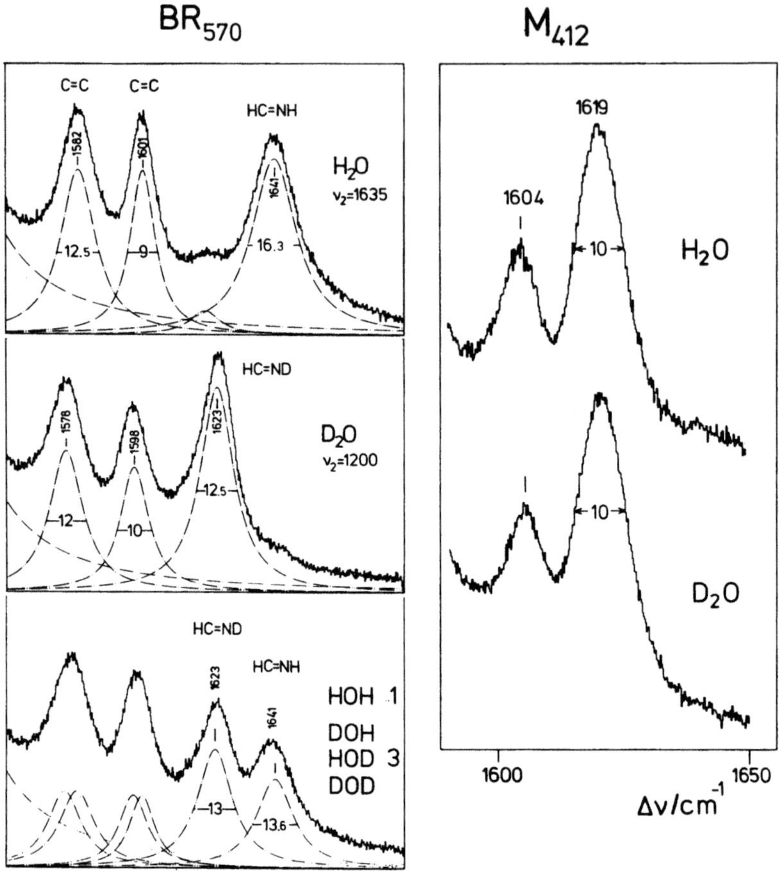

Fig. 4. RR bands of the C=N stretching vibrations of BR-570 and M-412 from PM suspensions in H_2O, D_2O and in a H_2O/D_2O mixture.

frequency down-shift of the C=ND stretch by 16 cm^{-1} the width of this band is strongly influenced. After a complete exchange the FWHH value decreases from 16.3 (100 % H_2O) to 12.5 cm^{-1} (100 % D_2O) while in a 1 : 1 mixture both bands have nearly the same widths (13.6 and 13.0 cm^{-1}). This means that the width of the band is independent of the fact if a proton or deuteron is attached to the SB group but is only affected by the solvent. Since the frequency of the C=NH stretch is in near resonance with the frequency of the bending mode of water (1635 cm^{-1}) but far off-resonance with the corresponding modes in D_2O or HOD,it is suggested that the broadening effect in H_2O is due to vibrational energy-transfer from the C=NH stretch to the H_2O bend. According to this interpretation,the low statistical weight of H_2O isomers in the mixture (25 %) only leads to a slight increase of both FWHH values with respect to the band width of the C=ND stretching mode in 100 % D_2O. Since the vibrational energy-exchange mechanism is a short-range effect [6] this implies that at least one water molecule must be closely attached to the SB group.

The regeneration of the native BR from the dehydrated PM is complete and very fast in the presence of small amounts of water. Thus,irreversible denaturation cannot be responsible for the observed effects. Furthermore, neutron diffraction [7] and IR absorption [3] experiments could not detect any changes of the protein backbone after dehydration. Therefore, it is suggested that the important modifications at the chromophoric site, e. g. all-trans - 13-cis isomerization, deprotonation of the SB, different photochemical behaviour, are due to the removal of those water molecules which are located in the immediate vicinity of the chromophore.

There is convincing evidence that the intact BR structure and function of the chromophore are largely determined by the ionic interactions between the positively charged SB group and a negatively charged counterion (carboxylate) [8]. Based on our own experimental findings,we suggest that the ion pair structure is stabilized by water molecules. These water molecules must be fixed to the protein environment at the binding site,so that they cannot hydrolyze the C=N bond. The absence of line-broadening effects on the C=N stretching mode in the M-412 intermediate (Fig. 3) can be interpreted by an increased distance between water molecules and the SB group. This interpretation is in line with the RR spectroscopic similarities between M-412 and M^{dh}-412,which indicate that no ionic interactions of the SB group water molecules with side groups of the protein,but rather the nonpolar interactions of the chromophore with the protein,are responsible for the chromophoric structure. This means that during the isomerization processes,which start in the primary photochemical step, the SB group migrates from a polar to an unpolar environment [9].

The different chromophoric structure in the hydrated and dehydrated PM also gives rise to a different photochemical behaviour. It is suggested that in BR^{dh}-530 a positively charged protonated side group of the protein is near the (unprotonated) SB - presumably linked via a hydrogen bond. This can account for the absorption maximum of 530 nm [10]. The first thermal intermediate BR^{dh}-630 has a protonated SB. Therefore it is very likely that proton-transfer is the primary photochemical step according to Sandorfy's model [11]. The new charge distribution in BR^{dh}-630 is stabilized by conformational changes of the retinal chain. The main structural and kinetic features of both photocycles are summarized in Fig. 2.

References

1 W. Stoeckenius, R. H. Lozier, and R. A. Bogomolni, Biochim. Biophys. Acta 505, 215 (1979)
2 R. Korenstein and B. Hess, Nature 270, 184 (1977)
3 Y. A. Lazarev and E. L. Terpugov, Biochim. Biophys. Acta 590, 324 (1980)
4 P. Hildebrandt and M. Stockburger, Biochemistry 23, 5539 (1984)
5 T. Alshuth and M. Stockburger, Ber. Bunsenges. Phys. Chem. 85, 484 (1981)
6 F. Legay, in: "Applications of Lasers in Chemistry and Biochemistry" (C. B. Moore, ed.) Vol. 2, p. 43 (1977), Academic Press
7 P. K. Rogan and G. Zaccai, J. Mol. Biol. 145, 281 (1981)
8 U. Fischer and D. Oesterhelt, Biophys. J. 28, 211 (1979)
9 A. Warshel and N. Barboy, J. Am. Chem. Soc. 104, 1470 (1982)
10 J. Favrot, J. M. Leclerq, R. Roberge, C. Sandorfy, and D. Vocelle, Photochem. Photobiol. 29, 99 (1979)

Picosecond and Nanosecond Resonance Raman Evidence for Structural Relaxation in Bacteriorhodopsin's Primary Photoproduct

D. Stern and R. Mathies

Department of Chemistry, University of California, Berkeley, CA 94720, USA

The protein bacteriorhodopsin (bR) pumps protons across the cell membrane of Halobacterium halobium, producing an electrochemical potential gradient that drives ATP synthesis [1]. Visible light absorbed by bacteriorhodopsin's protonated retinal Schiff base chromophore initiates the proton-pumping photocycle (Fig. 1). Formation of the primary photoproduct, K, involves chromophore isomerization from all-trans to 13-cis [2] and the storage of ~16 kcal/mole of potential energy [3]. The Schiff base nitrogen becomes deprotonated and reprotonated during subsequent events in the photocycle; presumably the Schiff base proton is "pumped." Charge-separation (displacement of the Schiff base nitrogen away from a protein counterion) is thought to be responsible for differences in the photointermediates' absorption maxima and probably plays a role in energy-storage [4]. To understand the mechanism of proton pumping in more detail, we are using time-resolved Raman spectroscopy to study the changes in chromophore structure that occur in picoseconds and nanoseconds.

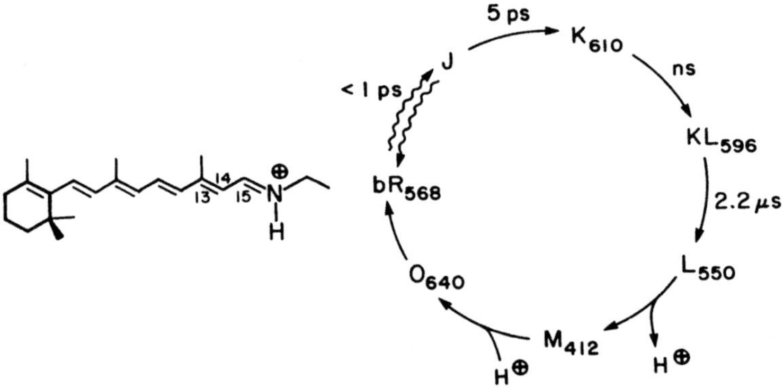

Figure 1. The proton-pumping photocycle of bacteriorhodopsin. Subscripts give absorption maxima of the intermediates in nm. We use the symbol K with no subscript to refer generically to J, K_{610}, and KL_{596}.

Time-resolved absorption spectroscopy shows that three photocycle intermediates appear on a picosecond to nanosecond time-scale at room temperature. The most red-shifted intermediate, J, forms in 430 fs and decays to K_{610} in 5 ps [5]. K_{610} decays to KL_{596} in several ns [6]. Time-resolved Raman methods having 40 ps to 100 ns time-resolution have been used to study the primary photochemistry. The first Raman spectrum of K at room temperature was reported by TERNER et al. [7], who showed that the Schiff base in K is protonated. HSIEH et al. [8,9] concluded that all-trans → 13-cis isomerization occurs within 40 ps and that some structural changes occur in the following nanoseconds. SMITH et al. [10] showed that K at room temperature is similar to K_{625}, the primary photoproduct trapped at liquid-nitrogen temperature. All of these Raman experiments were conducted before KL_{596} was identified as a distinct intermediate. Consequently, the structural differences between KL_{596} and K_{610}

have not yet been determined. Furthermore, no Raman experiment with sufficient time-resolution to see the J intermediate has yet been performed.

We have used pulsed laser excitation and multichannel detection to obtain spectra of K on 20 ns, 200 ps, and 7 ps time-scales. The pattern of frequencies and intensities in the 1100-1300 cm^{-1} fingerprint region implies that all-*trans* → 13-*cis* isomerization is complete within 7 ps. Time-evolution of the ethylenic stretching and C_{15}-H out-of-plane wag frequencies shows that significant structural relaxation of the Schiff base moiety occurs between 7 ps and 20 ns. This structural relaxation may involve movement of the Schiff base group relative to a protein counterion.

I. Experimental Methods

Time-resolved Raman spectra were obtained using a single laser pulse to both photolyze the sample and excite Raman scattering from the photoproduct. The photoalteration parameter, $F = 3.824 \times 10^{-21} E\varepsilon\phi/A$, is a useful measure of the extent of photolysis [11]. Here E is the number of photons per pulse, ε is the extinction coefficient, ϕ is the photoisomerization quantum yield, and A is the focused beam area. A pulse with a high photoalteration parameter ($F > 1$) gives a Raman spectrum of a mixture of bR_{568} and K. A pulse with a low photoalteration parameter ($F < 0.5$) gives a Raman spectrum of nearly pure bR_{568}. Because the quantum yield for the reaction K → bR_{568} is about twice the quantum yield for the reaction bR_{568} → K, at most ~30% of the sample can be converted to photoproducts. With 514 nm excitation the resonance enhancement for K is poor, and K contributes less than 30% of the intensity in the high-F Raman spectrum. Spectra of pure K must therefore be obtained by subtraction. The choice of subtraction parameter is a matter of judgment. Figure 2 shows that moderate over- or under-subtraction has little effect on the peak frequencies, although it changes the relative intensities of some lines.

The purple membrane suspension (30 μM in distilled water at 4°C) flows through a glass capillary fast enough that each high-F pulse illuminates a fresh portion of the sample. The flow-rate is 5 m/s for the 20 ns and 200 ps experiments, and 75 cm/s for the 7 ps experiment.

The 20 ns and 200 ps experiments make use of a Spectra-Physics argon ion laser operating at 514 nm. Cavity dumping produces 20 ns pulses at any desired repetition rate. The beam is focused onto the flowing sample to a diameter of 20 μm. The pulse energy and repetition rate are 160 nJ and 12 kHz for the high-F spectra, 8 nJ and 200 kHz for the low-F spectra. Mode-locking produces 200 ps pulses at a repetition rate of 82 MHz. Here the cavity dumper operates outside the laser cavity as a pulse selector to reduce the pulse energy and repetition rate to the desired values (4 nJ and 350 kHz for the high-F spectra, 0.5 nJ and 2.0 MHz for the low-F spectra). The focused beam diameter for the 200 ps experiment is 8 μm. The detection system consists of a PAR 1205A/1205D optical multichannel analyzer coupled to a modified Spex 1400 double monochromator [12].

For the 7 ps experiment we use a synchronously-pumped, extended-cavity Coherent 590 dye laser to produce nanojoule pulses at 576 nm. Autocorrelation (using a Spectra-Physics 409 autocorrelator) gives a double-sided exponential with a FWHM of 5 ps. The ratio of autocorrelation width to pulse duration depends on the pulse shape. MCDONALD *et al.* [13] show that the pulse is a noise burst with a Gaussian envelope even when the autocorrelation function looks like a double-sided exponential. The pulse duration, therefore, is probably ~7 ps. A three-stage amplifier (Quanta-Ray PDA-1) pumped by the second harmonic from a Nd:YAG laser (Quanta-Ray DCR-2A) amplifies 20 pulses per second to 0.6 mJ. The amplified pulses, attenuated to 0.15 mJ and focused to 2 mm diameter, give high-F spectra. The unamplified dye laser beam, attenuated to 5 mW and cylindrically focused to 2 mm x 20 μ, gives low-F spectra. The detection system for these experiments consists of a PAR 1218/1412 silicon photodiode array detector coupled to a Spex 1870 spectrograph.

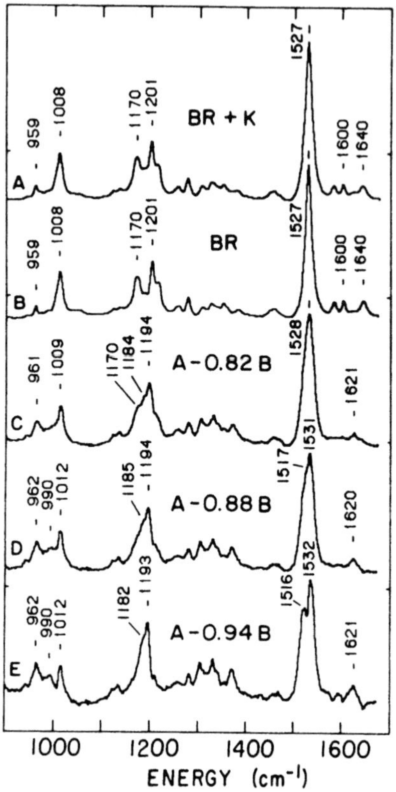

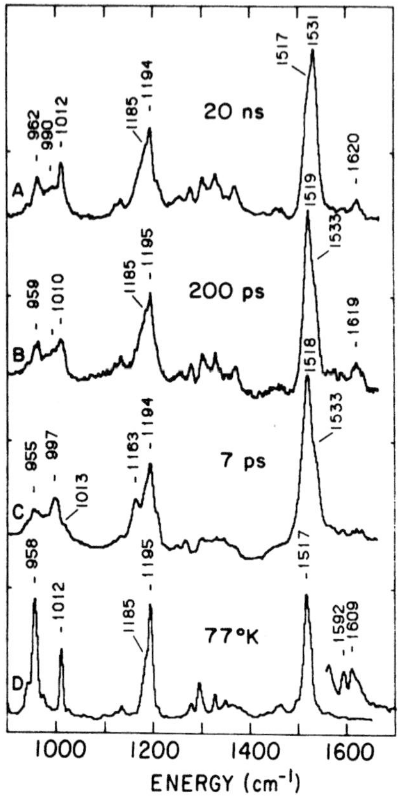

Figure 2 (left). An example of the subtraction procedure. A) high-F 20 ns spectrum B) low-F 20 ns spectrum C) under-subtracted, A − 0.82B D) correctly subtracted, A − 0.88B E) over-subtracted, A − 0.94B

Figure 3 (right). Time-resolved Raman spectra of bacteriorhodopsin's primary photoproduct. A) spectrum of KL_{596} taken with 20 ns pulses at 514 nm B) spectrum of K_{610} taken with 200 ps pulses at 514 nm C) spectrum of K_{610} taken with 7 ps pulses at 576 nm D) spectrum of K_{625} obtained using a low-temperature spinning cell [2] with a 514 nm pump beam and a 752 nm probe. The spectrum of the Schiff base region (insert) was obtained using a 647 nm pump and 514 nm probe.

II. Results

Time-resolved Raman spectra of K at 20 ns, 200 ps and 7 ps are shown in Fig. 3, along with a spectrum of K_{625} trapped at $77^{o}K$. In each spectrum, the most intense feature in the C-C stretching ("fingerprint") region is an asymmetric peak at ~1194 cm^{-1}, with a strong shoulder at 1185 cm^{-1} and a weaker shoulder at ~1208 cm^{-1}. The similarity of the fingerprints in our spectra shows that no major structural changes such as isomerization occur between 7 ps and 20 ns. The 1163 cm^{-1} line in the 7 ps spectrum is absent in the other spectra, but this probably arises because the 7 ps spectrum was obtained with longer-wavelength excitation than the 200 ps or 20 ns spectra (576 nm vs. 514 nm). A line in bR_{568} at a similar frequency is preferentially enhanced with red excitation [14]. The presence of a line at 1163 cm^{-1} in the 100 ns K spectrum obtained with 568 nm excitation by TERNER *et al.* [7] supports our argument.

Some temporal evolution in chromophore structure is apparent from the ethylenic stretching region of our spectra. The 20 ns spectrum has a peak at 1531 cm^{-1} with a shoulder at 1517 cm^{-1}. The 200 ps and 7 ps spectra each have a peak at ~1519 cm^{-1} with a shoulder at 1533 cm^{-1}. Probably the ~1531 cm^{-1} line belongs to KL_{596} and the ~1519 cm^{-1} line belongs to K_{610}. This assignment is qualitatively consistent with the known inverse correlation between ethylenic stretching frequency and absorption maximum [15]. The similarity of the ethylenic frequencies in the 200 ps and 7 ps spectra implies that 7 ps pulses are not short enough to resolve the $J \rightarrow K_{610}$ transition.

The C_{15}-hydrogen out-of-plane wag (HOOP) frequency in K seems to increase monotonically at shorter times. The line at ~990 cm^{-1} in the 20 ns and 200 ps spectra can be identified as the C_{15}-HOOP based on its characteristic frequency shifts in the $^{13}C_{15}$, C_{15}-D, and N-D isotopic derivatives [16]. In the 7 ps spectrum the C_{15}-HOOP appears at 997 cm^{-1}, where it mixes with and borrows intensity from the methyl rock at 1013 cm^{-1}. As the two lines approach each other in frequency, the amount of mixing increases; as a result the C_{15}-HOOP gains intensity and the methyl rock loses intensity. It is hard to determine the C_{15}-HOOP frequency accurately in the 20 ns and 200 ps spectra because the line is broad and not well resolved, but the reduced intensity of the methyl rock in the 200 ps spectrum suggests that the C_{15}-HOOP frequency is slightly higher at 200 ps than at 20 ns.

The Schiff base normal mode (predominantly C=N stretch mixed with N-H rock) appears at ~1620 cm^{-1} in the 20 ns and 200 ps spectra. The decrease in frequency observed upon $^{13}C_{15}$, N-D, and C_{15}-D isotopic substitution proves this assignment [16]. The frequency decrease upon N-D substitution measures the amount of coupling between the C=N stretch and N-H rock, which may be sensitive to the Schiff base hydrogen bonding environment. The N-D shift is 11 cm^{-1} at 20 ns, but only 7 cm^{-1} at 200 ps.

III. Discussion

The fingerprint region in our spectra resembles the fingerprint region in spectra of K_{625}, which BRAIMAN and MATHIES [2] have shown is 13-*cis*. The asymmetric band at ~1194 cm^{-1} in these spectra seems to be unique to K. No other photointermediate or retinal derivative has such a fingerprint. Since the ~1194 cm^{-1} band appears in even the 7 ps spectrum, we conclude that all-*trans* → 13-*cis* isomerization is complete within 7 ps.

It is reasonable to expect that the photoproduct trapped at low temperature might be almost identical to the first transient intermediate present at room temperature. We can test this idea by comparing the K_{625} spectrum in Fig. 3D to the room-temperature spectra in Fig. 3A-C. Using isotopic substitution we have assigned the 1609 cm^{-1} line in Fig. 3D as the Schiff base vibration [16], in agreement with recent FTIR results [17]. The Schiff base frequency in the 7 ps room-temperature intermediate is not yet known, but we believe it is ~1619 cm^{-1} based on the 200 ps and 20 ns spectra. The 10 cm^{-1} difference in Schiff base frequency indicates a significant difference in Schiff base environment. Additional evidence for structural differences between the room-temperature and low-temperature species comes from the C_{15}-HOOP frequency. The C_{15}-HOOP contributes intensity to lines at 958 cm^{-1} and 974 cm^{-1} in the K_{625} spectrum [10,18], but appears at 997 cm^{-1} in the 7 ps spectrum. Evidently, low-temperature trapping and time-resolved techniques do not capture the same pigment structure.

The 12 cm^{-1} increase in ethylenic frequency between 200 ps and 20 ns provides additional evidence that KL_{596} is an intermediate distinct from K_{610}. The increase in ethylenic frequency, and the blue shift in absorption observed on a similar time-scale [6], both suggest that the $K_{610} \rightarrow KL_{596}$ transition involves relaxation of the Schiff base nitrogen towards a protein counterion. Formation of a stronger hydrogen bond between the Schiff base proton and a counterion might accompany such a structural relaxation. Changes in hydrogen bonding can alter the coupling between C=N stretch and N-H rock [19], and indeed the shift in Schiff base frequency upon N-deuteration increases from 7 cm^{-1} at 200 ps to 11 cm^{-1} at 20 ns. Interactions between the retinal chromophore and negative charges in the binding site might also be responsible for the change in C_{15}-HOOP frequency from 997 cm^{-1} at 7 ps to 990 cm^{-1} at 20 ns. Attempts to calculate the effects of a negative charge near the Schiff base nitrogen on the Schiff base and C_{15}-HOOP frequencies are in progress.

Acknowledgment

This work was performed in collaboration with Prof. Johan Lugtenburg and his coworkers at the University of Leiden in the Netherlands. The K_{625} spectrum was obtained by Mark Braiman. The spectrum of the Schiff base region of K_{625} was obtained in collaboration with Steven O. Smith. Support from the NSF (CHE 8116042) and the NIH (EY-02051) is gratefully acknowledged.

References

1. W. Stoeckenius and R.A. Bogomolni: Annu. Rev. Biochem. 51, 587 (1982)
2. M. Braiman and R. Mathies: Proc. Natl. Acad. Sci. USA 79, 403 (1982)
3. R.R. Birge and T.M. Cooper: Biophys. J. 42, 61 (1983)
4. B. Honig, T. Ebrey, R.H. Callender, U. Dinur, and M. Ottolenghi: Proc. Natl. Acad. Sci. USA 76, 2503 (1979)
5. M.C. Nuss, W. Zinth, W. Kaiser, E. Kolling, and D. Oesterhelt: Chem. Phys. Lett. 117, 1 (1985)
6. Y. Shichida, S. Matuoka, Y. Hidaka, and T. Yoshizawa: Biochim. Biophys. Acta 723, 240 (1983)
7. J. Terner, C.-L. Hsieh, A.R. Burns, and M.A. El-Sayed: Proc. Natl. Acad. Sci. USA 76, 3046 (1979)
8. C.-L. Hsieh, M. Nagumo, M. Nicol, and M.A. El-Sayed: J. Phys. Chem. 85, 2714 (1981)
9. C.-L. Hsieh, M.A. El-Sayed, M. Nicol, N. Nagumo, and J.-H. Lee: Photochem. Photobiol. 38, 83 (1983)
10. S.O. Smith, M. Braiman, and R. Mathies: Time-Resolved Vibrational Spectroscopy, ed. by G. Atkinson (Academic, New York 1983), pp. 219-230
11. R. Mathies, A.R. Oseroff, and L. Stryer: Proc. Natl. Acad. Sci. USA 73, 1 (1976)
12. R. Mathies and N.-T. Yu: J. Raman Spectrosc. 7, 349 (1978)
13. D.B. McDonald, J.L. Rossel, and G.R. Fleming: IEEE J. Quantum Electron. 17, 1134 (1981)
14. A.B. Myers, R.A. Harris, and R.A. Mathies: J. Chem. Phys. 79, 603 (1983).
15. M.E. Heyde, D. Gill, R.G. Kilponen, and L. Rimai: J. Am. Chem. Soc. 93, 6776 (1971)
16. D. Stern, S.O. Smith, J. Lugtenburg, and R. Mathies: manuscript in preparation.
17. K.J. Rothschild, P. Roepe, J. Lugtenburg, and J.A. Pardoen: Biochemistry 23, 6103 (1984)
18. S.O. Smith: Ph.D. Thesis, University of California, Berkeley (1985)
19. A. Maeda, T. Ogurusu, T. Yoshizawa, and T. Kitagawa: Biochemistry 24, 2517 (1985)

Picosecond Time-Resolved Resonance Raman Spectroscopy of Bacteriorhodopsin Intermediates

G.H. Atkinson, I. Grieger, and G. Rumbles

Department of Chemistry, University of Arizona, Tucson, AZ 85721, USA

The value of bacteriorhodopsin (BR) as a model for the molecular dynamics associated with the retinal chromophore in visual pigments and for the associated proton and ion pumping mechanisms across membranes has been described extensively [1-3]. Although many aspects of both the BR dynamics and the structures of BR intermediates have been reported [2-3], significant parts of the molecular mechanism underlying its biochemical function remain either unknown or only partially characterized. One such area encompasses the initial molecular changes that occur upon radiative excitation of BR. This initial phase of the "BR photocycle" is followed by a complex series of interdependent changes in molecular bonding, conformation, and environment of both the retinal chromophore and the opsin. These changes drive proton pumping across the bacterial membrane, the function which underlies the biochemical activity of BR [1-3]. Since the chemical energy for this series of processes is thought to derive from these initial steps, an understanding of the specific molecular changes is of considerable importance.

Transient absorption measurements have suggested that these initial molecular changes occur on the picosecond time-scale for visual pigments containing retinal [4-6]. The absence of structural information in these picosecond transient absorption data, however, has limited the degree to which these molecular transformations have been characterized. Such characterization can be obtained from the vibrational spectra recorded in time-resolved resonance Raman (TR^3) spectroscopy [7-9]. The instrumental methods associated with recording picosecond TR^3 (PTR^3) spectra have recently been improved [10] to the extent that the BR photocycle can now be examined in considerable detail [11]. These instrumental techniques derive from a pump-probe configuration involving two tunable dye lasers operating with pulsewidths of less than 10 ps [10]. Early results have demonstrated that the laser intensities used to initiate and probe the BR photocycle by PTR^3 must be selected with considerable care. It is this specific topic which is addressed in the work presented here.

The processes normally associated with the BR photocycle occur on ground-state potential surfaces. The exception to this point involves the initial excitation of BR-570 and the subsequent photostationary equilibrium (PSE) established with K-590. This PSE, of course, involves the excited electronic states of both BR-570 and K-590. Other excited state (electronic and/or vibrational) processes can be initiated if the intensity of radiation to which the BR system is exposed exceeds well-defined thresholds. The optical conditions used to initiate and probe a photolabile system can be quantitatively measured by means of a photolability parameter, ℓ_i, defined as:

$$\ell_i = \phi_j \sigma(\lambda) I(\lambda, r) \tag{1}$$

where ϕ_j is the quantum yield for the photochemical process involving the formation of species j, $\sigma(\lambda)$ is the absorption cross-section of a particular species at λ, and $I(\lambda,r)$ is the intensity of the exciting radiation at λ in a beam of diameter 2r. This formulation has been suggested previously [12]. In the case of initiation of the BR photocycle from BR-570, ℓ_i becomes ℓ_o. When combined with the time during which the molecule of interest remains in the excitation beam, t, the value of ℓ_o defines the intensity threshold conditions for establishing a PSE (i.e., $\ell_o t$) [13].

Two examples of intensity-dependent processes were recently described involving the PSE between M-412 and M' caused by 420-440 nm radiation [14,15] and the direct optical (532 nm) conversion of BR-570 directly to M-412 within 10 ns [15]. These two reactions are examples of "photolytic interruptions" of the ground-state BR photocycle accessible at thermal (room temperature) energies. The molecular processes underlying such photolytic interruptions are of fundamental interest and are currently under examination. It is important to recognize the distinction between these excited-state processes and those which form the ground-state BR photocycle. In the latter case, optical excitation occurs only in the BR-570, and perhaps the K-590, species.

The role of laser intensity becomes particularly significant for the initial, picosecond processes associated with the PSE between BR-570 and K-590. The intensity selected to initiate the BR photocycle simultaneously determines the extent of the PSE between BR-570 and K-590 and thereby, their relative concentrations. The existence of a PSE insures that (1) a TR^3 spectrum of BR-570 alone is difficult to record and (2) K-590 will be present only in conjunction with BR-570. To record a TR^3 spectrum of BR-570 only, it is necessary to use very low laser intensities in order to minimize the degree to which the PSE becomes established. Low intensity, cw laser excitation at various wavelengths has been used to record RR spectra assigned to BR-570 [16-19]. For recording PTR^3 spectra, however, pulsed, picosecond laser excitation must be used to record the RR spectrum of BR-570 under experimental conditions,which can be compared directly to the resonance Raman spectra of the BR-570/K-590 mixtures. A well-characterized RR spectrum of BR-570 itself is essential for the correct interpretation of the PTR^3 spectra of mixtures including BR-570.

Data addressing this point are presented in Figures 1. Low intensity cw laser excitation at 514.5 nm was used to obtain RR spectra of BR-570 in H_2O and D_2O solvents [20]. The RR spectrum for BR-570 in H_2O (Figure 1B) with $\ell_o t < 0.1$ is in good agreement with other RR spectra assigned to BR-570 [16-19]. Specifically, the fingerprint region of 1150 cm^{-1} to 1400 cm^{-1} provides the clearest characterization of the configuration of the retinal chromophore. The RR bands between 1150 cm^{-1} and 1250 cm^{-1} are due to the C-C stretching vibrations,while those between 800 cm^{-1} and 1000 cm^{-1} have been assigned to the hydrogen out-of-plane (HOOP) vibrations [21]. The HOOP bands gain intensity when the retinal hydrocarbon chain undergoes significant conformational distortion [18]. The strongest RR band at 1529 cm^{-1} is assignable to the C=C stretching vibrations. The frequency position of this band correlates well with the Π-electron density distribution in the hydrocarbon chain and therefore, also correlates with the absorption maxima [22,23]. The protonation environment of the Schiff base (SB) linkage of retinal to the protein is characterized by RR bands near 1640 cm^{-1} associated with C=N stretching motions. These bands are sensitive to deuterium substitution since the N-H (D) inplane (ip) bending motion is coupled to the C=N stretching vibration [24]. Such a vibrational analysis demonstrates that the structure of the chromophore in BR-570 has an all-trans, 15-anti (trans), protonated SB linkage to the protein [19,25,26]. The configurational changes that occur in this structure of BR-570 on the picosecond time-scale are opened to examination by the PTR^3 spectroscopy.

The RR spectra of BR-570 in H_2O and D_2O suspensions also were recorded using pulsed (8 ps, 1MHz) laser radiation at 590 nm [11]. The PTR^3 spectrum of BR-570 in H_2O is presented in Figure 1A. The laser intensities were selected to be low enough ($\ell_o t \sim 0.12$; average power of 1.7 mW) to insure that scattering from primarily, if not only, BR-570 would be observed. A comparison of the respective RR spectra (compare Figure 1A with 1B) shows that in general the results are the same. Detailed examination, however, reveals, several small changes in frequency positions (typically 2-4 cm^{-1}) and in intensities for the weaker bands. An exception involves the 4 cm^{-1} shift of the strongest band in the RR spectrum, the C=C stretch near 1529 cm^{-1}. The only new band to appear in the PTR^3 spectrum is the band at 796 cm^{-1}.

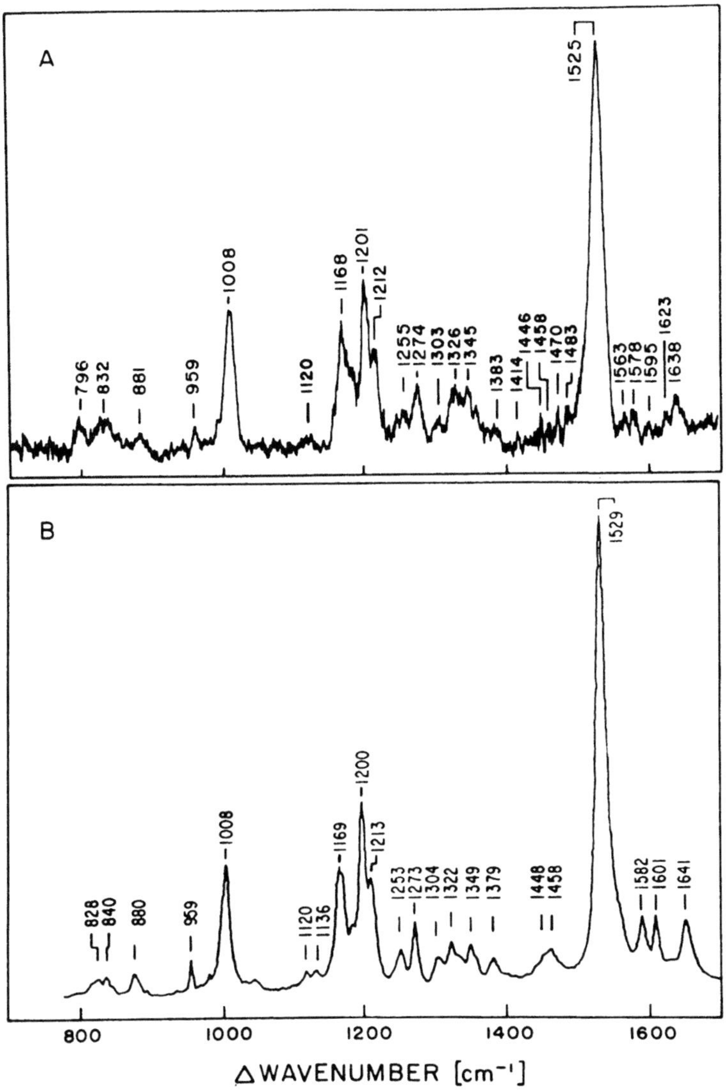

Figure 1 (A) PTR^3 spectrum of BR-570 in H_2O suspension (8 ps pulsewidth, 590 nm, average power = 1.7 mW, repetition rate = 1 MHz, $\ell_o t \sim 0.12$), (B) cw spectrum of BR-570 in H_2O suspension (514.5 nm, $\ell_o t < 0.1$)

These differences can be attributed to either one or both of two phenomena. (1) The change of excitation wavelength from 514.5 nm to 590 nm alters the resonance enhancement for Raman scattering and thereby, could cause shifts in the wavenumber displacements of bands and changes in their relative intensities. The 4 cm^{-1} shift of the C=C stretching band position can be associated with this effect,since the position of this vibrational mode correlates strongly with the absorption maxima of the BR species [22,23]. The remaining small wavenumber shifts and intensities changes may also arise from the difference in excitation wavelengths. (2) The photochemical formation of an intermediate must be considered as an alternative cause of these spectral differences. Specifically for 8 ps excitation, the formation of K-590 could contribute significantly to the spectrum in Figure 1A. Recent

PTR[3] studies have shown, however, that the RR spectrum of K-590 differs substantially from that of BR-570 in the HOOP and fingerprint regions [15]. Since neither the HOOP nor fingerprint regions of the PTR[3] spectrum contain the band structure assigned to K-590, it is apparent that essentially no photochemistry has occurred. The 8 ps spectrum, therefore, is considered to arise primarily, if not exclusively, from BR-570 and the differences between the spectra in Figure 1B and 1A are caused by changing the excitation wavelength from 514.5 nm to 590 nm.

Finally, it should be noted that the peak power of excitation in these two spectra differ by more than six orders of magnitude, and yet the $\ell_0 t$ formulation has provided an accurate measure of the photolability properties of the BR photocycle. Since it is now widely recognized that quantitative control over optical excitation and detection in TR[3] and transient absorption experiments is crucial to unraveling the molecular dynamics, the conditions under which formulations such as $\ell_0 t$ are valid have become of primary importance in designing experiments. This conclusion pertains not only to studies of the BR photocycle, but to the examination of other biochemical systems as well.

1. D. Oesterhelt and W. Stoeckenius: Proc. Natl. Acad. Sci. USA 70, 289 (1973)
2. W. Stoeckenius, R.H. Lozier, and R.A. Bogomolni: Biochim. Biophys. Acta 505, 215 (1979)
3. W. Stoeckenius and R.A. Bogomolni: Ann. Rev. Biochem. 51, 587 (1982)
4. K.J. Kaufmann, V. Sundstrom, T. Yamane, and P.M. Rentzepis Biophys. J. 22, 121 (1978)
5. M.L. Applebury, K.S. Peters, and P.M. Rentzepis Biophys. J. 23, 375 (1978)
6. E.P. Ippen, C.V. Shank, A. Lewis, and M.A. Marcus Science 200, 1279 (1978)
7. G.H. Atkinson: "Time-Resolved Raman Spectroscopy", in Advances in Infrared and Raman Spectroscopy Vol IX R.E. Hester and R.J.H. Clark eds. (North-Holland Publ. London) 1981, pp. 1-61
8. G.H. Atkinson: Time-Resolved Vibrational Spectroscopy (Academic Press, New York 1983)
9. J. Terner, C.-L. Hsieh, A.R. Burns, and M.A. El-Sayed Proc Natl. Acad. Sci. USA 76, 3026 (1979)
10. G. Rumbles and G.H. Atkinson (unpublished results)
11. I. Grieger, G. Rumbles, and G.H. Atkinson Proc Natl Acad. Sci. (submitted for publication)
12. M. Stockburger, W. Klusmann, H. Gattermann, G. Massig, and R. Peters Biochemistry 18, 4886 (1979)
13. T. Alshuth, I. Grieger, and M. Stockburger: in Time-Resolved Vibrational Spectroscoppy (G.H. Atkinson ed.) Academic Press, New York (1983) p. 231
14. I. Grieger and G.H. Atkinson: "Time-Resolved Resonance Raman Spectroscopy: Photolytic Excitation Processes in the Bacteriorhodopsin Photocycle" in Time-Resolved Raman Spectroscopy (D. Phillips and G.H. Atkinson, eds.) Gordon and Breach, New York (in press)
15. I. Grieger and G.H. Atkinson: Biochemistry (in press)
16. T. Alshuth and M. Stockburger Ber. Bunseuges. Phys. Chemie 85, 484 (1981)
17. G. Massig, M. Stockburger, W. Gartner, D. Oesterhelt, and P. Towner J. Raman Spectrosc 12, 287 (1982)
18. G. Eyring, B. Curry, A. Broek, J. Lugtenburg, and R. Mathies Biochemistry 21, 384 (1982)
19. S.O. Smith, A.B. Myers, J.A. Pardoen, C. Winkel, P.P.J. Munder, J. Lugtenburt, and R. Mathies Proc. Natl. Acad. Sci. U.S.A. 81, 2055 (1984)
20. I. Grieger, Ph.D. Thesis Georg-August-Universitat, Gottingen, F.R.G. (1981)
21. G. Eyring, B. Curry, R. Mathies, R. Fransen, I. Palings, and J. Lugtenburg Biochemistry 19, 2410 (1980)
22. M. Heyde, D. Gill, R. Kilponen, and L. Rimai J. Am. Chem. Soc. 93 6776 (1971)
23. B. Honig, T. Ebrey, R.H. Callender, V. Dinur, and M. Ottolenghi Proc. Natl. Acad. Sci. U.S.A. 76, 2503 (1976)
24. B. Aton, A.G. Doukas, D. Narva, R.H. Callender, V. Dinur, and B. Honig Biophys. J. 29, 79 (1980)
25. G. Orlandi and R. Schulten Chem. Phys. Lett. 64, 370 (1979)
26. G.S. Harbinson, S.O. Smith, J.A. Pardoen, C. Winkel, J. Lugtenburg, J. Herzfeld, R. Mathies, and R. Griffin Proc. Natl. Acad. USA 81, 1706 (1984)

Evidence from FTIR-Measurements for a Separation of the Protonated Schiff Base from the Counterion into a Less Polar Environment

K. Gerwert and F. Siebert*

Institut für Biophysik und Strahlenbiologie, Albertstraße 23, D-7800 Freiburg, F. R. G.

J.A. Pardoen, C. Winkel, and J. Lugtenburg

Department of Organic Chemistry, Leiden University, NL-2300 RA Leiden, The Netherlands

The chromoprotein Bacteriorhodopsin transduces light-energy into electrochemical energy by a light-driven protontransfer across the membrane 1). In order to elucidate the proton pump mechanism in molecular detail, we investigated the photocycle with low-temperature FTIR Difference-Spectroscopy.

In the photocycle, the photo product K decays thermally via L, M and O back to BR. At low temperature it was possible to freeze these intermediates. From the IR difference spectra between BR and the intermediates, we obtained information about changes of the chromophore and of the apoprotein.

In a previous publication we showed that four internal aspartic acids undergo protonation-changes during the photocycle 2). By these protonation changes we were able to explain the absorption-maxima of the different photo-intermediates. Nevertheless, because of lack of accurate structural data, the location of the aspartic acids was chosen arbitrarily.

To obtain information about the micro-environment of the retinal and the different isomerisation-states, we took difference spectra of BR(548)-BR(568), BR(568)-K(610), BR(568)-L(550), and BR(568)-M(412), using bacteriorhodopsin with $14^{13}C$, $15^{13}C$, $14\text{-}15^{13}C$, $10\text{-}11^{13}C$, 14D, 15D labelled retinal in H_2O and D_2O. Based on the observed shifts, we were able to identify the main in-plane and out-of-plane vibrations of the retinal.

The protonated Schiff base -the link between chromophore and protein- plays an essential role in the mechanism. It was, therefore, important to identify its stretching-vibration.

Results

The IR-difference spectra selected only absorption changes of the chromophore and the protein during the photocycle, under the backround absorption of the chromoprotein. In the IR-difference spectra, the negative bands are from BR568 and the positive bands are from the photo-intermediates. Because two species are present, bands could be obscured. For better understanding of the shifts caused by labelling, we first discuss the BR-M difference spectrum. Measurements of model compounds have shown that the absorption of a protonated schiff base retinal - as in BR568 - is much stronger than the absorption of an unprotonated Schiff base retinal as in M412.3) Therefore, the BR-M difference spectra mirrored mostly the BR bands for the chromophore bands. If the shifts of the BR568 bands are clear, the shifts of K bands can be seen more easily in the BR-K difference spectrum, in which both species have strong bands.

Schiff base:

In the region where the Schiff base frequency is expected, there are some experimental difficulties. First, there is a high background absorption caused by Amid I (1640 cm^{-1}), Amid II (1550 cm^{-1}) and the OH bending vibration from H_2O (1640 cm^{-1}). It is necessary to have at least 10 % transmission under the Amid I/OH bending vibration band. Otherwise baseline distortions would be present. Another problem for identifying the C-N band is caused by the presence of additional bands in this region. Most of these bands are not seen in the Resonance Raman spectra, which only mirrored chromophore bands. These additional bands are, therefore, mostly opsin bands. It is generally accepted that in BR(568) the C=N stretch is located at about 1640 cm^{-1}. (fig. 1)

For K there are differing assignments 4) 5). Our $15^{13}C$ label in H_2O und D_2O then

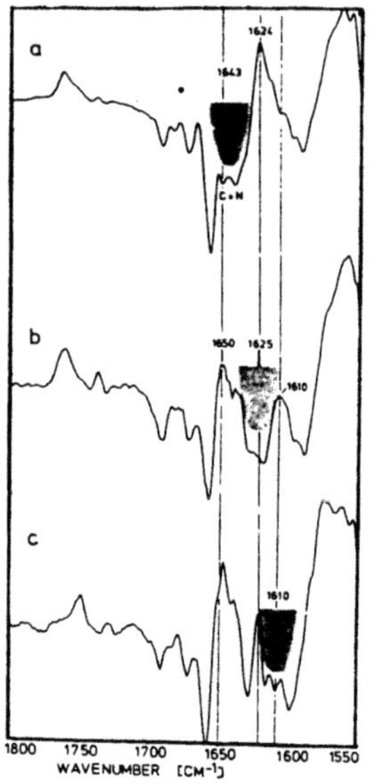

Fig. 1a shows the BR-M difference spectrum in the Schiff base region in H_2O, fig. 1b) 15-^{13}C labelled compounds in H_2O and fig 1c) 15-13C labelled compounds in D_2O. The mode arising from the Schiff base stretching vibration is assigned to a broad band at 1643 cm^{-1}, as indicated by the dotted area. This band has shifted to about 1625 cm^{-1} in 15 ^{13}C. This is indicated by a new positive band at 1650 cm^{-1} and the vanishing of the positive band at 1624 cm^{-1}. In 15 13 D_2O the band at 1624 cm^{-1} has risen again because the C=N band has shifted to about 1610 cm^{-1}, as indicated by the vanishing of the 1610 cm^{-1} band.

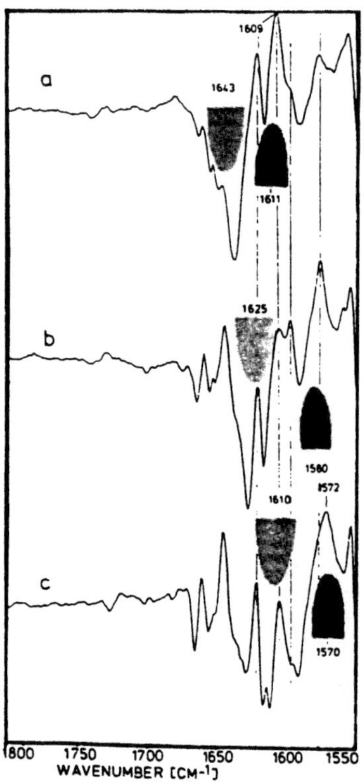

Fig. 2a shows the BR-K difference spectrum in the Schiff base region in H_2O, fig 1b) 15-13C labelled compounds in H_2O and fig. 1c) 15-13C labelled compounds in D_2O. For the assignment of the Schiff base frequency in K one has also to account for the shifts of the BR C=N stretch. The shifts of the Br bands are explained in fig. 1. The BR C=N band is indicated by the dotted area and the K C=N band is indicated by the shaded area. We computed differences of the difference spectra and saw additional shifts of a broad positive band at about 1611 cm^{-1} in the BR K difference spectrum. The computed differences of the difference spectra clearly demonstrated that the Schiff base band in K is not only the band at 1609 cm^{-1} but also a broader band at about 1611 cm^{-1}. This is confirmed by the fact that a band at 1609 is still present in 15 ^{13}C, H_2O. The shift of the C=N band to 1580 is indicated by a new positive band at 1575 cm^{-1}. This band has shifted to 1572 cm^{-1} in 15 ^{13}C D_2O.

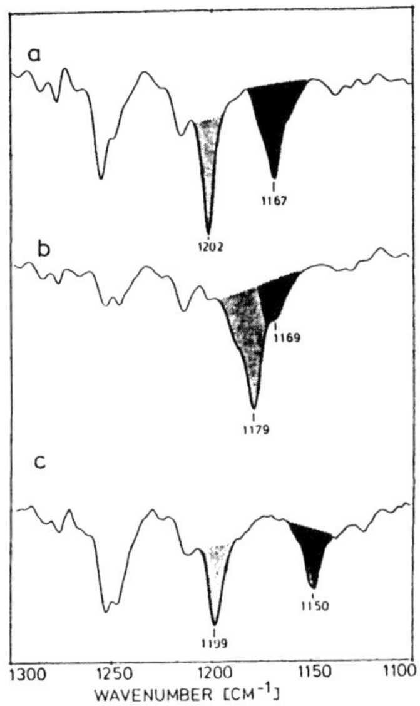

Fig. 3a shows the BR-M difference spectrum in the fingerprint region in H_2O, b) shows the 14-15 ^{13}C labelled compound in H_2O and c) 10-11 labelled compounds in H_2O. In 14-15 13C the band at 1202 cm^{-1} has shifted down to 1179 cm^{-1} and contains, therefore, mainly contribution from the C14-C15 stretching-vibration. Nevertheless, the 22 cm^{-1} downshift shows a contribution from other modes. For a pure stretching-vibration, a downshift of about 40 cm^{-1} would be expected. The band at 1167 cm^{-1} has shifted down to 1150 cm^{-1} in 10-11 13C. Therefore, the band at 1167 cm^{-1} contains an essential part of the C10-C11 stretching-vibration. The band at 1169 cm^{-1} contains also an essential part of 14 H bending vibration.

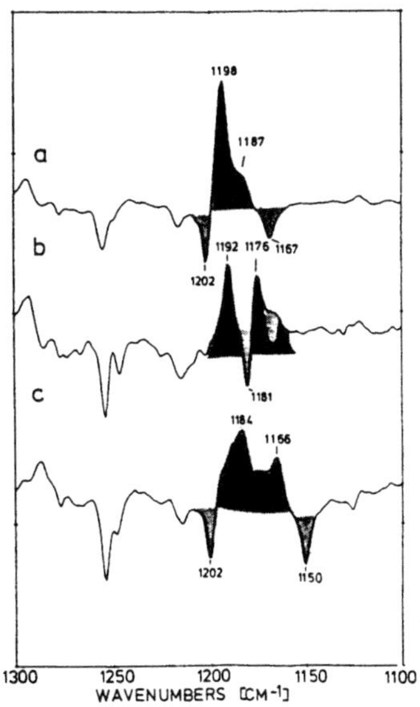

Fig. 4a shows the BRK difference spectrum in the fingerprint region in H_2O, b) 14-15^{13}C in H_2O and c) 10-11 ^{13}C in H_2O. In the BRK difference spectra one has also to account for the shifts of the BR bands. In 14-15 ^{13}C the band of the BR C14-C15 stretch has shifted down to 1181 cm^{-1} whilst the C10-C11 stretch remains at 1167 cm^{-1}. This is indicated by the dotted area. The positive K band at 1198 cm^{-1} with shoulder at 1187 cm^{-1} has shifted down to 1192 cm^{-1} and 1176 cm^{-1} in 14-15 ^{13}C. The true shift is obscured by the negative BR bands. The shifts are better seen in 10-11 ^{13}C where the K band has shifted down to 1184 cm^{-1}. Because both bands are shifted down in the same way in 10-11 ^{13}C and 14-15 ^{13}C, we have assigned the strong band at 1198 cm^{-1} with shoulder at 1187 cm^{-1} to a mixing of mainly C10-C11 stretch with C14-C15 stretch. This mixing could also explain the greater intensity of the 1198 cm^{-1} band.

clearly showed that the C=N vibration is a very broad band at about 1611 cm^{-1} at 77 K. (fig. 2). This is in contrast to a room-temperature RR measurement. They assigned a band at 1620 cm^{-1} to the C=N vibration 6). The discrepancies in the RR measurements could arise from temperature effects.

Fingerprint:

For BR(568), our 14-15 ^{13}C label showed that the band at 1202 cm^{-1} contains an essential part of the C14 - C15 stretching vibration, and 10-11 ^{13}C label an essential part of C10-C-11 stretching vibration in the band at 1167 cm^{-1}. (fig. 3)

For K(610), the fingerprint region has been completely changed due to 13 cis isomerisation. The 10-11 ^{13}C and 14-15 ^{13}C label showed that the band at 1198 cm^{-1} with the shoulder at 1186 cm^{-1} is a mixing of the C10-C11 and C14-C15 stretching-vibrations. (fig. 4)

Discussion

The experimental data showed a large downshift in frequency of the C=N stretching-vibration, an increase in frequency of the C10-C11 stretching-vibration and an unexpectedly low downshift in frequency of the C14-C15 stretching-vibration. For the 14-15 stretching- vibration, a large downshift in frequency is expected due to 13 cis isomerisation. 7)

These experimental results could be explained by a more delocalised pi-electron-distribution in K. The most likely explanation of these experimental results could be a charge-separation of the protonated Schiff base from the aspartic acid counterion caused by the retinal isomerisation into a less polar enviroment (fig.5) which cannot stabilize the positive charge at the terminal part of the retinal. This is in contrast to the 13 cis isomerisation leading to BR(548), where the Schiff base stretching- vibration is at about the same frequency as in BR(568) and the C14-C15 stretching-vibration has shifted down to about 1170 cm^{-1}. Therefore, we assume that the salt bridge is not broken by dark-adaptation. 8) Because dark-adapted BR does not pump protons across the membrane, we concluded that this movement into a less polar environment is an essential requirement for proton pumping.

Acknowledgement

We thank Mrs. Heppeler for her help in the preparation of the purple membrane and the Wissenschaftliche Gesellschaft in Freiburg im Breisgau for financial support to visit the TRVS meeting.

References

1. W. Stoekenius and R.A. Bogmolni, 1982 Ann.Rev. Biochem. 51, 587-616
2. M. Engelhardt, K. Gerwert, B. Hess, W. Kreutz and F. Siebert, Biochem. 1985, 24, 400
3. F. Siebert, W. Mäntele, Biophys. Struct. Mech. 6, 147-164 (1980)
4. Rothshild , P. Roep , J. Lugtenburg , J. Pardoen, Biochem. 1984, 23, 6103-6109
5. F. Siebert, W. Maentele 1983, Eur J. Bioch. 130, 565-573
6. Matties, talk, Intern. Conf. o. Photobiologie, Philadelphia 1984
7. B. Curry, A. Broek, J. Lugtenburg and R. Mathies, J.Am.Chem.Soz. 1982, 104, 5274-5286
8. S. Smith, A. Myers, J. Pardoen, Ch. Winkel, P. Mulder, J. Lugtenburg, R. Mathies, P.N.A.S. 81, 2055-2059 (1984).

Time-Resolved Infrared Spectroscopy Applied to Photobiological Systems

K. Gerwert, R. Rodriguez-Gonzalez, and F. Siebert

Institut für Biophysik und Strahlenbiologie, Albertstraße 23, D-7800 Freiburg, F.R.G.

The main interest in the study of photobiological systems is the role of the chromophore. Its structure and molecular changes during the photoreaction can be investigated by uv-vis and Resonance Raman Spectroscopy. However, there is a need to learn more about the role of the protein in the functioning of these systems. Especially, the interaction and interplay between specific amino acid residues and the chromophore are of great interest. In principle, infrared spectroscopy can be employed for such problems. But, since it is generally considered to be a relatively insensitive method, it appears, at first sight, questionable that it would enable one to detect molecular changes of a few functional groups against the large background absorption caused by the total system, which is a protein or even a biological membrane. With the development of new infrared detectors, however, which are fast and at least two orders of magnitude more sensitive than the usual thermopile detector, such an undertaking now appears feasible. In addition to acquiring the spectral information, it is also important to obtain information on the kinetics of the molecular changes. It appears interesting, for example, to ascertain whether or not the time-course of protein changes parallels that of chromophore changes. Therefore, the method of time-resolved infrared spectroscopy has become a natural extension of time-resolved uv-vis and Resonance Raman Spectroscopy for the investigation of photobiological systems. Since the time-resolved difference spectra reflect only those groups which undergo molecular transformations, their interpretation is facilitated. Nevertheless, additional techniques, such as chemical modifications and isotopic labelling influencing the difference spectra, may be required. Such methods would help to distinguish between molecular changes caused by the protein as apposed to the chromophore, and provide detailed information on the molecular events.

Figure 1 shows the schematics of our apparatus for time-resolved infrared spectroscopy. It resembles a conventional laser-flash

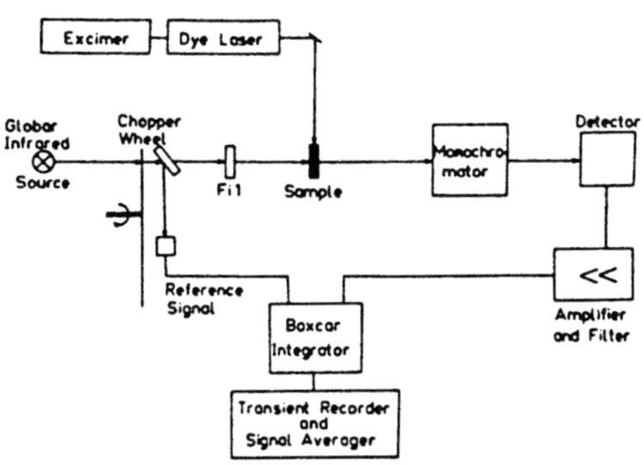

Fig. 1

Diagram of the apparatus for time-resolved infrared spectroscopy photolysis system. Source and monochromator are part of an infrared spectrophotometer. The thermopile detector was replaced by a HgCdTe-detector. Since its 1/f-noise, dominating below 2 kHz, disturbs the measurements of spectral changes with a time-constant of 1 ms and longer, the infrared beam is modulated by means of a mechanical chopper with a frequency of 10 kHz, transforming the low-frequency noise to high-frequency noise. For the measurement of faster spectral changes, the equipment is used without the chopper. The signal from the detector is digitized by the transient-recorder and transferred to a computer, where signal-averaging and the evaluation of the data is performed. The spectral-resolution of the system is depending on the background absorption, 5 cm^{-1} - 10 cm^{-1}. For small transmission changes, up to 1000 signals are averaged. The time-resolution is limited by the detector and amounts, for HgCdTe-detectors, to approx. 0.5 µs. This system was first described in (1). In the following, the application of this method to two biological systems, which can be triggered by light, is described. In the first example, the photo-dissociation of CO-myoglobin and the rebinding of CO to the binding site is investigated. The subject of the second example is the molecular changes of bacteriorhodopsin during the photoreaction.

CO-Myoglobin

This system is regarded as a model system for the study of the dynamics of proteins (2). Its Photo-dissociation has been studied by flash photolysis investigations (3) and by time-resolved Resonance Raman spectroscopy (4). Static infrared difference spectroscopy has demonstrated that several kinds of CO-myoglobin exist, differing in the Co-band of the bound Co (5). This has been interpreted in terms of several conformational sub-states of myoglobin, which fluctuate rapidly at room temperature. Here, the CO molecule in the bound and photo-dissociated states is monitored. In addition, the influence of CO on the protein is investigated. At room temperature, the photolysis expels CO from its binding-site to the aqueous phase and the time-limiting step for rebinding is the bi-molecular reaction between the dissolved CO and the protein. Therefore, the time-resolved infrared difference-spectrum will exhibit the changes occurring between CO-myoglobin and myoglobin + CO in an aqueous solution.

Fig. 2a

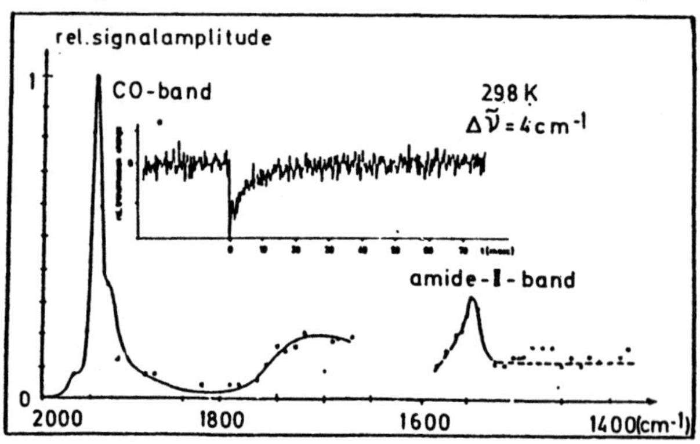

Fig. 2b

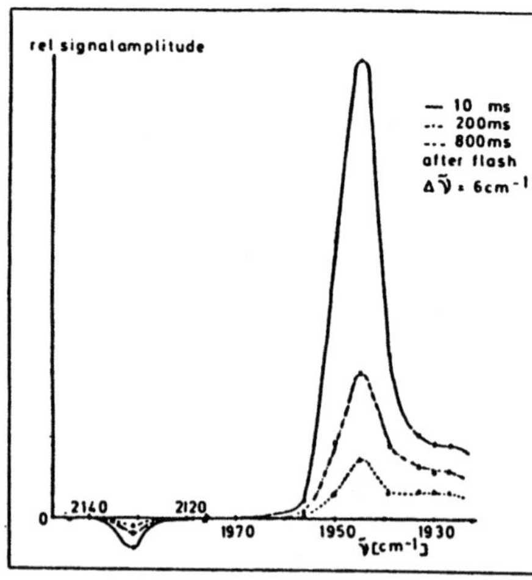

Photolysis of Co-myoglobin; a: difference spectrum at room temperature, insert shows signal at 945 cm^{-1}. b. difference spectra at 85 K for various times after the flash.

Since the interaction of the water molecules with CO broadens its absorption band to a large extent, dissolved CO cannot be detected. This difference-spectrum is shown in fig. 2a. The large structured-band around 1940 cm^{-1} represents the bound CO, the shoulders being caused by the sub-states. The continuous absorption change between 1800 cm^{-1} and 1680 cm^{-1} and below 1500 cm^{-1} is probably caused by the disturbance of the water caused by the dissolved CO. A distinct band can be discerned at 1550 cm^{-1}. This could represent a change of the amide II band of the protein, and indicate a structural change caused by the photo-dissociation.

To obtain more information on the internal steps of rebinding as proposed by Frauenfelder (3) and on the protein dynamics, similar experiments were performed at low temperature. Here, the Co molecule remains within the protein. Around 80 K the rebinding is essentially influenced by barrier I (3). The resulting difference-spectra obtained for various times after flash are shown in fig 2b. Again, the large structured-band caused by bound CO is present. The small negative band at 2130 cm^{-1} represents photolysed CO. This wavenumber is even lower than that of free Co, indicating an interaction with the protein. There is a one-to-one correspondence with respect to the different times between the band at 1945 cm^{-1} and the band at 2130 cm^{-1}. However, it is already evident from fig 2b that the relaxation at 1927 cm^{-1}, corresponding to another sub-state, proceeds with a slower time course. This is even more clear from fig. 3, in which a more detailed investigation of the temperature-dependence of the kinetics of rebinding at 1945 cm^{-1} and at 1927 cm^{-1} is represented. It can be concluded that rebinding does not proceed with a single exponential for both cases. The temperature-dependence of the rate-constant for the slow phase is depicted in fig. 3. The different behaviour for the two wavenumbers is striking. Whereas the activation energies are comparable, clear differences for the activation entropies are observed. In our opinion, these findings support the model as proposed by Frauenfelder, in which the sub-states influence the rebinding kinetics. At elevated temperature, the fluctuations are fast enough to yield an average rate-constant. At low temperature, however, the sub-states are frozen or their fluctuations sufficiently slowed down to now manifest their existence.

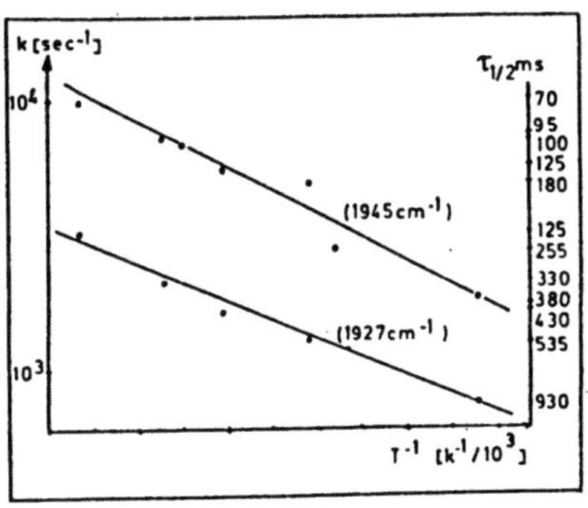

Fig. 3
Temperature-dependence of the slow phase rate-constants of rebinding at 85 K, measured at 1945 cm^{-1} and 1927 cm^{-1}.

Bacteriorhodopsin

Although Resonance Raman spectroscopy is able to provide detailed information on the chromophore during the photocycle, (some examples of this were presented during this meeting) the role of the protein remains elusive to investigations. With the development of the method of static FTIR difference spectroscopy, measuring difference spectra between bacteriorhodopsin and its intermediates trapped at low temperature (e.g.7 and references cited therein), information on

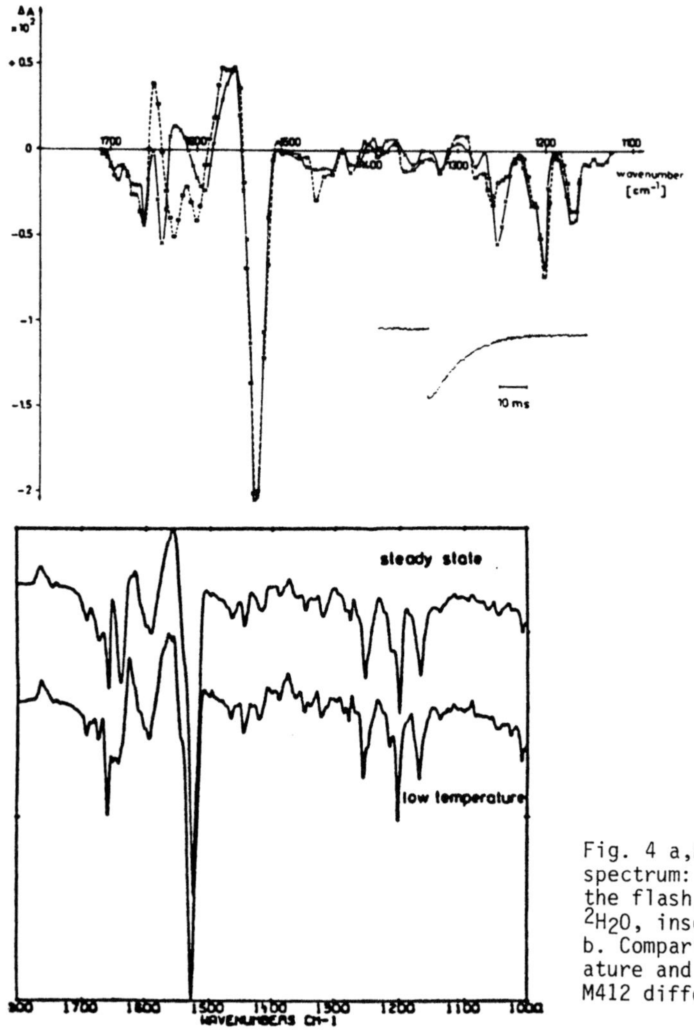

Fig. 4 a,b. BR568-M412 difference spectrum: a: evaluated 0.2 ms after the flash, ——— H_2O, ------- ^2H_2O, insert shows typical signal. b. Comparison of static low-temperature and static steady-state BR568-M412 difference spectra.

the protein could be obtained. However, there are indications that the photoreactions at low temperature differ from those at room temperature. Also, since it is a static method, no information in the time course of the spectral changes is obtained. Therefore, we applied our method of time-resolved infrared difference spectroscopy to the investigation of the photocycle of bacteriorhodopsin, thus avoiding the difficulties involved with static difference spectroscopy.

Fig. 4a shows the time-resolved difference spectrum, ca. 0.2 ms after the flash. It essentially reflects the molecular changes occurring between BR68 and the intermediate M412. For comparison, the static BR568-M412 FTIR-difference spectrum obtained at -60° C and the steady state FTIR difference spectrum are shown in fig. 4b. The latter was obtained by accumulating the M412 intermediate under continuous illumination at 0° C. From this it can be seen that the time-resolved difference spectrum resembles more closely the steady state spectrum. Especially in the region between 1600 cm^{-1} and 1700 cm^{-1}, the low-temperature spectrum exhibits deviations . A more detailed analysis has shown that these deviations are caused by molecular changes of the protein. Therefore, it can be concluded that regarding the chromophore, no differences are observed between the room-temperature and low-temperature measurements, but that special care has to be taken regarding the protein. It appears that some changes of the protein are frozen at low temperature. The spectra are dominated by bands of the chromophore. This indicates that the chromophore in bacteriorhodopsin is considerably more polar than most of the amino-acid residues. The polarity is induced by the protonation of the Schiff base linking the retinal to the protein.

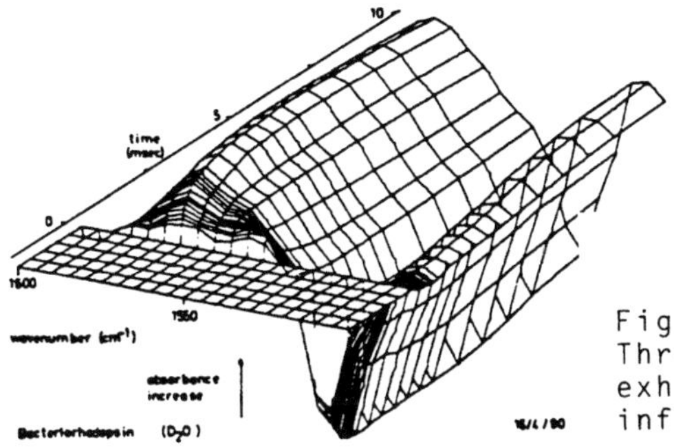

Fig. 5
Three-dimensional computer evaluation, exhibiting the spectral and temporal information.

Fig. 5 shows a three-dimensional plot of the time- and wavelength-dependence of the spectral changes. The large band at 1525 cm^{-1} is the C=C stretching band of the retinal. But protein changes, occurring about 1ms after the flash, can also be seen around 1570 cm^{-1}. It is clear that the information content of the time-resolved difference spectra is large. A computer will therefore be required for the evaluation. Fig. 6a shows a special case of protein changes. The difference spectrum reflects changes occurring between BR568 and the L550 intermediate. It can be deduced from the time course of the signals that the band decays with the rise of the M412 intermediate. This is especially evident from the kinetic isotope effect observed for measurements in ^2H_2O. In addition to the kinetic effect, ^2H_2O shifts the band at 1740 cm^{-1} about 10 cm^{-1} to lower frequencies. The observations indicate that with the formation of the L550 intermediate, one carboxylic acid becomes deprotonated and reprotonated with the formation of the M412 intermediate.
The same spectral range was investigated at later times after the flash. The corresponding BR568-M412 difference spectrum is shown in fig. 6b. Representative signals, from which the spectrum was deduced, are also given. From this spectrum it can be deduced that one carboxylic acid becomes protonated with the formation of M412 and an additional one at a somewhat later time. To identify the carboxylic acids with either glutamic or aspartic acids, bacteriorhodopsin was selectively labelled by the incorporation of (4-13C) aspartic acid. The resulting difference spectrum is also presented in fig. 6b. The complete band at 1760 cm^{-1} with the shoulder at 1750 cm^{-1} is shifted by 40 cm^{-1} to lower frequencies. Therefore, both carboxylic acids are aspartic acids. A similar investigation of the carboxylic acid of fig. 6a has also identified it with an aspartic acid. From these investigations a detailed picture of protonation changes in the neighbourhood of the chromophore could be obtained. A more detailed discussion of these points is given in (7). It is reasonable to assume that these protonation changes play an essential role in the protein pumping mechanism, and that they influence the absorption properties of bacteriorhodopsin intermediates. In (7), a model was proposed which

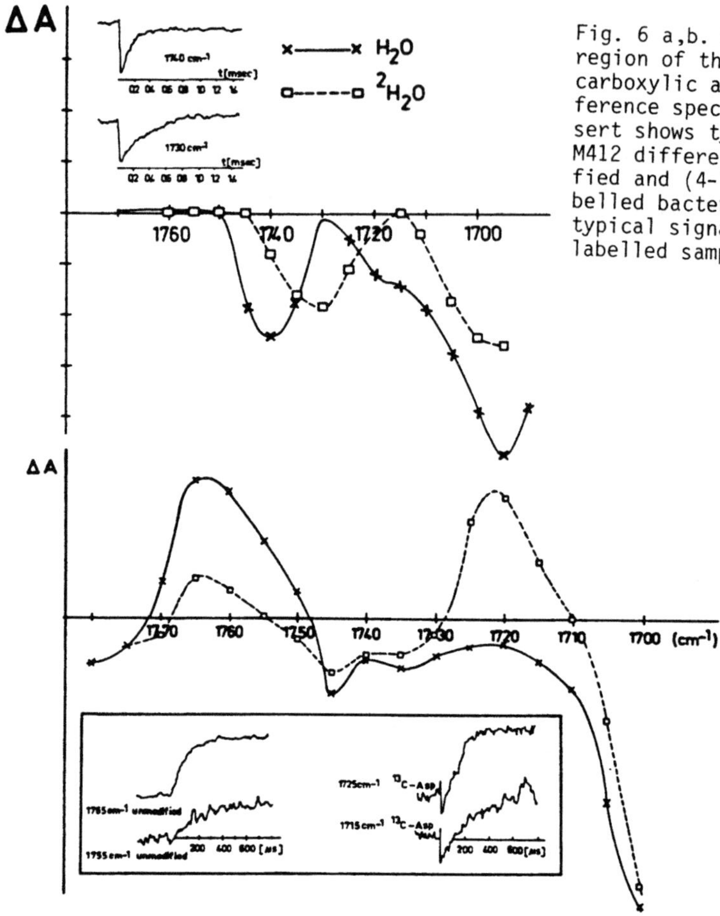

Fig. 6 a,b. Difference spectra in the region of the C=O band of protonated carboxylic acids; a: BR568-L550 difference spectrum for H2O and ^2H_2O, insert shows typical signals. b: BR568-M412 difference spectrum for unmodified and (4-^{13}C) aspartic acid labelled bacteriorhodopsin. Insert shows typical signals for unmodified and labelled samples.

took into account the absorption properties, the retinal isomerisation and the protonation changes. It could also explain the observation that two protons can be pumped per photocycle.
Summing up, the two examples have shown that time-resolved infrared difference spectroscopy, combining temporal and spectral information, is able to provide results which are difficult to obtain by other methods.

1. F. Siebert, W. Mäntele & W. Kreutz: Biophys. Stuct. Mech. 6, 139 (1975)
2. H. Frauenfelder, G.A. Petsko & D. Tsernoglu: Nature, 280, 558 (1979)
3. R.H. Austin, K.W. Besson, L. Eisenstein, H. Frauenfelder, I.C. Gunsalus: Biochemistry 14, 5355 (1975)
4. F.S. Parker: Application of Infrared, and Resonance Raman Spectroscopy in Biochemistry (Plenum Press,New York 1983) 254 ff.
5. ibid. 233 ff
6. J. Alben, D. Beece, S.F. Bowne, W. Doster, L. Eisenstein, H. Frauenfelder, D. Good, J.D. McDonald, M.C. Marden, P.P. Moh, L. Reinisch, H.H. Reynolds, E. Shymasunder, K.T. Yue: Proc. Natl. Acad. Sci. USA 79, 3744 (1982)
7. M. Engelhardt, K. Gerwert, B. Hess, W. Kreutz, and F. Siebert, Biochem. 24, 400 (1985)

Part VII

Theoretical Aspects

Electrontransfer and Vibrational Relaxation in Bridged Donor-Acceptor Systems

S.F. Fischer and I. Nußbaum

Physik-Department, Technische Universität München, D-8046 Garching, F. R. G.

1. Introduction

In recent years several groups /1-4/ have studied the electron-transfer from a donor to a rigidly linked acceptor. By small variations of the donor or the acceptor molecule the free energy change ΔE can be varied without effecting other parameters such as Stokes shift or the transfer interaction significantly. If the rate k for the electron-transfer is plotted versus the free-energy change, it seems that the experiments can be interpreted within the frame of existing theories: that is the classical description by MARCUS /5/, which predicts a quadratic dependence of $\ln k$ versus ΔE or even better the quantum treatments /6-8/, which predict for large energy changes an energy-gap law similar to the one known for nonradiative processes in large molecules. The maximum of the rate occurs for an energy change equal to the overall Stokes shift. These theories treat the electronic interaction as constant /9/ and concentrate on the Franck Condon factors for the vibrational degrees of freedom.

In this note we want to analyse the electronic coupling in more detail. We define the initial and final states by a selfconsistent treatment of the extension of the electronic wavefunction from the donor into the bridge up to the acceptor and vice versa, together with the reorganization of the nuclear equilibrium positions for internal degrees of freedom and the solvent. The latter will be treated in the classical high-temperature limit. This approach is closely related to the construction of solitonic states /10/ or the self-trapping of polarons /11/. As a result, we find an additional strong enhancement of the rate for free-energy changes close to the Stokes shift, which stem from the electronic coupling. In particular, we find that only for free-energy changes smaller than the Stokes shift a self-trapped state can stabilize, while for higher free energies an initial state with the electron largely localized at the donor site can be stabilized only as an excited charge-transfer state. That means the wavefunction has a node between the donor and the acceptor sites.

2. The Rate Expression

We choose a model hamiltonian of the following form:

$$H = \sum_n E_n |n\rangle\langle n| + \sum_n^{N-1} V_{n,n+1}(|n\rangle\langle n+1| + |n+1\rangle\langle n|) + \sum_n \Lambda_n Q_n |n\rangle\langle n|$$

$$\sum_k (\lambda_{k1} q_k |1\rangle\langle 1| + \lambda_{kN} q_k |N\rangle\langle N|) + \frac{1}{2} \sum_n (P_n^2 + \Omega^2 Q_n^2) + \frac{1}{2} \sum_k (p_k^2 + \omega_k^2 q_k^2) \quad (1)$$

Here E_n is the energy of the transferrable electron at site n, $V_{n,n+1}$ the transfer interaction and Λ_n the coupling constant of an internal mode Q_n with frequency Ω. The medium modes q_k are coupled only to the donor site $|1\rangle$ and to the acceptor site $|N\rangle$ with coupling constants λ_{k1} and λ_{kN} respectively.

Taking the ground state expectation value with an electronic wavefunction $\psi = \sum_n \varphi_n |n\rangle$ one can determine φ_n and the equilibrium position Q_n^o or q_k^o by a variational procedure which gives for example /12/

$$Q_n^o = - \frac{\Lambda_n}{\Omega^2} |\varphi_n|^2 \tag{2}$$

By substituting the equilibrium values into the ground-state expectation value of H for the coordinates, one arrives at a nonlinear functional in φ_n

$$H = \sum_n E_n |\varphi_n|^2 + 2 \sum_n^{N-1} V_{n,n+1} \varphi_n \varphi_{n+1} - \frac{1}{2} \sum_n \frac{1}{\Omega^2} \Lambda_n^2 |\varphi_n|^4$$
$$- \frac{1}{2} \left(\sum_k \frac{1}{\omega_k^2} \lambda_{k1}^2 |\varphi_1|^4 + \sum_k \frac{1}{\omega_k^2} \lambda_{kN}^2 |\varphi_N|^4 \right) \tag{3}$$

The equation can be solved for φ_n. Let us assume that the acceptor has a lower energy than the donor. In this case one can always find a state localized at the acceptor site. Its energy is lowered with regard to E_n by solvent and internal reorganization. $\Delta E_{med}^N = \sum_k \frac{\lambda_{kN}^2}{2\omega_k^2}$ and $\Delta E_{int}^N = \frac{\Lambda_N^2}{2\Omega^2}$, respectively.

It is also possible to find a self-trapped state at the donor site, as long as the energy difference of the unrelaxed donor and acceptor state is smaller than twice the reorganization energy of the donor state

$$E_1 - E_N < 2\Delta E_{med}^1 + 2\Delta E_{int}^1$$

For larger energy differences, a state largely localized at the donor site can only be stabilized if it is an excited state, that means its wavefunction must have a node. We call this an excited charge-transfer state. Between these two regions there exists a small regime of the order of $2\sqrt{2}\,\Lambda$ where the excited state loses its character of charge localization.

It is possible to calculate the electron-transfer from either the self-trapped donor state or the excited charge-transfer state $\{\varphi_n^i\}$ to the relaxed ground state $\{\varphi_n^f\}$. The interaction is given by

$$V^i = \sum_n \gamma_n \Lambda_n (Q_n - Q_n^o) + \sum_k \gamma_1 \lambda_{1k} (q_k - q_{k1}^o) + \sum_k \gamma_N \lambda_{Nk} (q_k - q_{kN}^o) \tag{4}$$

$$\text{with } \gamma_n = \varphi_n^i \varphi_n^f - |\varphi_n^i|^2 \sum_{n'} \varphi_{n'}^i \varphi_{n'}^f \tag{5}$$

The last term in Eq. 5 reflects the fact that the initial and the final electronic states are not orthogonal. For strong localization we get $\gamma_1 \cong \gamma_N \cong \varphi_N^i$ and $\gamma_{n\neq 1,N} \ll \gamma_1$. The important result is that φ_N^i shows a resonance-like behaviour close to $E_i - E_f = 2\Delta E_{med}^1$. For the relaxed states $\tilde{E}_i = E_i - \Delta E_{med}^1 - \Delta E_{int}^1$ and $\tilde{E}_f = E_f - \Delta E_{med}^N - \Delta E_{int}^N$ it reads $\tilde{E}_i - \tilde{E}_f = \Delta E = \Delta E_{med}^1 + \Delta E_{int}^1 + \Delta E_{med}^N + \Delta E_{int}^N$.

The transfer rate has the approximate form /13/

$$k_{if} = (4\pi kT\lambda)^{-\frac{1}{2}} \exp(-G_1^2 - G_N^2) \sum_{m_1 m_N} \frac{G_1^{2m_1} G_N^{2m_N}}{m_1!\, m_N!} \exp\left(-\frac{(\Delta E - \hbar\Omega(m_1 + m_N))^2}{4kT\lambda}\right)$$

$$\left(\left(\gamma_1^2 \frac{(m_1 - G_1^2)^2}{G_1^2} + \gamma_N^2 \frac{(m_N - G_N^2)^2}{G_N^2}\right) \Delta E_{int}\Omega + (\gamma_1^2 \Delta E^1_{med} + \gamma_N^2 \Delta E^N_{med}) kT\right) \tag{6}$$

$$\text{with} \quad G_n^2 = \frac{\Delta E_{int}}{\hbar\Omega} (\varphi_n^{i2} - \varphi_n^{f2})^2 \tag{7}$$

and λ is the solvent stokes shift. Contributions originating from coupling to the bridging sites γ_n are neglected. The last term in brackets is the effective interaction V^2_{eff}, which consist of contributions describing transitions induced by the internal modes and those induced by the medium modes. The latter become more important at higher temperatures.

3. Discussion of the Results

Figure 1 shows the rate as function of the energy-change $\Delta E = \tilde{E}_i - \tilde{E}_f$. It can be seen that the rate increases by two orders of magnitude as the value of the Stokes shift is approached. The resonance-type increase results from the electronic coupling. So it is a typical non-Condon effect. The transition from the self-trapped to the charge-transfer state is discontinuous. It shows that electron-transfer between two different ground states (wavefunctions without a node)exist only in the normal regime where an activated process takes place. In the abnormal regime for $\Delta E > \Delta E^1_{med} + \Delta E^N_{med} + \Delta E^1_{int} + \Delta E^N_{int}$ an energy-gap law similar to the one for radiationless transitions induced by promoting modes applies. It should be noted, that the initial and the final state differ in the equilibrium position of the coupled modes,and this implies that the corresponding electronic wavefunctions are not orthogonal.

Comparison with experiments is difficult since the energy-difference ΔE can not be varied continuously,and since there exists no self-trapped state with charge-localization in the center of the resonance. The existing experimental results are certainly not in conflict with these predictions.

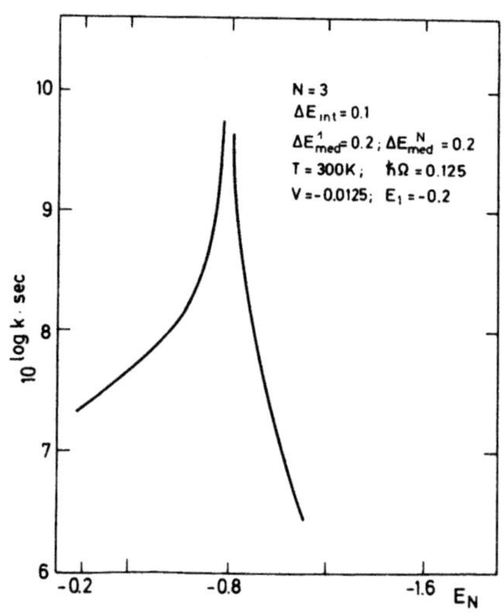

Figure 1

The transition rate as a function of the free energy change

Acknowledgement

The authors like to thank Dr. G. Venzl for his contributions to the developement of this theory /13/. This work has been supported by the Deutsche Forschungsgemeinschaft SFB 143 C2.

References

1. A. D. Joran, B. A. Leland, G. G. Geller, J. J. Hopfield, and P. B. Dervan: J. Am. Chem. Soc. 10b, 6090 (1984)
2. J. S. Lindsey and D. C. Manzerall: J. Am. Chem. Soc. 105, 6528 (1983)
3. M. R. Wasielewski and M. P. Niemczyk: J. Am. Chem. Soc. 106, 5043 (1984)
4. M. E. Michel-Beyerle and H. Heitele: J. Am. Chem. Soc. in press
5. R. A. Marcus: J. Chem. Phys. 43, 679 (1965)
6. V. G. Levich and R. R. Dogonadze: Coll. Czech. Chem. Commun. 26, 193 (1961)
7. R. P. Van Duyne and S. F. Fischer: Chem. Phys. 5, 183 (1974)
8. J. Ulstrup and J. Jortner: J. Chem. Phys. 63, 4358 (1975)
9. D. N. Beratan and J. J. Hopfield: J. Am. Chem. Soc. 106, 1584 (1984)
10. A. S. Davydov: Physica 3D, 1 (1981)
11. D. Emin and T. Holstein: Phys. Rev. Lett. 36, 323 (1976)
12. G. Venzl and S. F. Fischer: Phys. Rev. B, in press
13. S. F. Fischer and G. Venzl: Photoreaktive Festkörper (M. Wahl-Verlag, Karlsruhe 1984, Ed. H. Sixl, pp. 723)

Theory of Luminescence in the Three-Level System with Extension to the Multistate Problem by Transform Methods

A.C. Albrecht

Department of Chemistry, Cornell University, Ithaca, NY 14853, USA

In this paper we briefly address two current issues in resonance Raman theory. One concerns the general question of the "redistribution" of emission that can drastically alter the lineshape of Raman emission even in a simple system, depending on experimental conditions. The other concerns frequency domain transform methods (or their time-domain equivalents, including wave-packet propagation) which apply only to electronically resonant vibrational Raman scattering at a (quasi) adiabatic level and in the absence of any elastic (pure-dephasing) system-bath interactions.

A. "Redistribution of Emission" in Raman Scattering

When an incident quasi-monochromatic light field of frequency ω_L drives a material system to produce (incoherent) light at a different frequency, ω_R, either a spontaneous Raman process is said to have occurred, or some other form of emission such as fluorescence has taken place. The system itself will have passed from its initial state, g, say, to a final state, f. Formally, the rate for this transformation is given by the time derivative of the modulus square of the projection of the final state, f, onto the state of the system that has time-evolved according to Schroedinger's equation. Usually at this point perturbation theory is resorted to, under the assumption that the relevant Rabi frequencies of the problem are small compared to the damping constants of the material system. (This assumption, however, is acquiring increasingly critical attention as more intense laser fields are used in the laboratory, and their influence cannot be treated perturbatively.) The passage of the system from state g to state f must call for the evolution in time of both the bra vector and the ket vector from g to f. At the perturbative level, this implies intervention of light fields upon the system at four points in time. The four fields are the two conjugate pairs $E_L(\omega_L)$ and $E^*_L(\omega'_L)$ from the incident quasimonochromatic laser beam and $E_R(\omega_R)$ and $E^*_R(\omega'_R)$ from the emitted beam. (If the emission is spontaneous, these fields may be taken as black-body zero-point fields.) The bra and ket evolutions each pass through intermediate conditions, $\{m\}$, the virtual states of the problem, before finally reaching condition f. There are, in principle, a large number of distinct time evolutions that together achieve the overall g to f transition. However, for the condition of resonance, or near resonance, with a subset of intermediate states, m, then there are only three (plus their complex conjugates) important time evolutions to consider. In two of these, the driving fields E_L and E^*_L act at the first two time interventions. As a result <u>both</u> bra and ket are driven to condition m, leading to an actual population of <u>state</u> m. In the third time evolution pattern, one of the emissive fields E_R (or E^*_R) intervenes, following the initial perturbation by the laser field. In this history only coherent superposition states appear, but never is an excited state m, populated. The lineshape of the emission (the spectrum in ω_R) is derived from a superposition of the contributions from these two classes of time evolution, and can vary greatly, depending on the conditions of the experiment. The evolutions which produce real intermediate states, m, lead to resonance fluorescence-type emission (even when ω_L is not resonant on ω_{mg}), while the coherent superposition history produces a complex lineshape which carries a sharp "Raman-like" positive feature that "tracks" with ω_L, but also a negative going component that always destructively interferes with the resonance fluorescence. The relative weight of these two contributions can be a sensitive function of experimental conditions, particularly if these involve pulsed excitation, elastic material/bath interactions, or phase incoherence in the excitation pulse. As a matter of fact, any pure coherence-loss processes (within the system or within the driving light fields) weakens the "Raman-like" channel, and thus exposes the pure fluorescence signal (the emission is said to be "redistributed"). Furthermore, should the incident light be pulsed and have a time-width less than or comparable to reciprocal dephasing frequencies then, again, the Raman-type emission is quenched to expose the fluorescence component. In this case, the quenching mechanism follows directly from the need in the superposition time evolution for the insertion of an E_R (or E^*_R) between the E_L and E^*_L interventions. As the driving pulse is shortened, the time available between E_L and E^*_L can be reduced to the point where little or no Raman-type superposition channel survives.

On the other hand, when all sources of pure-dephasing are removed, the originating state has infinite lifetime, and the system is driven with cw ω_L the Raman-type superposition channel is at its strongest. Now, its negative component precisely cancels the fluorescence term, leaving only the Raman-type lineshape that tracks with ω_L. There is no net fluorescence. In this special case the emission is not redistributed.

The theory of steady state spectral redistribution must be carried out within the density matrix formalism, and it has now been addressed in considerable detail in the literature (we mention here only examples [1,2,3,4]). Spectral redistribution has not yet had a practical impact in resonance Raman studies of polyatomic molecules, rather one has been hard put to show examples of this phenomenon. However, it may well become important as Raman studies of ultrashort-lived states begin to appear. Here, the redistribution of emission away from Raman and towards fluorescence must take place, both because the initial state in the scattering process is short-lived (especially if its lifetime compares to dephasing times in general) and because of the ultrashort light pulse that would likely be used to probe the transient state.

In a recent paper [4] model calculations are presented for ω_R emission from a three level (g,m,f) system using a variety of light pulses and the presence or absence of Markovian pure-dephasing system-bath interaction. Furthermore, the emitted fields are filtered both in time and in frequency to produce "observable" frequency-time spectra in ω_R. We can offer here only one example. In Fig. 1 the ω_R is plotted as a function of time after exciting the three-level system by a quasi-monochromatic Gaussian pulse that is shorter than the dephasing times of the system. (The filtered ω_R is called ω_D and, in fact, the frequency coordinate is $\Delta_D=\omega_{mf}-\omega_D$, the detuning of emission from the fluorescence peak ($\Delta_D=0$. The frequency of the pulse, ω_L, is detuned from ω_{mg} by 20 THz which is about three times the transform-limited frequency spread of the incident pulse and about ten times the various damping parameters being used.

The most conspicuous result is that even with all pure coherence loss parameters absent, the pulsed emission has heavily redistributed towards fluorescence (the signal seen around zero detuning, $\Delta_R=0$). The Raman signal has been clearly quenched by the above time-interval argument. A second important point is that the Raman component at $\Delta_D \approx 20$ THz, weak as it is, possesses a lifetime distinctly greater than the driving pulse duration. It is not instantaneous but is sensitive to the g,f superposition coherence loss-rate. It is to be emphasized that this identical system, if excited with a cw laser, will give no redistributed emission. In that case one obtains only the familiar (Lorentzian) Raman spike at Δ_D = -20THz. The fluorescence is fully quenched.

B. Transform Methods in Resonance Raman

Transform techniques in resonance Raman scattering (RRS) establish a connection between a Raman excitation profile (REP) for a given vibration and the linear ab-

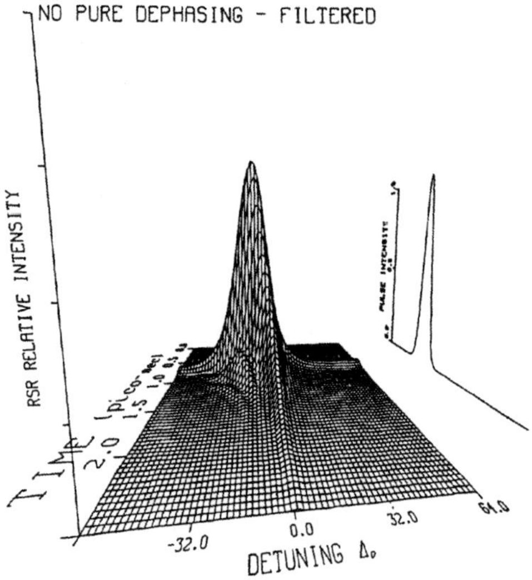

Fig. 1. The time-frequency ω_D (filtered ω_R) spectrum (as $\Delta_D = \omega_{mf} - \omega_D$) from a three-level system excited by an ω_L pulse of 165 fsec duration. Here $\omega_L - \omega_{mg} = 20$ THz. The damping parameters γ_{gm}, γ_{gf}, γ_{mf}, and γ_m are 2.0, 0.5, 2.5, and 4.0 Thz respectively. See [4] for details.

sorption spectrum (ABS) of the full electronic band responsible for the RRS [5]. The REP/ABS link appears within the adiabatic approximation, initially at the Condon level, though now it has been extended to include non-Condon effects as well as quadratic corrections involving small force constant changes upon electronic excitation. No transform relation has yet been uncovered which can incorporate elastic system/bath interactions (pure dephasing) or that goes fully beyond the adiabatic (electronic/vibrational) picture of the system eigenstates.

The centerpiece of all transform methods is the complex polarizability, whose real and imaginary parts form a Hilbert pair (a Hilbert transform relates one to the other). Any modeling of the complex polarizability will quickly lead to a determination of both ABS and the REP for any chosen vibration. On the other hand, an experimental ABS can produce the complex polarizability under a specified set of approximations, which then predicts an REP (the forward transform). The reverse procedure (inverse transform), though more difficult, is also possible and in principle can identify the active ABS underlying a possibly complicated absorption region.

The transform methods can be equally well expressed in the language of time correlators, as was originally done by Hizhnyakov and Tehver [6] and greatly elaborated upon by Page and his coworkers [7], or directly in frequency space as the appropriate Fourier transform of the time correlator [8] - the Kramers-Heisenberg formula. Mukamel speaks of "eigenstate-free" theory by treating the well-defined time correlator perturbatively in the difference operator of the excited and ground-state potential [9]. Alternatively, Heller and coworkers [10] observe that the time correlator contains the propagation of the initial vibrational state (modified by non-Condon terms if necessary) upon the excited state potential surface. They proceed to approximate this as the classical trajectory of a Gaussian wave packet on the upper surface. What is important to realize is that all of the methods are basically derived from the same conventional second-order time dependent perturbation theory. They cannot recognize elastic system-bath interaction responsible for pure-dephasing processes. In fact, normally, only a single population loss (T_1) damping parameter is assigned to all vibronic levels of the upper surface. Mukamel [11] addresses the formal generalization to the system-bath situation, but very quickly the problem becomes a many-parameter one.

In our own contributions [8] we have noted that in fact the transform can be formulated outside of a full adiabatic treatment. It is only necessary to adiabatically remove the scattered mode from the molecular eigenstate basis to achieve a transform law. This automatically permits variable (and unspecified) damping parameters among the remaining molecular eigenstates, as long as they are immune to the degree of excitation of the factored mode.

In passing, we note that transform relations can be discerned in other electronically resonant four-wave mixing spectroscopies. We are currently studying these for electronically resonant CARS for example [12].

While model calculations of multimode polyatomic spectra (ABS and REP) can be quickly accomplished and abound, the applications of transform methods to experiment are not many because of insufficient REP data. We have found transform successes in the Soret band of the ferri- and ferro-cytochrome-c with the absolute determination of several vibronic coupling energies [13]. Very recently, a most complete (and successful) experimental testing of refined transform methods has been carried out on REP's observed in the S_1 state of azulene [14]. We have just completed several careful REP studies for certain fundamentals, overtones and combinations in the intense S_4 electronic transition in azulene [15]. The results are disturbing. Good transform predictions are achieved for only one normal mode (674 cm^{-1}) and fair success is seen for a combination (674 cm^{-1} + 821 cm^{-1}). The REP's of four other scattered modes cannot be predicted by the appropriate transform of ABS. In Fig. 2 we show the REP data for the 674 cm^{-1} fundamental in the right panel, including a smooth curve through them. This is inverse transformed [16,17] to produce the dashed ABS in the left panel, where the observed ABS is also plotted. The forward transform of ABS' essentially reproduces the smoothed REP. This represents a fairly good, though not perfect, confirmation of simple transform theory. In Fig. 3 similar display is shown for the 1264 cm^{-1} mode. Simple transform theory clearly fails in this case, though the tech-

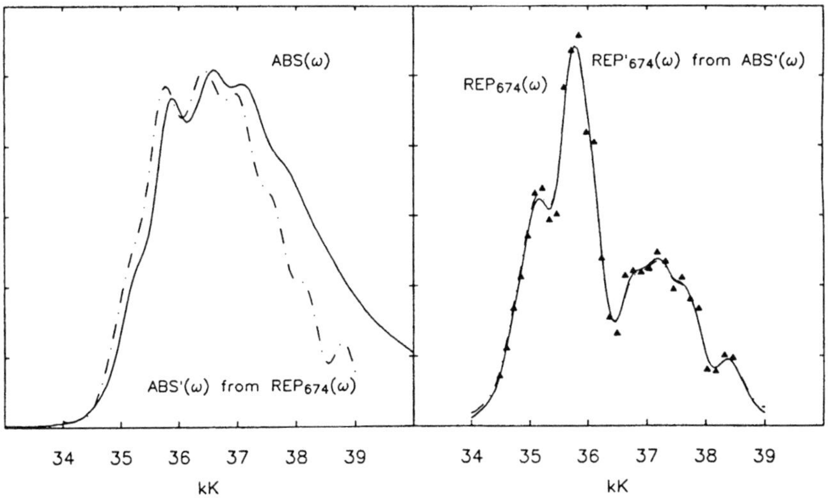

Fig. 2. The REP for the 674 cm^{-1} fundamental in the S_4 band of azulene (data points in right panel). The smoothed REP curve is inverse transformed to produce the dashed ABS' in left panel. The observed ABS is also given. ABS' is forward transformed to essentially match the smoothed curve in right panel (See [15]) for details.

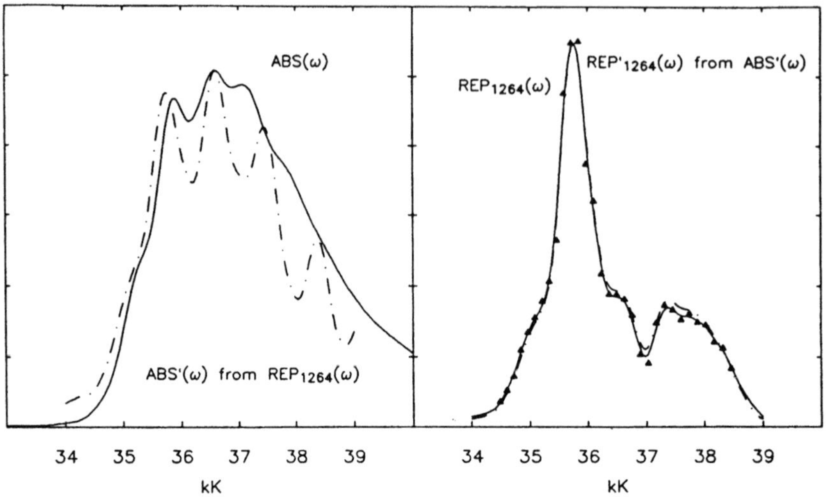

Fig. 3. Same as in Fig. 2 but for the 1264 cm^{-1} mode.

nique is upheld through the clear forward-inverse reversibility. In this case we have tentatively concluded that most modes are heavily Duschinsky mixed in the S_4 potential to disallow the simple transform theory. The fact that the 674 cm^{-1} mode is essentially successful seems to support the theoretical observation [8] that provided one mode can be factored from the molecular eigenstates it is eligible for the transform approach, no matter how complicated or confused the remaining space is.

Acknowledgements

Support through NSF Grant CHE-80-16526 and the Materials Science Center at Cornell University is gratefully acknowledged. Gratitude is also expressed to the CNRS Quantum Optics Laboratory of the Ecole Polytechnique (Palaiseau) for support and stimulus during a current extended visit by ACA.

References

1. T. Takagahara, E. Hanamura, R. Kubo: J. Phys. Soc. Japan 43, 802, 811 (1977) (for steady-state illumination).
2. T. Takagahara, E. Hanamura, R. Kubo: J. Phys. Soc. Japan 43, 1522 (1977) (for rectangular pulse excitation).
3. D. Lee and A. C. Albrecht" "Advances in Infrared and Raman Spectroscopy", Vol. 12, edited by R. J. Clark and R. E. Hester, 1985, Wiley-Heyden. (This deals with the steady-state problem in general with many references.)
4. J. Melinger and A. C. Albrecht: submitted to J. Chem. Phys.
5. For a review see P. M. Champion and A. C. Albrecht: Ann. Rev. Phys. Chem. 33, 353 (1982).
6. V. Hižhnyakov and I. Tehver: Phys. Stat. Sol. 21, 755 (1967).
7. D. L. Tonks and J. B. Page: Chem. Phys. Lett. 66, 449 (1979). See also several papers since then which can be traced through reference [14] below.
8. B. R. Stallard, P. M. Champion, P. R. Callis, and A. C. Albrecht: J. Chem. Phys. 78, 712 (1983) (see also reference [5] above).
9. S. Mukamel and James Sue: J. Chem. Phys. 82, 5291 (1985).
10. See reference [10] above. Some more recent references include D. J. Tannor and E. J. Heller: J. Chem. Phys. 77, 202 (1982); E. J. Heller, R. L.

Sundberg, D. J. Tannor: J. Phys. Chem. Chem. 86, 1822 (1982); D. J. Tannor: Ph.D. Thesis, UCLA, 1984.
11. S. Mukamel: Phys. Rep. 93, 1 (1982).
12. J. Melinger and A. C. Albrecht: manuscript in preparation. Also see M. Pfeiffer, A. Lau and W. Wernke: J. Raman Spectry. 15, 20 (1984) for comment on transform-type relationships in electronically resonant CARS.
13. B. R. Stallard, P. R. Callis, P. M. Champion, and A. C. Albrecht: J. Chem. Phys. 80, 70 (1984).
14. C. K. Chan, J. B. Page, D. L. Tonks, O. Brafman, B. Khodadoost, and C. T. Walker: J. Chem. Phys. 82, 4813 (1985).
15. J. Cable and A. C. Albrecht: manuscript in preparation.
16. D. Lee, B. R. Stallard, P. M. Champion, and A. C. Albrecht: J. Phys. Chem. 88, 6693 (1984).
17. J. Cable and A. C. Albrecht: manuscript in preparation (Figs. 2 and 3 are from this manuscript).

Application of Time-Dependent Raman Theory to the Analysis of Inorganic Photosystems*

D.E. Morris and W.H. Woodruff

Los Alamos National Laboratory, University of California,
Los Alamos, NM 87545, USA

1 Introduction

Recent developments by HELLER and coworkers [1] in the time-dependent theory of Raman scattering are of enormous potential value in determining structural and vibrational parameters in excited electronic states. Ground-state Raman intensities of fundamentals, overtones and combinations under resonance or near-resonance conditions can be used to determine normal coordinate displacements, force-constant changes and Duschinsky rotation in the excited state. This approach to excited-state characterization is particularly attractive because it does not require the direct observation of excited-state vibrational frequencies by TRVS or resolved vibronic spectroscopies. Thus, it is not hampered by short excited-state lifetimes and/or stringent experimental conditions. Furthermore, for systems where TRVS or vibronic data are available, Heller's theory affords an additional test of the accuracy of calculated excited-state parameters.

Continuing interest in elucidating the structure in the reactive excited state(s) of inorganic photosystems has prompted a re-examination of Heller's theory [2]. The aim has been to extend existing mathematical treatments to the more general (and commonly encountered) case in which coordinate displacements, force-constant changes and Duschinsky rotation occur simultaneously in the excited states. The most significant result of this work is the discovery of additional terms which were not explicitly developed in earlier treatments [1]. These new terms can make significant contributions to the scattering intensities of overtones and combinations. However, the broader implication is that attempts to obtain accurate excited-state structural information from ground-state Raman intensities will require consideration of all terms which can contribute to these intensities.

As demonstrated here, application of this more general formulation of Heller's theory to the analysis of real (and realistically obtainable) data is straightforward. As examples of the ease and utility of such analyses, attention is focused on some transition metal dimers whose photochemical and photophysical properties continue to be of considerable interest.

* Performed at Los Alamos National Laboratory under the auspices of the U.S. Dept. of Energy and at the University of Texas at Austin under NSF Grant CHE 84-03836

2 Theory, Results and Discussion

The time-dependent expression for the polarizability is [1]

$$\alpha_{f,i}(\omega_I) = \int_0^\infty \exp [i(\omega_I + E_i)t - \Gamma t] < \Psi_f \mid \Psi_i(t) > dt + NRT. \quad (1)$$

The central problem thus lies in calculating the time-dependent overlap between the final scattering state, Ψ_f, and the wave packet, $\Psi_i(t)$, propagating on the excited-state potential energy surface. Assuming harmonic ground- and excited-state surfaces and a gaussian form for $\Psi_i(t)$, this problem can be treated classically for short propagation times. This leads to expressions for the polarizabilities which are complex polynomials in time containing the desired excited-state structural parameters (vide infra). All terms in these polynomials are complex, so that modulus squaring of $\alpha_{f,i}$ to obtain the scattering intensity does not result in any cancellations of cross-terms. These cross-terms are the ones not explicitly developed previously [1] but which can significantly influence the total intensity of overtones and combinations.

The complete expressions for the fundamental (2) and first overtone (3) in mode k are;

$$|\alpha_{1,0}|^2 = (2\omega_{0k})^{-1} (V_k^2 / \sigma^4) \varepsilon_1(\beta) \quad (2)$$

$$|\alpha_{2,0}|^2 = (8\omega_{0k}^2)^{-1} [(V_k^4 / \sigma^6) \varepsilon_2(\beta) + ((\omega_{0k}^2 - V_{kk})^2 / \sigma^4) \varepsilon_1(\beta) + (V_k^2 (\omega_{0k}^2 - V_{kk}) / \sigma^5) \varepsilon'_{2,1}(\beta)] \quad (3)$$

where V_k is the first derivative with respect to q_k of the excited-state surface in the Franck-Condon region and V_{kk} is the second derivative corresponding to the excited-state force constant. Thus $\omega_{0k}^2 - V_{kk}$ is the change in force constant between ground and excited states. The $\varepsilon(\beta)$ are integral expressions describing the evolution of the wave packet (thus the overlap) on the excited-state surface at a point in the absorption envelope given by $\beta=(\omega_I - E)/\sigma$ and $2\sigma^2$ corresponds to the vibronic contribution to the absorption linewidth from totally symmetric modes. The integral expressions are plotted in Fig.1. The contribution from $\varepsilon'_{2,1}$ results from the cross-term described above. It is apparent that as excitation is detuned from exact resonance this new term will significantly effect the overtone intensity in totally symmetric modes for which there is a change in force constant in the excited state. Note also that the influence of this term is strongly excitation wavelength-dependent. An expression similar to (3) results for combination scattering intensity,but involves the Duschinsky rotation term in place of the force-constant change [2].

Equations (2) and (3) reveal the means for calculating the desired excited-state parameters. If fundamental and first overtone Raman intensities are available, $2\sigma^2$ can be estimated from the absorption spectrum and the excited-state force constant, V_{kk}, and displacement, $\Delta_k=V_k/V_{kk}$, can be obtained. Alternatively, TRVS or vibronic data can be used in place of overtone intensities to estimate V_{kk}. However, as illustrated in Fig. 1, if overtone intensities are used, the cross-term contribution cannot be neglected or erroneous V_{kk} values result. In addition, estimates of Δ_k

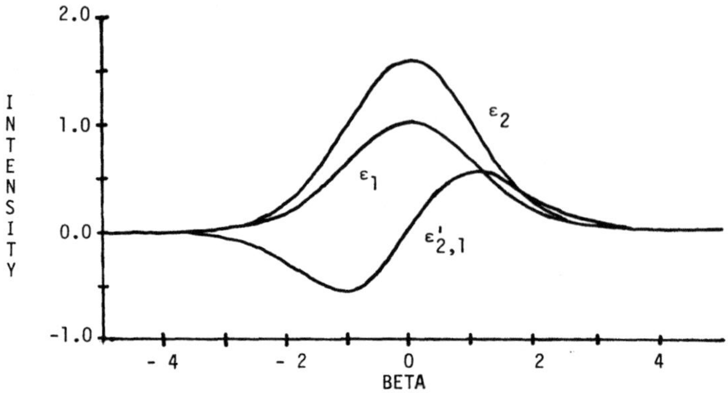

Figure 1. Time-dependent integral expressions in the generalized Heller theory

based solely on fundamental intensities will be in error when any appreciable change in force constant occurs.

To illustrate these points, calculations have been made for a number of quadruply bonded metal dimers. The characteristic features in these systems are a lowest-lying metal-localized δ-δ^* electronic transition which reduces the bond order to three in the excited state and a ground-state vibrational mode comprised almost entirely of metal-metal stretching which is strongly resonance enhanced in the δ-δ^* region. The results are summarized in Table 1. In all cases the vibrational frequency decreases and the bond length increases (3-8% elongation) in the excited state, both expected results for systems with a decreased excited-state bond order. The result for $Mo_2(O_2CCF_3)_4$ warrant further consideration, because the excited-state displacement has been determined previously by alternate means. Most notably, a simplified Heller theory analysis, in which the change in force constant was neglected, yielded Δ=0.045Å [7]. Explicit inclusion of this term (Table 1) results in a 50% increase in the value of the displacement, clearly illustrating the importance of considering all factors when calculating excited-state parameters. A Frank-Condon analysis of the vibronic spectrum [3] gave Δ=0.1Å for the acetate dimer, suggesting that some discrepancies between time-dependent theory and FC analyses exist and remain to be resolved.

Table 1. Excited-state displacements in the metal stretching mode in metal dimers

	$\bar{\omega}_{gs}$ [cm^{-1}]	$\bar{\omega}_{es}$ [cm^{-1}]	Δ [Å]
$Mo_2(O_2CCF_3)_4$	395	355[a]	0.07
$Mo_2Cl_8^{4-}$	346	336[b]	0.15
$Re_2Cl_8^{2-}$	274	248[c], 258[e]	0.13 , 0.12
$Re_2Br_8^{2-}$	275	255[d]	0.17

From vibronic absorption spectrum; [a]Ref.3 [b]Ref.4 [c]Ref.5 [d]Ref.6. [e]From Heller theory analysis of the overtone intensities.

3 Conclusions

The more general formulation of Heller's theory outlined here and in more detail elsewhere [2] should make it possible to characterize the excited states of real and chemically interesting systems with a high degree of reliability, provided care is excercised in considering the influence of all factors in determining the scattering dynamics. The reliance of this method on ground-state Raman intensities makes it particularly convenient and a valuable complement to the more direct but often experimentally prohibitive methods such as TRVS. While further testing of the predictions of this formalism is required, preliminary results as presented here are encouraging.

4 References

1 E.J.Heller, R.L.Sundberg and D.J.Tannor : J. Phys. Chem. 86, 1822 (1982) and references therein

2 D.E.Morris and W.H.Woodruff : J.Phys. Chem. submitted 5/85

3 W.C.Trogler, E.I.Solomon, I.Trabjerg, C.J.Ballhausen and H.B.Gray : Inorg. Chem. 16, 628 (1977)

4 P.E.Fanwick, D.S.Martin, F.A.Cotton and T.R.Webb : Inorg. Chem. 16, 2103 (1977)

5 W.C.Trogler, C.D.Cowman, H.B.Gray and F.A.Cotton : J. Am. Chem. Soc. 99, 2993 (1977)

6 C.D.Cowman and H.B.Gray : J. Am. Chem. Soc. 95, 8177 (1973)

7 C.S.Yoo and J.I. Zink : Inorg. Chem. 22, 2474 (1983)

Time-Dependent Quantum Dynamics of the Picosecond Vibrational IR-Excitation of Polyatomic Molecules

M. Quack, J. Stohner, and E. Sutcliffe

Laboratorium für Physikalische Chemie der ETH Zürich (Zentrum), CH-8092 Zürich, Switzerland

The picosecond vibrational excitation of polyatomic molecules is discussed in terms of solutions of the quantum dynamical equations of motion for simple model systems and realistic spectral structures of an asymmetric top molecule.

1. Introduction

The picosecond vibrational dynamics of polyatomic molecules irradiated with intense, coherent radiation from pulsed infrared lasers is of fundamental importance for our understanding of IR-photochemistry[1-3]. We have, in previous publications, already addressed such problems as quantum interference effects between different pathways[4], mode selectivity[5] and state selectivity[6] of IR-excitation. In the present paper, we shall report some instructive results on wavepacket dynamics and elaborate upon some of the consequences for nonlinear picosecond excitation of ozone.

2. Picosecond Quantum Statistical Trajectories of the Harmonic Oscillator

We consider the time-dependent motion according to the Schrödinger equation

$$i \frac{h}{2\pi} \frac{\partial \psi}{\partial t} = \hat{H}(t)\psi \tag{1}$$

In the basis of eigenstates ϕ of the molecular hamiltonian $\hat{H}_{mol}$ with coherent monochromatic excitation (frequency ω) this is equivalent to [1]

$$\psi(t) = \sum_k b_k \phi_k \tag{2}$$

$$\dot{\underline{b}} = \{\underline{W} + \underline{V}\cos(\omega t)\}\underline{b} \tag{3}$$

$$\underline{b}(t) = \underline{U}(t)\ \underline{b}(0)$$

$\underline{U}$ is the time evolution matrix, which also solves the Liouville von Neumann equation for the density matrix $\underline{P}$.

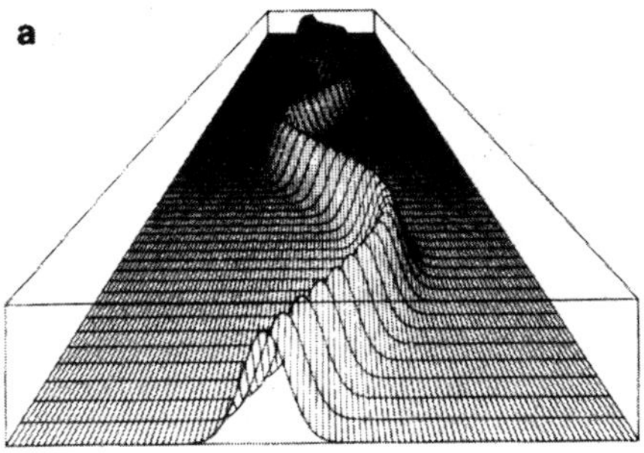

Fig. 1a: Harmonic oscillator coherent motion after excitation from the ground state

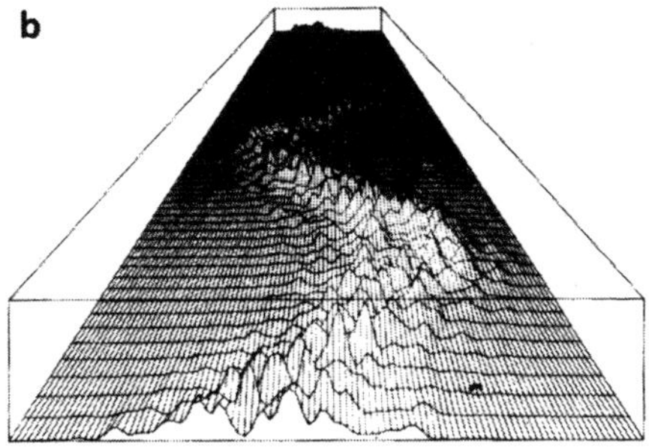

Fig. 1b: Harmonic oscillator motion after coherent excitation randomly drawn from a thermal (10'000 K) initial state (5 trajectories superposed)

$$\underline{P}(t) = \underline{U}(t)\ \underline{P}(0)\ \underline{U}^{\dagger}(t) \tag{5}$$

Details of the basic theory have been reviewed before [2]. We shall consider here some quantum mechanical and quantum statistical trajectories for the harmonic oscillator. Figure 1a shows the wavepacket motion of a harmonic oscillator initially in its ground-state after 16.5 ps slightly off-resonant excitation (resonance frequency 1000 cm^{-1}, laser frequency 999 cm^{-1}). The vertical axis in the perspective drawing is the probability density $|\psi(r,t)|^2$, the horizontal axis is the coordinate r and the time axis goes to the back in 1 fs steps. One finds the expected periodic motion of the gaussian wavepacket with the classical oscillator frequency.

Figure 1b is a little more interesting. Here we have drawn at random an initial state vector from a thermal ensemble at 10'000 K. Otherwise the conditions are the same as in Fig. 1a. Now one has a wavepacket, which is spread out and shows two types of motion: The forced harmonic oscillator motion and a random, Brownian-like motion. The figure shows in fact the average over five trajectories. As is well known [2], the average over infinitely many random quantum trajectories will give the same result as the density matrix equation (5) with the appropriate initial conditions. In the case of the harmonic oscillator this should be a broad gaussian. This is illustrated in Fig. 2 (the conditions are otherwise the same as for Fig. 1). What do these rather didactic calculations teach us about coherent IR-multiphoton excitation? The excitation of a harmonic oscillator at resonance or

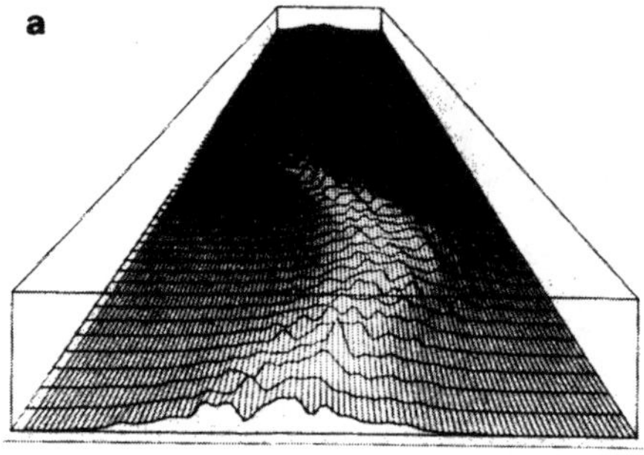

<u>Fig. 2a</u>: Average over 100 quantum trajectories drawn from a 10'000 K ensemble

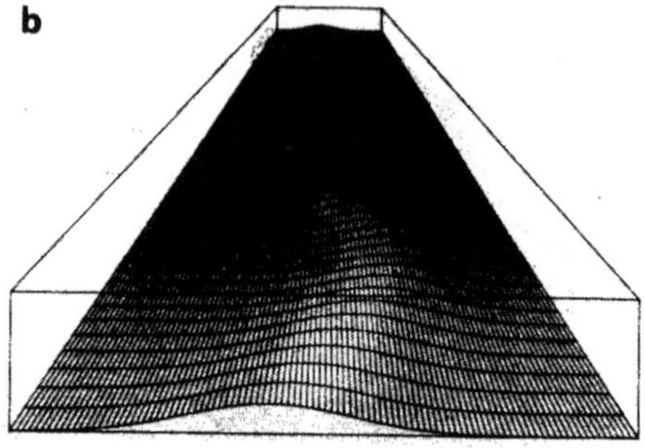

<u>Fig. 2b</u>: Coherent excitation of thermal initial ensemble with density matrix at 10'000 K (other conditions as Fig. 1)

off-resonance lead to coherent oscillation at the oscillator frequency, not to state selective excitation. The motion is quasiclassical with a broadened gaussian in the thermal situation. Individual trajectories drawn from thermal ensembles give in addition a Brownian motion superimposed on the oscillation. The degree of convergence with many trajectories is shown in Fig. 3. The gaussian is reasonably approached only with about 1000 trajectories. We can contrast this behaviour with the multiphoton excitation of the anharmonic oscillator [7]. In this case, highly state selective excitation was observed with non-classical motion and pronounced multiphoton resonances. Neither of these models is, of course, a realistic representation of the excitation of polyatomic molecules. They do provide, however, some insight into the primary dynamics.

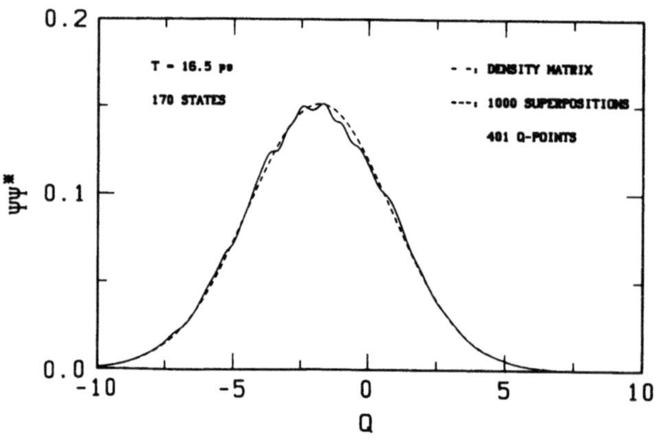

<u>Fig. 3</u>: Convergence to a thermal gaussian (full line) after 1000 trajectories (dashed line)

3. Nonlinear Picosecond Rovibrational Excitation of Ozone

Ozone, O_3 , is a nonlinear asymmetric top triatomic molecule with two anharmonic stretching vibrations, strongly coupled by Coriolis and Darling Dennison resonances, both falling in the range for CO_2-laser excitation. The bending vibration seems to be largely decoupled and almost harmonic. Its frequency near 700 cm^{-1} falls close to the H_2-Raman shifted CO_2-laser [5]. We have presented elsewhere extensive three-dimensional quantum dynamical calculations on the IR-multiphoton excitation of ozone to dissociation threshold [4,5,6,7,8]. We shall summarize here only the main results: (i) The excitation is quasiresonant stepwise. Direct multiphoton transitions are irrelevant up to intensities of 100 GW cm^{-2} and the quasiresonant approximation [1] is quantitatively valid [8]. The Floquet Liapunov [1,2] solutions are unnecessary. (ii) The excitation at moderate intensities of about 100 MW cm^{-2} is highly state selective and may be used to produce an IR-multiphoton pumped laser [6]. (iii) The excitation at high intensities beyond 100 GW cm^{-2} may be mode selective to near dissociation threshold [5]. (iv) In spite of the small size of the molecules, rotational averaging and rovibrational couplings render a quantitative statistical treatment possible [9]. The coarse-grained motion is effectively incoherent even for coherent excitation, quite different from the harmonic oscillator case [4,9]. (v) In spite of the step-wise nature of the process, the excitation may be highly nonlinear [4]. The nonlinearity increases with tuning off-resonance, and this theoretical finding is in agreement with experiments on another molecule (CF_3I [10]).

Only a brief review of our work was possible here, but we hope that some insight was provided into the intriguing phenomena of nonlinear coherent IR-multiphoton excitation.

References

1 M. Quack, J. Chem. Phys. 69, 1282 (1978)

2 M. Quack, Adv. Chem. Phys. 50, 395 (1982)

3 K. von Puttkamer, H.R. Dübal and M. Quack, Faraday Disc. Chem. Soc. 75, 197, 263 (1983)

4 M. Quack and E. Sutcliffe, Chem. Phys. Lett. 99, 167 (1983)

5 M. Quack and E. Sutcliffe, Chem. Phys. Lett. 105, 147 (1984)

6 M. Quack and E. Sutcliffe, Isr. J. Chem. 24, 204 (1984)
7 M. Quack and E. Sutcliffe, Infrared Phys. 25, 163 (1985)
8 M. Quack and E. Sutcliffe, J. Chem. Phys. 00, 000 (1985)
9 M. Quack and E. Sutcliffe, to be published
10 M. Quack and G. Seyfang, Chem. Phys. Lett. 93, 442 (1982)

Decay Law in Conformational Relaxation

A.A. Villaeys

Centre de Recherches Nucléaires et Université Louis Pasteur,
23, rue du Loess, F-67037 Strasbourg Cedex, France

1. Introduction

The problem of large amplitude motion is quite relevant in a number of fields in physics. With the recent developments of picosecond and subpicosecond experiments, data are now available concerning the effect of solvent viscosity on the dynamics of molecules relaxing through conformational relaxation in liquids. From the theoretical point of view, a large number of steady-state descriptions of these processes exists. Most of them are classical, and are based on the KRAMERS' works [1] which is a purely adiabatic approach. They can be classified into two categories, depending on whether an intramolecular potential barrier exists or not, to motion on the excited state surface.

More recently, we have developed a theory from a different point of view to analyze the effects of solvent viscosity on non-adiabatic photoisomerization or electronic relaxation of molecules in liquids [2-3]. Only the case with a sizeable potential barrier has been studied. In this quantum and non-adiabatic model, the rate constant has been evaluated in the overdamped and underdamped cases for which solutions are analytically tractable. The viscosity was introduced through the vibrational correlation function, which has been evaluated classically by using the Langevin equation. The viscosity dependences are expressed as the product of a constant w_0 times an universal function of the viscosity

$$W(\eta^*) = w_0\eta^* \exp(\eta^{*2}) \sum_{p=0}^{\infty} (-)^p(\eta^*)^{2p}[p!(\eta^{*2}+p)]^{-1} \quad , \quad \eta^* = \eta/\eta_0 \tag{1}$$

for the overdamped regime and

$$W(\eta^*) = w_0' \frac{1 - \exp(-\eta^*)\cos(2\pi\bar{\omega}_{ab}/\omega)}{1 + \exp(-2\eta^*) - 2\exp(-\eta^*)\cos(2\pi\bar{\omega}_{ab}/\omega)} \tag{2}$$

for the underdamped regime, where η is the viscosity and $\bar{\omega}_{ab}$, η_0, w_0' are constants explicited in references [2-3].

Later on, time-dependent descriptions have tried to disentangle the complicated role of solvent viscosity on time-resolved situations. In the present work, by adopting the same approach previously developed, we give a quantum and non-adiabatic time-dependent description. Although the formalism is general and not restricted to the overdamped regime, only this last case will be presented here.

2. Model

The molecule of interest here is described by two electronic configurations b and a where b is the upper excited one. From the adiabatic approximation, the molecular Hamiltonian H_m is the sum of the electronic H_e, vibrational H_v and interaction part H_{ev} which represents the perturbation responsible for the conformational relaxation

$$H_m = H_e + H_v + H_{ev} \tag{3}$$

The electronic potential surfaces are usually different in both configurations, so we modelize the vibrational structure by a set of displaced harmonic oscillators. The vibrational Hamiltonians take a form which depends on the electronic state

$$H_{vb} = \sum_j [\frac{1}{2} P_j^2 + \frac{1}{2} \omega_j^2 Q_j^2]$$
$$H_{va} = H_{vb} + V_{ab} \tag{4}$$

where the residual interaction V_{ab} is given by

$$V_{ab} = \sum_j \omega_j^2 Q_j d_j + \frac{1}{2} \sum_j \omega_j^2 d_j^2 \tag{5}$$

if d_j is the displacement of the j mode. From the zero-order Hamiltonian $H_0 = H_e + H_v$ the adiabatic Born Oppenheimer states correspond to $|av\rangle$ and $|bv'\rangle$.

3. Formalism

Systems exhibiting statistical properties are conveniently described in the density matrix formalism [4]. If H is the Hamiltonian of the total system, then

$$H = H_m + H_b + H_{mb} \tag{6}$$

if H_b and H_{mb} correspond respectively to the purely dissipation bath and its interaction with the molecular system. By using the same notation and partition, the total Liouvillian $L = [H, \,]/\hbar$ is decomposed as

$$L = L_e + L_v + L_{ev} + L_b + L_{mb}. \tag{7}$$

From the factorization of the reduced density matrices at the initial time t=0, the reduced density matrix of the molecular system, $\chi_m(t) = Tr_b\, \varrho(t)$, is governed by the equation

$$\frac{\partial \chi_m(t)}{\partial t} = -iL_0 \chi_m(t) - \int_0^t d\tau\, Tr_b[L^1_{mb}(t)\exp[-iL_0(t-\tau)]L^1_{mb}(\tau)]\exp[iL_0(t-\tau)]\chi_m(t) \tag{8}$$

if $Tr_b[L_{mb}(t)] = 0$ is assumed. Also we have

$$L_0 = L_e + L_v + L_{ev}$$
$$L^1_{mb}(t) = \exp(iL_b t)L_{mb} \exp(-iL_b t). \tag{9}$$

Here $\varrho(t)$ stands for the total density matrix, Tr_b for the trace over the bath states and a second-order cumulant expansion has been used. By assuming the Markoff approximation, a damping operator Γ_{ev} can be defined which reduces the density matrix equation (8) to

$$\frac{\partial \chi_m(t)}{\partial t} = -i(L_0 - i\Gamma_{ev})\chi_m(t). \tag{10}$$

In addition, we trace over the vibrational degrees of freedom to introduce the electronic reduced density matrix $\sigma_e(t) = Tr_v[\chi_m(t)]$. Following the same procedure, we obtain

$$\frac{\partial \sigma_e(t)}{\partial t} = -iL_e\sigma_e(t) - i\exp(-iL_e t)Tr_v\{[\tilde{L}_{ev}(t) - i\tilde{\Gamma}_{ev}(t)]\sigma_v(0)\}\exp(iL_e t)\,\sigma_e(t)$$
$$- \int_0^t d\tau\, Tr_v\{\exp(iL_v t)[L_{ev} - i\Gamma_{ev}]\exp(-i(L_e+L_v)(t-\tau))[L_{ev} - i\Gamma_{ev}]$$

$$\times \exp\,(-iL_v\tau)\sigma_v(o)\}\ \exp\,(iL_e(t-\tau))\sigma_e(t) \tag{11}$$

where

$$\tilde{L}_{ev}(t) = \exp(i(L_e+L_v)t)\ L_{ev}\ \exp(-i(L_e+L_v)t) \tag{12}$$

and similarly for $\tilde{\Gamma}_{ev}(t)$. By assuming the statistical average of L_{ev} to be zero and neglecting the Γ_{ev}^2 term, we finally get the equation

$$\frac{\partial\sigma_e(t)}{\partial t} = -i\bar{L}_e\sigma_e(t) - \int_0^t d\tau\ Tr_v[L_{ev}^1(t)\exp(-iL_e(t-\tau))L_{ev}^1(\tau)\sigma_v(0)]\exp(iL_e(t-\tau))\sigma_e(t) \tag{13}$$

where $\bar{L}_e$ defines an effective electronic Liouvillian

$$\bar{L}_e = L_e - i\ Tr_v[\exp(iL_vt)\Gamma_{ev}\ \exp(-iL_vt)\sigma_v(0)] \tag{14}$$

which is assumed time-independent here. In fact, this implies strictly $[\Gamma_{ev},L_v]=0$. Also, as previously mentioned, we have

$$L_{ev}^1(t) = \exp(iL_vt)L_{ev}\ \exp(-iL_vt) \tag{15}$$

At this stage, it is convenient to introduce the probability conservation, that is to say

$$\langle\langle bb|\sigma_e(t)\rangle\rangle + \langle\langle aa|\sigma_e(t)\rangle\rangle = 1 \tag{16}$$

Then, assuming the single line approximation valid, that is to say neglecting the overlapping resonances, the equation for $\sigma_e(t)$ becomes

$$\frac{\partial}{\partial t}\langle\langle bb|\sigma_e(t)\rangle\rangle = a(t)\ \langle\langle bb|\sigma_e(t)\rangle\rangle + b(t) \tag{17}$$

where, using $\langle\langle ij|L_e|kl\rangle\rangle = E_{ij}\ \delta_{ik}\ \delta_{jl}$, we have

$$\begin{aligned} a(t) &= -i\bar{E}_{bb} - 2b(t) \\ b(t) &= 2\hbar^{-2}\mathrm{Reel}\{\int_0^t d\tau\ \exp[-iE_{ab}(t-\tau)]Tr_v[\sigma_v(0)H'_{ab}(t)H'_{ba}(\tau)]\} \end{aligned} \tag{18}$$

if $H'_{ij} = \langle\phi_i|H_{ev}|\phi_j\rangle$ is the electronic matrix element. Then, all that we need is the evaluation of the vibrational correlation function. Its expression has been evaluated previously. If we assume H'_{ab} independent of the nuclear coordinates and introducing a second-order cumulant expansion, we get

$$\begin{aligned} G(t,\tau) &= Tr_v[\sigma_v(o)H'_{ab}(t)H'_{ba}(\tau)] = |H'_{ba}|^2\exp[\gamma_{ab}(t-\tau)] \\ \gamma_{ab}(t) &= -\frac{it}{2\hbar}\sum_j \omega_j^2\ d_j^2 - \sum_j \frac{\omega_j^4 d_j^2}{\hbar^2}\int_0^t dt_1\int_0^{t_1} dt_2\ \langle\langle Q_j(t_1)Q_j(t_2)\rangle\rangle \end{aligned} \tag{19}$$

The correlation functions are evaluated as usual for all the modes except for the mode coupled to the solvent. For this last mode, because of its statistical behavior, it is deduced from the Langevin equation [2-3]. Then, assuming $d_j \approx 0$ for the quantum modes, $G(t,\tau)$ can be approximated, in the strong overdamped case, by

$$G(t,\tau) = |H'_{ba}|^2\ \exp[\{-(i\omega^2d^2/2\hbar)-(kTd^2\beta/\hbar^2)\}(t-\tau)] \tag{20}$$

where T and ω are the temperature and the frequency of the classical mode and k the Boltzmann constant.

4. Decay Law

From the previous results, we are now able to establish the expression of the decay law $P_b(t) = \langle\langle bb|\sigma_e(t)\rangle\rangle$. Although the formalism is general, only the strong overdamped case is considered here. In the case of a zero-energy gap $\bar{\omega}_{ab} = \omega_{ab}+\omega^2 d^2/2\hbar$ between the electronic configurations, $\bar{E}_{ab} = -i\Gamma_b$ and the solution is analytically tractable. We obtain the final result

$$P_b(t) = \{\exp[-\Gamma_b t+(4|H'_{ba}|^2/\hbar^2 K\beta)t - (4|H'_{ba}|^2/\hbar^2 K^2\beta^2)(\exp(-K\beta t)-1)]\}$$

$$\times \{1 + (2|H'_{ba}|^2/\hbar^2 K\beta) \sum_{n=0}^{\infty} \frac{1}{n!} \left(\frac{4|H'_{ba}|^2}{\hbar^2 K^2\beta^2}\right)^n \left[\frac{\exp[\Gamma_b t-(4|H'_{ba}|^2/\hbar^2 K\beta)t-K\beta(1+n)t]-1}{\Gamma_b-(4|H'_{ba}|^2/\hbar^2 K\beta)-K\beta(1+n)}\right.$$

$$\left. - \frac{\exp[\Gamma_b t-(4|H'_{ba}|^2/\hbar^2 K\beta)t-Kn\beta t]-1}{\Gamma_b-(4|H'_{ba}|^2/\hbar^2 K\beta)-Kn\beta}\right]\}, \quad K = kT\, d^2/\hbar^2 \qquad (21)$$

This expression emphasizes the complicated role of the solvent viscosity on the time evolution of conformational change. Of course, less extreme situations need to be treated to disentangle experimental results. This is currently under investigation.

References

1. H.A. Kramers, Physica 7, 284 (1940)
2. A.A. Villaeys, A. Boeglin and S.H. Lin, Chem. Phys. Lett. 116 (1985) 210
3. A.A. Villaeys, A. Boeglin and S.H. Lin, J. Chem. Phys. 82 (1985) 4040
4. S. Mukamel, I. Oppenheimer and J. Ross, Phys. Rev. A 17 (1978) 1988

Stochastic Theory of Vibrational Dephasing and Transient CARS Lineshapes

R.F. Loring and S. Mukamel[†]

University of Rochester, Rochester, NY 14627, USA

Considerable progress has been made in recent years in studies of the dynamics of vibrational excitations in liquids[1]. Before the advent of picosecond spectroscopic methods, vibrational dephasing was studied through measurements of the isotropic Raman lineshape $I_0(\omega)$. This lineshape is proportional to the Fourier transform of the vibrational correlation function $\langle\hat{q}(t)\hat{q}(0)\rangle$, where $\hat{q}$ is the coordinate of the relevant vibrational mode, and $\langle\cdots\rangle$ indicate a trace over the equilibrium density matrix[2]. The distinction is traditionally made between homogeneous and inhomogeneous contributions to spectral lineshapes. Although these two limits are well defined, the classification of realistic line broadening mechanisms into these two categories is not always possible. Consider a molecule coupled to a bath with time scale $1/\Lambda$. If Λ is much smaller than the observed line width, the line is inhomogeneously broadened. As Λ is increased, the lineshape will change, and only when Λ is much larger than the observed linewidth, will a Lorentzian, homogeneously broadened line emerge. For intermediate Λ, the line cannot be classified as either homogeneously or inhomogeneously broadened. Stochastic[2] and microscopic[3] models, which are valid for arbitrary Λ and interpolate between these limits, allow a more realistic calculation of spectral lineshapes in liquids. A fundamental question concerning the time-resolved CARS spectroscopy in liquids is, under what conditions does the experiment provide more microscopic information than is contained in $\hat{I}_0(\omega)$. Particular attention has been given to whether coherent transient CARS spectroscopy can distinguish between homogeneous and inhomogeneous broadening mechanisms[1,4]. In this article, we analyze the information content of coherent CARS spectra for a general line-broadening mechanism with an arbitrary correlation time $1/\Lambda$. Our model[5,6] interpolates continuously between the limits of homogeneous and inhomogeneous line broadening[7]. We model the molecular liquid as a collection of three-level systems of the type depicted in Fig. 1. $|1\rangle$ and $|2\rangle$ denote vibronic states belonging to the manifold of the ground state, whereas the vibronic state $|3\rangle$ belongs to an electronically excited state. The energy difference between $|2\rangle$ and $|1\rangle$ is $\hbar\Omega_2 + \hbar\Delta(t)$. $\Delta(t)$ is a stochastic process with zero mean arising from the random force exerted on the absorber molecule by its fluctuating environment. The effect of this fluctuating energy is to broaden the spontaneous Raman lineshape for the $|1\rangle$ to $|2\rangle$ transition (proper dephasing). We shall consider two pulse sequences denoted Raman free induction decay (RFID) and Raman echo (RE). The experiments both begin with a pair of time-coincident pulses with frequencies ω_L and ω_S ($\Omega_2=\omega_L-\omega_S$) that excite a coherence between $|1\rangle$ and $|2\rangle$ by stimulated Stokes scattering. There follows a delay period of duration t_D. In the RFID, the system is then irradiated by a probe pulse with frequency ω_{L2} and emission is detected at frequency $\omega_{L2}-\Omega_2$ (Stokes) or frequency $\omega_{L2}+\Omega_2$ (anti-Stokes). In the RE, the system is irradiated with another pair of excitation pulses. After another delay period of duration t, the system is probed with a pulse with frequency ω_{L2}, and the coherent Stokes or anti-Stokes signal is detected. The RFID consists of one pair of excitation pulses, one delay period, and a probe pulse, while the RE consists of two pairs of excitation pulses, two delay periods, and a probe pulse. The pulse sequences are shown in Fig. 2. In both cases, the coherent sig-

[†]Camille and Henry Dreyfus Teacher-Scholar

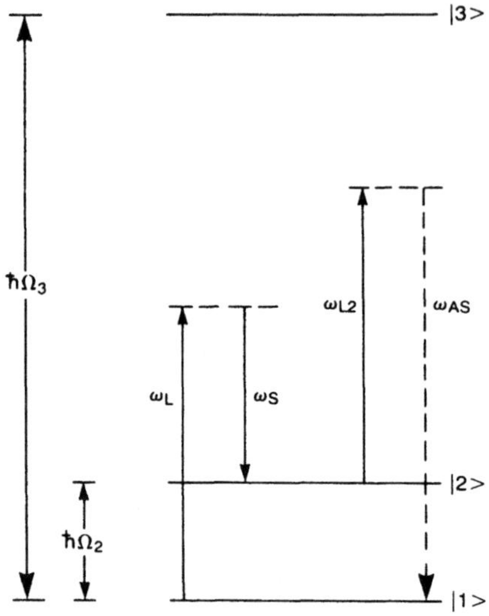

Fig. 1

Energy-level diagram for our model system

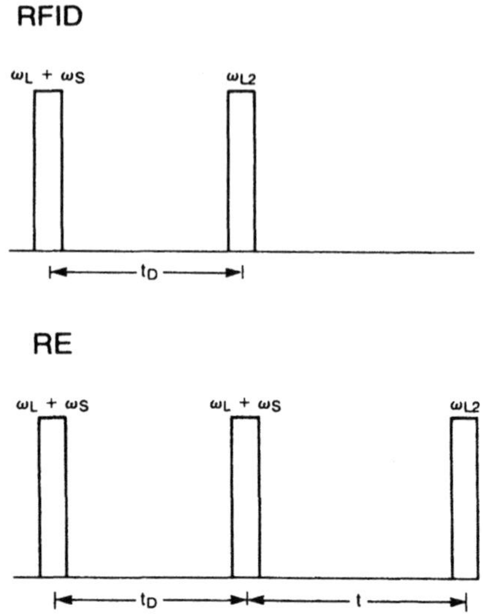

Fig. 2

Pulse sequences for Raman free induction decay and Raman echo experiments.

nal is observed at the direction $k_L-k_S+k_{L2}$. We have calculated the signal for both experiments using the tetradic (Liouville space) scattering formalism[3]. If the pulses are very short (compared with all detuning frequences and relaxation rates), the signals are[6]:

$$S_{RFID}(t_D) \propto |\chi_{RFID}(t_D)|^2 \tag{1}$$

$$\chi_{RFID}(t_D) = \langle\hat{q}(t_D)\hat{q}(0)\rangle \tag{2}$$

and

$$S_{RE}(t,t_D) \propto |\chi_{RE}(t,t_D)|^2 \tag{3}$$

$$\chi_{RE}(t,t_D) = \langle\hat{q}(-t)\hat{q}(0)\hat{q}(t_D)\hat{q}(0)\rangle. \tag{4}$$

The averages over the fluctuating energy of state $|2\rangle$ in (2) and (4) can be carried out in a straightforward manner if $\Delta(t)$ is taken to be a Gaussian variable[2,6]. In this case χ_{RFID} and χ_{RE} are given by

$$\chi_{RFID}(t_D) = \exp\,[-g(t_D)] \tag{5}$$

$$\chi_{RE}(t_1t_D) = \exp\,[-2g(t) - 2g(t_D) + g(t+t_D)], \tag{6}$$

where

$$g(t) = \int_0^t d\tau_1 \int_0^{\tau_1} d\tau_2 \; \langle\Delta(\tau_1-\tau_2)\,\Delta\,(0)\rangle. \tag{7}$$

For illustrative purposes, we shall consider dephasing that is induced by two independent processes that modulate the energy of state $|2\rangle$.

$$\langle\Delta(t)\Delta(0)\rangle = D_1^2\exp(-\Lambda_1|t|) + D_2^2\,\exp(-\Lambda_2|t|). \tag{8}$$

D_n and Λ_n are respectively the magnitude and inverse time-scale of process n. Substitution of (8) and (7) into (5) and (6) will give general expressions for the RE

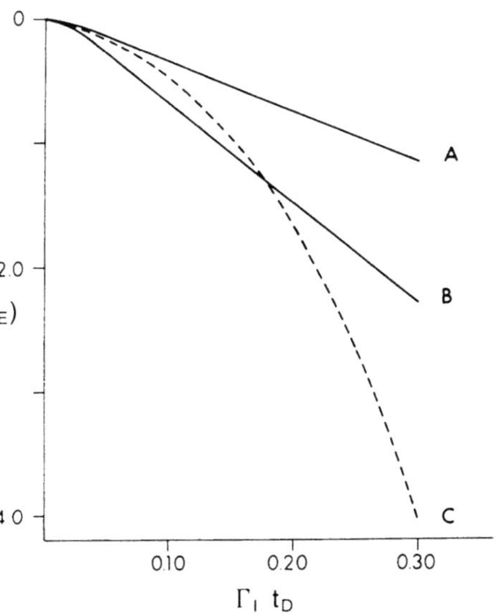

Fig. 3

Raman echo signal, calculated from (6)-(8).
$\Lambda_1=10^2$, $D_1=D_2=10$. $\Gamma_1=D_1^2/\Lambda_1$.
A) $\Lambda_2=10^{-6}$. B) $\Lambda_2=10^{-2}$.
C) $\Lambda_2=1$. Exponential RE signals will be observed from a system with two dephasing mechanisms if both processes are fast on the experimental time-scale (A), or if one process is fast, and the other is slow on this time-scale (B). Otherwise, the decay will be nonexponential. (C).

and RFID observables for a system with two dephasing mechanisms of arbitrary time-scale and magnitude.

Figure 3 shows logarithmic plots of the RE signal, calculated from (6) - (8). All plots are calculated for $\Lambda_1=100$, $D_1=D_2=10$. Plots A, B, and C are calculated respectively for $\Lambda_2=10^{-6}$, 100, 1. In all cases, process 1 causes homogeneous line broadening ($\Lambda_1/D_1 \gg 1$). In case A, $\Lambda_2/D_2 \ll 1$. In this case, the RE signal is nonexponential at very short times but becomes exponential with decay constant $4\Gamma_1$. ($\Gamma_1 \equiv D_1^2/\Lambda_1$). In case B, process 2 has a time-scale identical to that of process 1 ($\Lambda_1=\Lambda_2$). In this case, the RE signal is nonexponential at very short times but becomes exponential with decay constant $4(\Gamma_1+\Gamma_2)=8\Gamma_1$. In case C, process 2 has a time-scale that is comparable in magnitude to the time-scale of observation. In this case, the RE signal is <u>nonexponential</u> on the time-scale of observation. Fig. 3 shows that if vibrational dephasing is caused by two different processes with arbitrary time-scales, the observed RE signal is, in general, nonexponential. The signal will be exponential (except at very short times) in two limiting cases. The first such case holds, when the two processes occur on very different time-scales. In this case, one process is "dynamic" on the time-scale of observation, and the other is "static", relative to this time-scale. The line broadening in such a system has homogeneous and inhomogeneous components. The second limiting case that leads to an exponential RE signal occurs, when both processes are "dynamic" on the time-scale of observation. Such a system has a homogeneously broadened line.

In conclusion, we note the following: (i) (2) shows that for weak and short pulses the RFID cannot provide information about the material that is not also present in the low power lineshape function $I_0(\omega)$. This conclusion differs from earlier treatments of this problem [1],[8] (ii) George and Harris [4] have argued that the RFID experiment can be used to measure homogeneous dephasing times of a system with an inhomogeneously broadened Raman lineshape, provided that the initial laser pump pulse is significantly depleted. We have made the distinction between laser pulse depletion, which occurs when the sample is optically dense, and saturation of the material transition, which occurs when the fields are strong. The Bloch-Maxwell equations considered by George and Harris include laser depletion, but they do not include material saturation. We have shown [6] that the experimental observable calculated from these equations cannot show selectivity. Selectivity is defined as the capacity to provide more information than an ordinary lineshape measurement. If these equations are expanded to include material saturation, the information content of the experimental observable will no longer be identical to that of a low-power lineshape measurement. There are two experimental conditions under which the RFID experiment is not selective: short pulses of arbitrary strength and weak pulses of arbitrary duration. If the experiment is carried out with long, strong pulses, the signal will undoubtedly be characterized by some degree of selectivity. The signal, under these conditions, can be calculated numerically from the complete Bloch-Maxwell equations. (iii) The Raman echo experiment, on the other hand, can be used to go beyond the spontaneous Raman lineshape in obtaining dynamical information, since it is a probe for a four-point correlation function of the dipole operator (similar to the photon echo) [5],[6],[12]. The theory of the Raman echo for a system with an inhomogeneously broadened lineshape was originally developed by Hartmann[9]. Our present treatment generalizes that theory to an arbitrary time-scale of the bath and is valid from the homogeneous to the inhomogeneous broadening limits. Raman echo experiments have been carried out in gases [10] and solids [11] but have not yet been applied to liquids. The Raman echo experiment has the capability of providing significant new information on the dynamics of vibrations in liquids.

Acknowledgements

The support of the National Science Foundation, The Office of Naval Research, The Army Research Office, and the Petroleum Research Fund, administered by the American Chemical Society, is gratefully acknowledged.

References

1. A. Laubereau and W. Kaiser: Rev. Mod. Phys. 50, 607 (1978)
2. R. Kubo: Adv. Chem. Phys. 15, 101 (1969)
3. S. Mukamel: Phys. Rep. 93, 1 (1982); Phys. Rev. A 26, 617 (1982); Ibid 28, 3480 (1983); D. Grimbert and S. Mukamel: J. Chem. Phys. 76, 834 (1982)
4. S. George and C. B. Harris: Phys. Rev. A 28, 863 (1983); S. M. George, A. L. Harris, M. Berg, and C. B. Harris: J. Chem. Phys. 80, 83 (1984)
5. R. F. Loring and S. Mukamel: Chem. Phys. Lett. 114, 426 (1985)
6. R. F. Loring and S. Mukamel: J. Chem. Phys. (to be published)
7. E. Hanamura: J. Phys. Soc. Japan 52, 2258 (1983); H. Tsunetsugu, T. Taniguchi, and E. Hanamura: Solid State Comm. 52, 663 (1984)
8. D. W. Oxtoby: Adv. Chem. Phys. 40, 1 (1979); J. Chem. Phys. 74, 5371 (1981)
9. S. R. Hartmann: IEEE J. Quantum Electron. 4, 802 (1968)
10. K. P. Leung, T. W. Mossberg, and S. R. Hartmann: Phys. Rev. A 25, 3097 (1982)
11. P. Hu, S. Geschwind, and T. M. Jedju: Phys. Rev. Lett. 37, 1357 (1967)
12. J. L. Skinner, H. C. Andersen, and M. D. Fayer: Phys. Rev. A 24, 1994 (1981)

On the Lineshape of Molecular Aggregate Spectra

P.O.J. Scherer and S.F. Fischer

Physik-Department der Technischen Universität München, Boltzmannstraße, D-8046 Garching, F. R. G.

1. Introduction

Frequently visible absorption spectra of molecular aggregates exhibit vibronic structure which is usually interpreted in comparison with theoretical line spectra.The calculated line spectra do not make any predictions about the lineshape. For small intermolecular coupling, strength changes of the lineshape due to the aggregation are of minor importance. For larger intermolecular coupling, however , the lineshapes of different vibronic transitions differ from each other. For example,we point at the spectra of the dye 1,1'-diethyl-N,N'-isocyaninchloride (PIC) . Absorption spectra of PIC aggregates [1,2,3]contain an intense narrow line at 17450 cm^{-1}, the so called J-band,which is absent in monomer and dimer spectra[3]. Simulations of monomer and dimer spectra yield all parameters needed for model calculations of an aggregate line spectrum [3,4]. The resulting positions and intensities correlate well with the experimental PIC aggregate spectrum.

Earlier attempts [5,6]to calculate continuous aggregate spectra treat the vibronic structure as part of the inhomogeneous broadening. Thereby the details of the vibronic structure of the spectrum are lost. In our approach,vibronic structure and inhomogeneous broadening are treated separately. In a first approximation each vibronic transition of the line spectrum is replaced by an appropriate line profile which is obtained from a transformation of the gaussian monomer line profile. Only this procedure gives good agreement with the experimental data [7] .

The theoretical absorption spectrum is given by the groundstate average of the aggregate Green's function. For room temperature a thermal average over thermally populated groundstate vibrations is performed. This is important for the understanding of the long wavelength tail of the lowest absorption maximum.

2. The Model Hamiltonian

We consider the following model Hamiltonian for a one-dimensional molecular aggregate

$$\hat{H} = \sum_{n=o}^{N-1} |n><n| \{ \sum_{m=o}^{N-1} \hbar\omega_o b_m^* b_m + \hbar\omega_o \lambda(b_n^* + b_n) + \hbar\omega_o\lambda^2 + \hat{h}_n \}$$
$$+ V \sum_{n=o}^{N-2} \{ |n+1><n| + |n><n+1| \} \quad (1)$$

where $\hat{h}_n$ is an operator acting on the internal coordinates of all molecules in the aggregate. We assume that the N molecules are equivalent and in an equivalent environment. Taking cyclic boundary conditions (i.e. monomers 0 and N are identical) a canonical transformation of the Hamiltonian leads to a sum of N independent Hamiltonians.

$$\bar{H} = \sum_k |k><k| \{ H_k^{(o)} + \hat{h}_o \} \qquad k=0,\ 2\pi/N\ ,\ldots,(N-1)2\pi/N$$
$$H_k^{(o)} = \hbar\omega_o \sum_{n=o}^{N-1} b_n^* b_n + \hbar\omega_o \lambda(b_o^* + b_o) + \hbar\omega_o\lambda^2 + e^{ik}\hat{G}V + e^{-ik}\hat{G}^{-1}V \quad (2)$$

The operator $\hat{G}$ performs a cyclic permutation of the N monomers.

3. Approximate Calculation of the Spectrum

The eigenfunctions and eigenvalues of (2) can be computed numerically [4,7] and will be used as a starting point. The contribution of the mode with wavevector k to the absorption spectrum is given by the imaginary part of the spectral function.

$$I_k(E) = \sum_{s,s'} <\Omega\ \Phi_{ks}^{(o)}><\Phi_{ks'}^{(o)} \quad [\ E - <\Phi_{ks}^{(o)}\ H_k\ \Phi_{ks'}^{(o)}>\ ]^{-1}\Omega > \quad (3)$$
$$H_k^{(o)}\ \Phi_{ks}^{(o)} = E_{ks}\ \Phi_{ks}^{(o)}$$

For an aggregate with two molecules in its elementary cell, the spectrum consists mainly of transitions from the groundstate to the band edges k=0 and k= π .However, at frequencies below the lowest absorpion maximum transitions from thermally populated initial states with phonon momentum Q_i and energy E_i become important.

$$I(E) = \sum_i \exp(-\beta E_i)\ [\frac{1+\cos\alpha}{2}\ I_{k=Q_i}(E) + \frac{1-\cos\alpha}{2} I_{k=Q_i+\pi}(E)\ ] \quad (4)$$

Here α is the angle between the dipole transition moments of the two molecules in the elementary cell.

As a first approximation, we take only the diagonal part of (3). Employing a single-site approximation [5] we get the result

$$I_k(E) = \sum_s |<\Omega\ \Phi_{ks}^{(o)}>|^2\ I^{(o)}(E-E_{ks}+\Delta_{ks})[1-\Delta_{ks}I^{(o)}(E-E_{ks}+\Delta_{ks})]^{-1}$$

$$\Delta_{ks} = Ve^{ik} < \Phi_{ks}^{(o)}\ \hat{G}\ \Phi_{ks}^{(o)} > + \text{c.c.} \quad (5)$$

where $I^{(o)}$ is the spectral function of a single monomer line. In this approximation the result is the line spectrum of $H_k^{(o)}$ where now each line is replaced by a transformed line profile.

The off-diagonal terms in (3) redistribute intensity among the vibronic transitions and change the line shape. From an expansion of the inverse matrix corrections to the spectrum can be computed. Careful investigations showed that in the framework of the single-site approximation these corrections are generally small. This is plausible, as in the single-site approximation the groundstate averaged off-diagonal matrix elements in (3) vanish.

4. Application to PIC

For PIC all the relevant parameters are well known from the monomer and dimer spectra. The only new parameter is a shift of the whole spectrum to longer wavelengths of 470 cm^{-1}. Figure 1 shows the experimental absorption spectrum (dashed curve) of highly concentrated PIC solution in water [3] together with theoretical results for a 6-mer with cyclic boundary conditions (solid curve). The angle between the transition moments was changed with respect to the dimer value [3] from 70° to 90° to give better correlation with the experimental data.

The long wavelength tail of the lowest transition shows a characteristic behaviour which is of great interest. The experimental data [3] are shown in Fig.2 (dashed curve). After a rapid decay we observe an

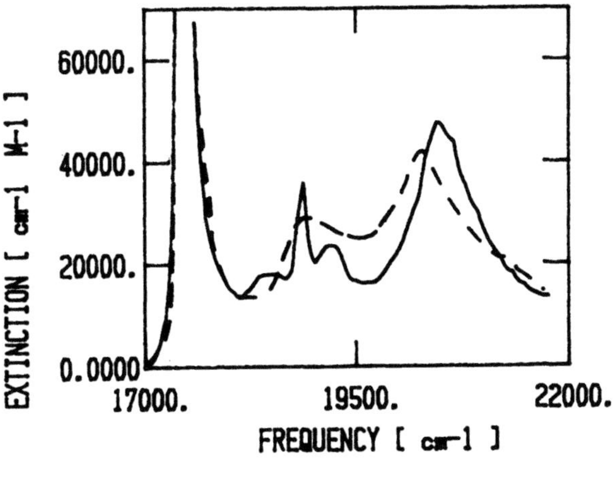

Figure 1

Experimental extinction spectrum of PIC-aggregates from [3] (dashed line)

Theoretical results for a 6-mer with cyclic boundary conditions (solid line)

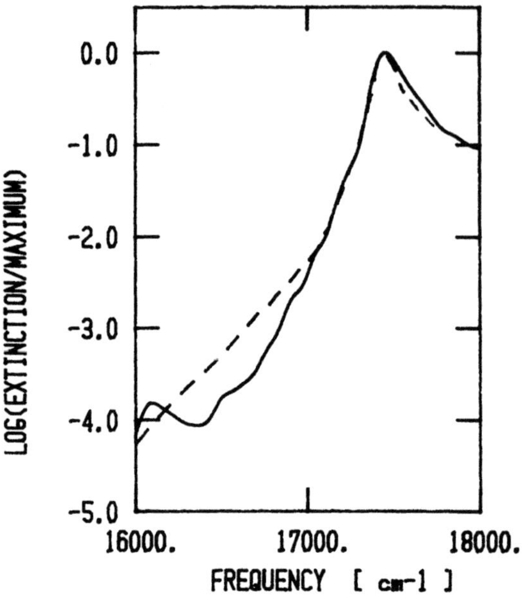

Figure 2

The long wavelength tail of the lowest absorption band of PIC is depicted logarithmically. The experimental data (dashed curve) are compared with theoretical results (solid curve) for a 6-mer with cyclic boundary conditions.

exponential decrease of the absorption with increasing frequency.This exponential decay is also observed in the monomer spectrum and is believed to reflect the thermal population of vibrations in the electronic groundstate [3] .The theoretical spectrum which is also shown in Fig.2 (solid curve) agrees with the experimental data. As in Fig.1 the theoretical spectrum tends to show more structure than the experimental data.Especially a hot band is observed at about $16000cm^{-1}$ which is absent in the experimental spectrum.

References

1 G.Scheibe,Angew.Chem. 49(1936) 563
2 E.E. Jelley, Nature 138 (1936) 1009
3 B.Kopainsky,J.K.Hallermeier,and W.Kaiser,Chem.Phys.Lett. 83 (1981) 498 , 87 (1982) 7
4 P.O.J.Scherer and S.F.Fischer,Chem.Phys. 86(1984)269
5 J.S.Briggs and A.Herzenberg,J.Phys. 3(1970)1663
6 R.P.Hemenger,J.Chem.Phys. 66 (1977) 1795
7 E.W.Knapp,P.O.J.Scherer and S.F. Fischer,Chem.Phys.Lett. 111(1984)481

Index of Contributors

Springer Proceedings in Physics

Volume 1

Fluctuations and Sensitivity in Nonequilibrium Systems

Proceedings of an International Conference, University of Texas, Austin, Texas, March 12-16, 1984

Editors: **W. Horsthemke, D. K. Kondepudi**

1984. 108 figures. IX, 273 pages. ISBN 3-540-13736-X

Contents: Basic Theory. - Pattern Formation and Selection. - Bistable Systems. - Response to Stochastic and Periodic Forcing. - Noise and Deterministic Chaos. - Sensitivity in Nonequilibrium Systems. - Contributed Papers and Posters. - Index of Contributors.

Volume 2

EXAFS and Near Edge Structure III

Proceedings of an International Conference, Stanford, CA, July 16-20, 1984

Editors: **K. O. Hodgson, B. Hedman, J. E. Penner-Hahn**

1984. 392 figures. XV, 533 pages. ISBN 3-540-15013-7

Contents: Fundamental Aspects of EXAFS and XANES. - EXAFS Data Analysis. - Biological Systems. - Catalytic Systems and Small Metal Clusters. - Surface Structure. - Amorphous Materials and Glasses. - Geology and Geochemistry, and High Pressure. - Other Applications. - Related Techniques and Instrumentation. - Index of Contributors.

GPSR Compliance

The European Union's (EU) General Product Safety Regulation (GPSR) is a set of rules that requires consumer products to be safe and our obligations to ensure this.

If you have any concerns about our products, you can contact us on ProductSafety@springernature.com

In case Publisher is established outside the EU, the EU authorized representative is:

Springer Nature Customer Service Center GmbH
Europaplatz 3
69115 Heidelberg, Germany

Batch number: 10620881

Printed by Printforce, the Netherlands